NEUROMETHODS

Series Editor
Wolfgang Walz
University of Saskatchewan
Saskatoon, SK, Canada

For further volumes:
http://www.springer.com/series/7657

Neuromethods publishes cutting-edge methods and protocols in all areas of neuroscience as well as translational neurological and mental research. Each volume in the series offers tested laboratory protocols, step-by-step methods for reproducible lab experiments and addresses methodological controversies and pitfalls in order to aid neuroscientists in experimentation. *Neuromethods* focuses on traditional and emerging topics with wide-ranging implications to brain function, such as electrophysiology, neuroimaging, behavioral analysis, genomics, neurodegeneration, translational research and clinical trials. *Neuromethods* provides investigators and trainees with highly useful compendiums of key strategies and approaches for successful research in animal and human brain function including translational "bench to bedside" approaches to mental and neurological diseases.

Identification, Characterization, and Manipulation of Neuronal Ensembles

Edited by

Luis Carrillo-Reid

Instituto de Neurobiología, Universidad Nacional Autónoma de México, Juriquilla, Mexico

Humana Press

Editor
Luis Carrillo-Reid
Instituto de Neurobiología
Universidad Nacional Autónoma de México
Juriquilla, Mexico

ISSN 0893-2336 ISSN 1940-6045 (electronic)
Neuromethods
ISBN 978-1-0716-4207-8 ISBN 978-1-0716-4208-5 (eBook)
https://doi.org/10.1007/978-1-0716-4208-5

This Humana imprint is published by the registered company Springer Science+Business Media, LLC, part of Springer Nature.
The registered company address is: 1 New York Plaza, New York, NY 10004, U.S.A.

Preface to the Series

Experimental life sciences have two basic foundations: concepts and tools. The Neuromethods series focuses on the tools and techniques unique to the investigation of the nervous system and excitable cells. It will not, however, shortchange the concept side of things as care has been taken to integrate these tools within the context of the concepts and questions under investigation. In this way, the series is unique in that it not only collects protocols but also includes theoretical background information and critiques which led to the methods and their development. Thus, it gives the reader a better understanding of the origin of the techniques and their potential future development. The Neuromethods publishing program strikes a balance between recent and exciting developments like those concerning new animal models of disease, imaging, in vivo methods, and more established techniques, including, for example, immunocytochemistry and electrophysiological technologies. New trainees in neurosciences still need a sound footing in these older methods in order to apply a critical approach to their results.

Under the guidance of its founders, Alan Boulton and Glen Baker, the Neuromethods series has been a success since its first volume was published through Humana Press in 1985. The series continues to flourish through many changes over the years. It is now published under the umbrella of Springer Protocols. While methods involving brain research have changed a lot since the series started, the publishing environment and technology have changed even more radically. Neuromethods has the distinct layout and style of the Springer Protocols program, designed specifically for readability and ease of reference in a laboratory setting.

The careful application of methods is potentially the most important step in the process of scientific inquiry. In the past, new methodologies led the way in developing new disciplines in the biological and medical sciences. For example, Physiology emerged out of Anatomy in the nineteenth century by harnessing new methods based on the newly discovered phenomenon of electricity. Nowadays, the relationships between disciplines and methods are more complex. Methods are now widely shared between disciplines and research areas. New developments in electronic publishing make it possible for scientists that encounter new methods to quickly find sources of information electronically. The design of individual volumes and chapters in this series takes this new access technology into account. Springer Protocols makes it possible to download single protocols separately. In addition, Springer makes its print-on-demand technology available globally. A print copy can therefore be acquired quickly and for a competitive price anywhere in the world.

Saskatoon, SK, USA *Wolfgang Walz*

Preface to the Series

Preface

Neuronal ensembles are groups of neurons with coordinated activity that underly perception, movement, memory, and cognitive states. Recent technological advances have allowed the recording and manipulation of hundreds of neurons opening the possibility to investigate the causal relationship between neuronal activity and complex behaviors. However, the standardization of methods to identify, characterize, and manipulate neuronal ensembles is still lacking.

The understanding of functional neuronal microcircuits requires the combination of methodological tools to close the bridge between transcriptomics, biophysics, network analysis, and dynamical systems.

The scope of this volume is to provide conceptual intuition and practical knowledge underlying experimental and analytical methods that have been used to study neuronal ensembles and their role in brain processes, so the readers can choose and compare different protocols to find the best pipeline to describe their own data.

This book is targeted to researchers interested in delving into neuronal population dynamics. Chapter 1 presents a comprehensive approach to implement multisensory experiments including the basics of head fixation, behavioral considerations, and analysis of laminar electrophysiology. Chapters 2 and 3 further extend the conceptual understanding of dimensionality reduction algorithms and network analysis applied to high-density electrophysiological recordings.

Additionally, the book contains analytical methods that have been used for neuronal ensemble analysis of optical recordings, starting with a protocol for fast segmentation of two-photon calcium imaging data (Chap. 4) and followed by dimensionality reduction methods to identify neuronal ensembles from large imaging population recordings where the points in multidimensional space are represented by population vectors (Chap. 5) or single neurons (Chap. 6). Chapters 7 and 8 leverage the use of graph theory to detect groups of neurons in optical recordings with high levels of coactivity. Chapter 9 introduces the use of regression and probabilistic graphical models to characterize the changes in information processing of neuronal ensembles from longitudinal calcium imaging recordings. In Chap. 10, considerations to compare neuronal ensemble analyses between control and pathological animals are discussed.

The implementation of tagging and manipulation techniques using optogenetic and chemogenetic approaches to study fear engrams is presented in Chap. 11.

Finally, Chap. 12 describes a potential method to characterize molecular signatures of neuronal ensembles using spatially defined transcriptomics.

Assessing the similarities, advantages, and disadvantages between different methodological approaches could lead to the design of interventional experiments aiming to reprogram brain functions in pathological conditions.

Lastly, I would like to voice my deepest gratitude to all the authors and colleagues for their contributions to the book and their comments. I hope that this volume will be a useful starting point for future generations of neuroscientists.

Juriquilla, Mexico *Luis Carrillo-Reid*

Contents

Contributors

MASAKAZU AGETSUMA • *Quantum Regenerative and Biomedical Engineering Team, Institute for Quantum Life Science, National Institutes for Quantum Science and Technology (QST), Chiba, Japan; Division of Homeostatic Development, National Institute for Physiological Sciences, Okazaki, Aichi, Japan; Japan Science and Technology Agency, Kawaguchi, Saitama, Japan; SANKEN (The Institute of Scientific and Industrial Research), Osaka University, IbarakiOsaka, Japan*

JOSE BARGAS • *División de Neurociencias, Instituto de Fisiología Celular, Universidad Nacional Autónoma de México, Mexico City, Mexico*

MARCO BRONDI • *Optical Approaches to Brain Function Laboratory, Istituto Italiano di Tecnologia, Genoa, Italy; Veneto Institute of Molecular Medicine (VIMM), Padova, Italy; Neuroscience Institute CNR, Padova, Italy*

LUIS CARRILLO-REID • *Instituto de Neurobiología, Universidad Nacional Autónoma de Mé xico, Juriquilla, Mexico*

RODRIGO COFRE • *Paris-Saclay Institute of Neuroscience (NeuroPSI), CNRS, Paris-Saclay University, Saclay, France*

SEBASTIANO CURRELI • *Optical Approaches to Brain Function Laboratory, Istituto Italiano di Tecnologia, Genoa, Italy*

ERIC DIJKEMA • *Cognitive and Systems Neuroscience Group, Swammerdam Institute for Life Sciences, University of Amsterdam, Amsterdam, The Netherlands*

MARIA JOSE ESCOBAR • *Departamento de Electronica, Universidad Técnica Federico Santa María, Valparaíso, Chile*

TOMMASO FELLIN • *Optical Approaches to Brain Function Laboratory, Istituto Italiano di Tecnologia, Genoa, Italy*

PAUL W. FRANKLAND • *Program in Neurosciences & Mental Health, Hospital for Sick Children, Toronto, ON, Canada; Institute of Medical Sciences, University of Toronto, Toronto, ON, Canada; Department of Psychology, University of Toronto, Toronto, ON, Canada; Child & Brain Development Program, Canadian Institute for Advanced Research, Toronto, ON, Canada*

GEOFFREY J. GOODHILL • *Departments of Developmental Biology and Neuroscience, Washington University School of Medicine, St. Louis, MO, USA*

JORDAN P. HAMM • *Neuroscience Institute, Georgia State University, Atlanta, GA, USA; Department of Psychiatry and Neuroscience Institute, New York University Grossman School of Medicine, New York, NY, USA; Emotional Brain Institute, Nathan S. Kline Institute for Psychiatric Research, Orangeburg, NY, USA*

RUBÉN HERZOG-AMUNÁTEGUI • *Institute du Cerveau, Paris, France*

SHEENA A. JOSSELYN • *Program in Neurosciences & Mental Health, Hospital for Sick Children, Toronto, ON, Canada; Institute of Medical Sciences, University of Toronto, Toronto, ON, Canada; Department of Psychology, University of Toronto, Toronto, ON, Canada*

YINQING LI • *School of Pharmaceutical Sciences, IDG/McGovern Institute for Brain Research, Center for Synthetic and Systems Biology, Tsinghua University, Beijing, P. R. China; McGovern Institute for Brain Research and the Department of Brain and Cognitive Sciences, Massachusetts Institute of Technology, Cambridge, MA, USA*

Violeta Gisselle Lopez-Huerta • *Institute of Cellular Physiology, National Autonomous University of Mexico, Mexico City, Mexico*

Gabriel Makdah • *Institut de Neurosciences des Systèmes (INS) – INSERM UMR1106, Aix-Marseille Université, Marseille, France; CHU Henri Mondor, Assistance Publique – Hôpitaux de Paris (AP-HP), Créteil, France*

Jean-Gabriel Minonzio • *Escuela de Ingenieria Informatica, Universidad de Valparaíso, Valparaíso, Chile*

Jan Mölter • *Department of Mathematics, School of Computation, Information and Technology, Technical University of Munich, Garching bei München, Germany*

Luca Montelisciani • *Cognitive and Systems Neuroscience Group, Swammerdam Institute for Life Sciences, University of Amsterdam, Amsterdam, The Netherlands*

Soraya Mora • *Centro Interdisciplinario de Neurociencia de Valparaíso, Universidad de Valparaíso, Valparaíso, Chile*

Umberto Olcese • *Cognitive and Systems Neuroscience Group, Swammerdam Institute for Life Sciences, University of Amsterdam, Amsterdam, The Netherlands*

Adrián G. Palacios • *Centro Interdisciplinario de Neurociencia de Valparaíso, Universidad de Valparaíso, Valparaíso, Chile; Instituto de Sistemas Complejos de Valparaíso, Valparaíso, Chile*

Marco N. Pompili • *Institut de Neurosciences des Systèmes (INS) – INSERM UMR1106, Aix-Marseille Université, Marseille, France*

Garance Prada • *Aix-Marseille Université, Marseille, France*

Asim J. Rashid • *Program in Neurosciences & Mental Health, Hospital for Sick Children, Toronto, ON, Canada; Institute of Medical Sciences, University of Toronto, Toronto, ON, Canada; Department of Psychology, University of Toronto, Toronto, ON, Canada*

Sebastián A. Romano • *Instituto de Investigación en Biomedicina de Buenos Aires (IBioBA)-CONICET-Partner Institute of the Max Planck Society, Buenos Aires, Argentina*

Issei Sato • *Department of Computer Science, Graduate School of Information Science and Technology, The University of Tokyo, Bunkyo-ku, Tokyo, Japan*

Miguel Serrano-Reyes • *División de Neurociencias, Instituto de Fisiología Celular, Universidad Nacional Autónoma de México, Mexico City, Mexico; Departamento de Ingeniería en Sistemas Biomédicos, Centro de Ingeniería Avanzada, Facultad de Ingeniería, Universidad Nacional Autónoma de México, Mexico City, Mexico; Licenciatura en Neurociencias, Facultad de Medicina, Universidad Nacional Autónoma de México, Mexico City, Mexico*

Luca Sità • *Optical Approaches to Brain Function Laboratory, Istituto Italiano di Tecnologia, Genoa, Italy*

Germán Sumbre • *Institut de biologie de l'Ecole normale supérieure (IBENS), Ecole normale supérieure, CNRS, INSERM, PSL Université Paris, Paris, France*

Yasuhiro R. Tanaka • *Brain Science Institute, Tamagawa University, Machida, Tokyo, Japan*

Ricardo Velazquez-Contreras • *Instituto de Neurobiología, Universidad Nacional Autónoma de México, Juriquilla, Mexico*

Sidney I. Wiener • *Centre Interdisciplinaire de Recherche en Biologie (CIRB) – CNRS UMR 7241 – INSERM U1050, Collège de France, Université PSL, Paris, France*

Marie-Theres Wittmann • *Program in Neurosciences & Mental Health, Hospital for Sick Children, Toronto, ON, Canada*

Chapter 1

Single-Neuron and Population Methods to Study the Circuit-Level Cortical Mechanisms of Multisensory Processing

Luca Montelisciani, Eric Dijkema, and Umberto Olcese

Abstract

In the past decade, the study of the circuit-level mechanisms of cortical multisensory processing has made great progress, thanks to the development of methods to record and manipulate the activity of neuronal ensembles in head-fixed rodent models, in which the controlled delivery of sensory stimuli can be associated to behavioral tasks. However, compared to unisensory processing, the field is still in its infancy, and our understanding of how multiple sensory modalities are integrated and compared in the neocortex keeps being disrupted by novel findings. In this context, the emergence of tools to record and analyze the activity of large populations of neurons localized across multiple brain regions has emerged as a key strategy to link multivariate patterns of spiking activity to sensory stimuli and the associated behavioral responses. This chapter aims to provide a manual for researchers aiming to apply these techniques. Specifically, we provide an overview of the main experimental steps required to study multisensory processing in head-fixed mice with laminar electrophysiology. Furthermore, we discuss in depth the main preprocessing steps and the most relevant techniques used to analyze how single neurons and neuronal populations process sensory information.

Key words Multisensory processing, Electrophysiology, Ensemble activity, Population activity

1 Introduction

Our perception is intrinsically multisensory. When we drink our morning coffee, for example, we do not only taste it but also smell it, see it, and feel the warmth of the cup in our hands. Furthermore, flavor perception itself is also modulated by the other senses, for example, the color of what we eat or drink modifies the taste that we perceive [1, 2]. Many cognitive functions are also strongly influenced by multisensory factors. For instance, performance in a visual task is affected by auditory stimuli: a low-volume white noise enhances visual discrimination performance [3], while strong sounds typically impair visual detection [4]. Analogous effects are

Luis Carrillo-Reid (ed.), *Identification, Characterization, and Manipulation of Neuronal Ensembles*, Neuromethods, vol. 215, https://doi.org/10.1007/978-1-0716-4208-5_1,

observed for visual influences in auditory tasks [5]. In spite of the pervasiveness of multisensory processing, it is therefore surprising that the study of the underlying neuron-level mechanisms in the cortex is still in its infancy, compared to the study of unisensory processing [6]. In fact, most of our knowledge about the way in which neurons encode multisensory information historically either came from studies performed in the superior colliculus [7] or—for cortical processing—was largely limited to the study of cue integration [8, 9]. This is surprising for two reasons. First, in mammals, most cognitive functions influenced by—or based upon—multisensory factors reside in areas other than the superior colliculus [10]. Second, multisensory processing entails more than just cue integration and also includes—among other factors—the segregation and prioritization of sensory stimuli [6, 11–14].

Nevertheless, in the past decade, the study of the circuit-level mechanisms of cortical multisensory processing has made great advances thanks to (1) the development of strategies to train rodents, in particular mice, for various types of multisensory tasks [6, 15], and (2) the availability of methods to record and manipulate neuronal activity with unprecedented precision across areas [16]. Such tools, corroborated by transgenic animal models, include optogenetics, two-photon calcium imaging, and, crucially, high-channel-count silicon probes. This latter approach, in particular following the advent of Neuropixels probes [17], allows to record the activity of up to hundreds of neurons located in multiple areas, across both superficial and deep regions, and with a temporal resolution capable of resolving the shape of single action potentials. In the context of multisensory processing, these are essential features. First, multisensory processing necessarily involves multiple areas and, for what pertains to cortical regions, a complex interplay between different layers [4, 18, 19]. Second, neuron-level interactions between signals related to stimuli coming from different sensory modalities occur at timescales that preclude the use of techniques with slower temporal resolutions. For example, interactions between visual and auditory or tactile signals significantly change based on whether the stimulus onset asynchronicity of the different stimuli varies by only 50 ms [18, 20]; the major implications of the fast timescales involved in multisensory processing are exemplified by the renown double flash illusion [21]. Third, it has recently been shown that not all neural responses underlie multisensory processing. For example, auditory responses in the primary visual cortex are actually, for a large fraction, corollary discharges evoked by stereotypical, involuntary motor reactions to auditory stimuli [22, 23]. In order to disentangle these components, it is essential to have recording techniques with high temporal resolution, so that the various sources of neuronal activity—that can differ by only tens of milliseconds [23]—can be disambiguated. Finally, while multisensory processing in the superior colliculus has traditionally been studied in terms of single-neuron responses [7], cortical multisensory processing (including cue integration) is a phenomenon thought to involve population-level dynamics [9, 24, 25], for whose investigation large-scale recordings are necessary.

These factors also apply to other fields of study (as is the case, for example, for what pertains the influence of spontaneous motor activity on the activity of cortical areas [26] or on visual processing and decision making [27, 28]) but are especially relevant for multisensory processing. In particular, several methods for the analysis of single-neuron and population-level activity have become very commonly applied to silicon probe recordings in the context of studying cortical multisensory processing, as they allow to both capture ongoing dynamics and disentangle distinct sources of activity. This chapter will therefore focus on providing an overview of such analytical methods. First, we will discuss the material and methodological requirements to study multisensory processing in head-fixed mice. This approach has become especially relevant because it allows to combine controlled delivery of sensory stimuli, monitoring of behavioral responses and internal state (e.g., arousal), and simultaneous application of multi-area recordings and optogenetics [13, 27]. A proper implementation of head-fixed recordings is essential to obtain high-quality data, as is the application of proper preprocessing techniques such as spike sorting. Spike sorting is essential to assign individual action potentials to single neurons, which is a prerequisite to study the neurophysiology of sensory processing. For this reason, a section of the chapter will be devoted to this topic. We will then focus on single-neuron methods for the analysis of sensory-evoked responses: classical peristimulus time histograms (PSTHs) and more advanced approaches to decode experimental variables from single-neuron activity or study neuronal encoding (e.g., generalized linear models). These approaches are essential to go beyond the level of firing rates and disentangle the various components embedded in spike trains. Finally, we will focus on population-level methods, for what pertains both dimensionality reduction approaches [29] as well as decoding methods. These techniques exploit the high-dimensional nature of large-scale recordings to uncover low-dimensional latent variables jointly encoded by neuronal ensembles [30]. Overall, this chapter will hopefully provide a useful resource for studying the circuit-level mechanisms of multisensory processing.

2 Materials and Experimental Paradigm

The following section is a step-by-step guide to perform an in vivo multisensory research paradigm in mice. It may function as a reference for those who wish to perform similar experiments or as a template for in vivo systems neuroscience approaches, in general. Methods are described in chronological order of execution, encompassing all general steps from animal habituation to electrophysiological recordings. For applied examples of these methods, the reader is referred to Oude Lohuis et al. (2022) [13]. Due to their diverse nature and complexity, neuronal manipulations such as chemo- and optogenetics are beyond the scope of this chapter.

2.1 Habituation

Mice should be at least 8 weeks of age before starting any kind of behavioral or surgical procedure that does not concern developmental phenomena. This ensures that all sensory systems are mature and that any subsequent procedure is not affected by remaining body growth.

The first step of any experimental protocol should be habituation: animals need to get comfortable with being picked up and manipulated by experimenters. It is advisable to do so in a gradual manner, via which each animal is exposed to a gradually increasing extent of handling each day over a period of 1–2 weeks. Habituation can start with a researcher simply placing their hand in the home cage and letting mice get accustomed. Due to their exploratory nature, mice will approach the hand and sniff it, which is quickly followed by a retraction and return to their nest. Animals may be lured to approach by offering a sunflower seed or drop of milk. After 2–3 days, the handler will observe a change in behavior and animals will not hide from their hand anymore. The next step involves picking up the mice. Here, our preferred method is handling them with the use of a tube. Handling tubes are placed in each cage, so that animals will consider it a part of their environment and a safe place to hide. Often, tubes of red plastic are used, since they are translucent to the handler but not the animal, which sees the tube as a secluded and safe shelter. Once a mouse enters the tube, it may be removed from the cage. Simply lifting the tube to a distance where mice can still comfortably leave it (approximately 5 cm) to return to the cage is a good first step. Subsequent habituation steps include placing the tube on the arm of a handler, all the while keeping their hand in the cage so mice may traverse the arm to return to the home cage. Mice may be considered "habituated" once they can be placed on the arm or hands of a handler for a minute without showing signs of distress such as hyperactivity, aggression, or jumping.

2.2 Headbars for Head Restriction and Electrophysiology

In most (multi-)sensory research, controlling stimulus delivery is of prime importance [15]. This is achieved by head-fixation of subjects and requires the implantation of a so-called headbar. These headbars will almost always include a recording chamber—an opening that allows access to the skull without requiring any additional surgery. Headbars come in many shapes and sizes depending on the

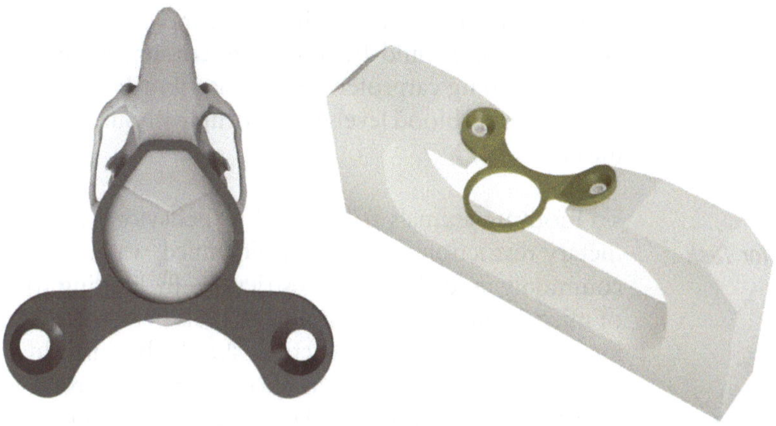

Fig. 1 Headbar example. Left: Rendering of a headbar superimposed on a mouse skull. This specific design allows simultaneous implantation of probes into posterior and anterior cortical areas. Right: Rendering of a similar headbar as it attaches to its complementary holder. "Ears" function as screw holes. Exposed skull is covered with a silicon elastomer and metal plate to prevent damage

animal species and experimental requirements such as area(s) of interest (see, e.g., Fig. 1).

Headbar surgery is performed under anesthesia and analgesia. A preoperative intraperitoneal injection of carprofen (10 mg/kg)—or another analgesic—is administered, before anesthesia is induced, typically with 3–5% of isoflurane. The animal is then moved to a nose clamp to ensure stability, where anesthesia is maintained with 1–1.5% isoflurane. Other anesthetics can of course be used. When deep anesthesia is achieved, confirmed by the absence of a pain reflex to a toe-pinch, an incision is made on the scalp using a tweezer and surgical scissors. Then, the periosteum is separated from the skull using a cotton swab. When a large headbar is implanted, it may be necessary to detach muscles such as pseudotemporalis profundus from the skull using a scalpel. When the skull's surface has been cleaned and dried, the headbar is attached to the skull using a glue (either regular super glue or glue for dental implants), after which a permanent protective layer is applied, usually in the form of dental cement. The bond between the skull and headbar will have to be able to withstand significant mechanical strain, so proper adhesion is crucial. Detachment of a headbar is considered a humane endpoint and thus requires the immediate sacrifice of the animal. It is therefore very important to ensure that the entire skull-implant interface is filled with cement. Following implantation, a small layer of cyanoacrylate glue is applied to the exposed skull, which is crucial to preserve the bone and prevent infection. The recording chamber is then sealed with a silicon elastomer and covered with a metal plate to avoid scratching. Subjects should be allowed to recover for at least 72 h after surgery, and ad libitum water, food pellets, and soft food should be available.

During this time, animals should be monitored for welfare issues, and analgesia should be administered appropriately. This can be done by adding carprofen to the water bottle, as it has been shown to keep steady blood levels of the drug without the need for invasive injections [31].

2.3 Dietary Restrictions for Task Performance

To motivate animals to respond to behavioral trials, some type of dietary restriction is necessary. For mice, water restriction is more commonly used than food restriction. By limiting the water intake of mice, both the total number of trials performed and the hit rate increase [32]. Water restriction should be introduced gradually, with the availability of hydration decreasing from ad libitum to approximately 1 mL per day for a typical adult mouse (with a minimum of 0,025 mL/g of body weight [33]) over the course of a week. At 1 mL per day, body weight is generally stable without animals showing signs of distress or discomfort. A convenient alternative to liquid water, whose administration is hard to control, is hydrogels. Hydrogels are materials that can contain up to 99% of water by volume, and gels for rodent consumption are commercially available. During the initial stages of water restriction. Hydrogel can be used to habituate animals to water restriction, supplement water during days with no training, or when animals did not collect their minimum daily hydration during training (approximately 1 mL). When animals are not being trained (i.e., during the weekend), an alternative to giving hydrogel is citric acid water. By adding 2–3% citric acid to water bottles, mice self-regulate their water intake without impacting behavioral performance [34].

2.4 Audiovisual Change Detection Task

There exists a rich variety of rodent behavioral tasks, ranging from freely moving to head-restricted, detection- to memory-based, spanning all sensory modalities. The scope of this section is confined to an audiovisual change detection task that we developed for mice [13, 14, 23], but the described principles will apply to many other tasks as well [6, 15]. Mice are trained on an audiovisual change detection task, where a change in one sensory domain is paired with licking a left lick spout and changes in the other domain to the right side. Before training, animals are habituated to head restriction in a gradual manner, increasing the duration of head restriction over the period of approximately 5 days. During the habituation phase, no stimuli are presented but licking the spout does result in the delivery of a small reward, familiarizing animals with collecting their daily hydration during head restriction. While restricted, the body of the animal is either in a plastic tube to align them to the screen or alternatively they can be placed on a floating Styrofoam ball to allow a certain degree of mobility. The latter approach is gaining popularity, especially in virtual reality experiments, where the movement of the ball is tracked and used to navigate virtual environments [35–37]. After habituation, animals learn two separate unimodal change detection tasks before combining them into the multimodal task. By doing so, we strive to teach the task in small and easily learnable increments. Only when they perform well on the hardest version of a unimodal task do they progress to the next modality. The initial stages of training involve passive reward delivery paired with each stimulus change. When animals have learned to lick after a large fraction of the stimulus changes (~80%), typically after 2–3 days, rewards are only delivered following a correct lick. The d-prime (d'), a measure adapted from signal detection theory, is used as a metric for behavioral performance [38]. Animals progress between training stages when some criterium is met (e.g., d' > 0.5 for 3 consecutive sessions), in addition to exhibiting a sigmoid psychometric curve to confirm they are in fact responding to stimuli as a function of saliency (Fig. 2). Generally, mice will learn auditory and tactile tasks quickly and take longer to learn visual tasks.

Whenever a behavioral task involves multiple stimulus modalities and choices, possible biases toward specific modalities need to be accounted for. During multimodal training, a preference toward one of the stimulus sensory modalities may appear, resulting in the animal predominantly reacting to one modality. It is crucial to

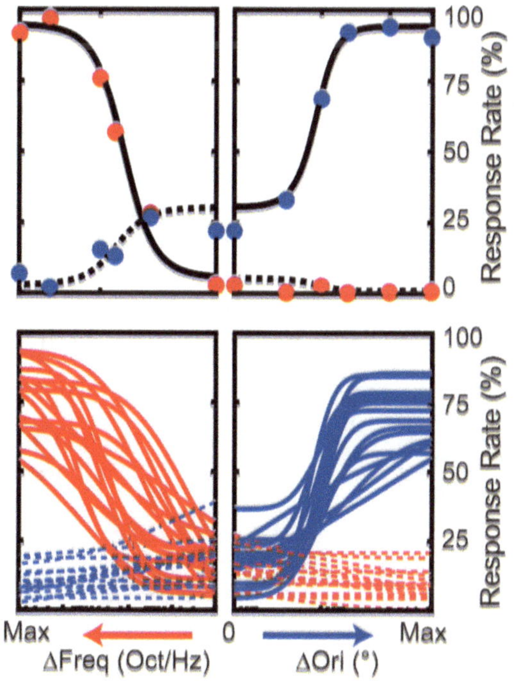

Fig. 2 Audiovisual change detection task psychometrics. Top panel shows response rates (dots) and model fit (line) for an example session. Bottom panel shows average psychometrics for multiple mice. (Adapted with permission from Ref. [13])

address such imbalances, for example, by changing stimulus saliency, lick detector distance, reward size, or the number of trials in a session for each modality. Additionally, individual subjects differ in their performance on each modality, so some type of normalization is necessary to compare between animals. We do so by determining a threshold stimulus, that is the stimulus at which 50% of trials are responded to correctly. During the experiment, each animal is presented with its respective threshold stimulus and a maximum saliency stimulus, e.g., a 90° orientation change of a drifting grating.

Analyses of behavioral data related to multisensory tasks also need to take biases into account. This is aided by the inclusion of probe trials into the task, i.e., trials where no stimulus is presented (Fig. 3). By comparing the animals' choices between stimulus- and probe trial, a better assessment of task performance can be made than simply looking at hit rates. The previously mentioned d' is a convenient measure of an observer's sensitivity to a signal but is limited to binary yes/no decisions in its original form. However, a modified version that allows the analyses multivariate decision tasks has been proposed [39].

2.5 Experimental Setup

This section will discuss several aspects related to experimental setups used in electrophysiological recordings in behaving rodents. Properly engineering and maintaining experimental setups is crucial to collecting high-quality, repeatable data. This is especially important because failure of a single device may result in the invalidation of the entire experimental session. For example, in the case of in vivo electrophysiology in behaving animals, failure to register the timestamps of stimuli on the same clock as the probe signal renders the data practically useless.

When it comes to choice detection, head-restricted rodent experiments often use lick detectors; animals lick to give their response to a trial, which is detected by a specialized device. Two common mechanisms used for lick detection are capacitive and

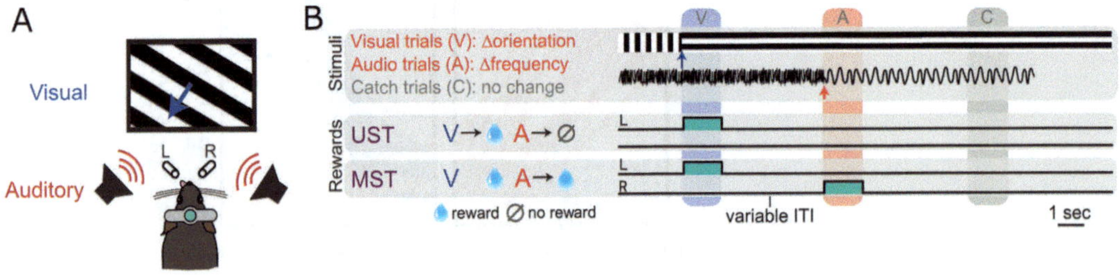

Fig. 3 Task setup. (**a**): Schematic overview of an audiovisual change detection task setup. Mice observe a drifting grating and auditory tone and are trained to detect changes in either one or both modalities. (**b**): Trial structure of visual, auditory, and catch trials. In visual trials, licks to, e.g., the left side are rewarded, whereas auditory trials are rewarded for right licks. During catch trials, no stimulus change is presented and animals are expected not to lick. Unisensory trained animals (UST) are only rewarded for one modality. (Adapted with permission from Ref. [13])

piezoelectric sensing. Capacitive detectors measure the capacitance of a conductive material, in this case, a lickspout. Whenever another conductive medium (e.g., a mouse) touches the spout, the capacitance changes and is registered as a lick. Capacitive lick detectors are robust and easy to use but not suitable for electrophysiology experiments, because each time the spout is licked, a tiny current runs through the animal which can be picked up as an electrical artifact. For this reason, sensors leveraging the piezo-electric effect are often used during experiments—lick spouts are attached to piezo-electric discs that generate a voltage whenever they are deformed. However, these devices are much more fragile and sensitive and thus inconvenient for daily training use.

A primary concern of all electrophysiologists is the elimination of electromagnetic interference from sources external to the brain. To isolate the subject from noise, experiments are usually performed inside a faraday cage. It is also of utmost importance to keep a proper ground tree, meaning all ground wires used in the setup lead to a single junction, instead of forming loops. Small currents can start to run in ground loops and introduce significant noise. Loaded power lines and AC-powered devices can cause noise several orders of magnitude larger than the signals of interest, which are typically in the micro- to millivolt range. These sources can easily be recognized by their 50/60 Hz waveform. After eliminating the 50/60 Hz noise component, residual noise sources may be isolated by carefully covering devices with a grounded material or placing them further away from the electrodes. Some devices, such as transformers and automatic switching devices (i.e., containing relays) can cause particularly large noise, and their use should be avoided if possible. In case digitization of neural signals does not happen at the headstage, it may also prove worthwhile to cover headstages in grounded aluminum foil and to use coaxed cables to connect to the acquisition system.

Another source of signal artifacts is animal motion. Motion artifacts are generated by a combination of electrical, magnetic, and ionic mechanisms [40], but all find their origin in the relative movement of the implanted probe with respect to the brain. For this type of noise, as with others, the best cure is prevention. Primarily, headbars should be securely attached to the holder so that body movement does not translate into head movement. Critical factors related to the probe are its leeway and the flexibility of the holder. Any movement of the probe may be disastrous to the integrity of the brain, as well as the probe itself. Ideally, both the animal platform and probe stand are attached to the same platform, for example, through magnetic clamps. Only very rigid materials should be used as a probe stand, so that touching probe stands doesn't result in movement. Relative movement between probes and the brain, so-called drift, may also be corrected post hoc during spike sorting. This method is discussed in more detail in this chapter's section on spike sorting.

2.6 Localization of Recording Sites: Mapping Retinotopy in Visual Cortex

A key aspect of multisensory research is to perform recordings in the correct brain regions to be studied. This, for large areas, can often be done via the use of stereotactic coordinates. However, when small regions need to be targeted or when subdivision of cortical areas (e.g., responding to a portion of the visual field) has to be reached, some form of functional mapping is required. For instance, to record cells that are tuned to visual stimuli, it is necessary to implant probes into the correct retinotopic location of V1. This can be guided by some type of retinotopic mapping, and in the context of this chapter, we will use this as a paradigmatic example of functional mapping. The most common options for retinotopic mapping are intrinsic optical signal (IOS, often also abbreviated as ISI) imaging or calcium imaging. Calcium imaging potentially provides a better spatiotemporal resolution and higher-quality mapping than IOS but requires the expression of genetically encoded calcium indicators, such as GCaMP6. IOS imaging utilizes the difference in red light absorption of blood as a function of oxygenation and is thus more widely applicable. For an extensive state-of-the-art retinotopic mapping method utilizing calcium imaging, the reader is referred to Zhuang et al. [41].

IOS imaging can be performed through the skull, without the use of an implanted cranial window. The procedure involves the careful thinning of the skull overlaying the area of interest, with the exception of animals young enough to image through the intact skull. The entire procedure is performed under anesthesia. When the bone is sufficiently thin, it becomes flexible to touch and translucent upon wetting. At this point, IOS imaging is possible. Stimuli equal to the experimental paradigm are presented, during which the cortex is illuminated by a red laser and frames are acquired by a camera. There are several factors that will determine the quality of the obtained signal:

- *Depth of anesthesia*—It is advisable to use low concentrations of anesthetics during imaging, e.g., 1% isoflurane. Deep anesthesia drastically reduces neural responses.
- *Smoothness of thinned skull*—If the thinned skull contains bumps or dents, light may scatter and introduce artifacts in the optical signal. Using larger drill bits can help to avoid this problem.
- *Body temperature*—It is pertinent to keep the animal at physiological body temperature, both for the well-being of the animal and the quality of the signal. A 1 °C deviation from homeostasis is enough to degrade the signal.
- *Camera focal point*—The camera must be focused on the tissue of interest. This is best achieved by using an animal platform with an adjustable z-axis. First, the platform is moved to a

position where the skull is in focus. Then, the platform should be moved up approximately 0.5–1 mm, depending on the thickness of the skull. This should place the focal plane of the camera within the cortex.

After acquisition, the hemodynamic response can be visualized by averaging each sequential frame over trials and then subtracting a no-stimulus baseline. A full description of an IOS experiment is beyond the scope of this chapter but is provided by Juavinett et al. [42]. When the area responsive to visual stimuli has been identified, a craniotomy is cut at that location using a scalpel. Animals should recover for at least 24 h before initiating awake experiments.

2.7 Laminar Electrophysiology

Once the area of interest has been identified, a common technique to measure in vivo neural activity is laminar-probe electrophysiology. Silicon probes can achieve very high spatial and temporal resolution recordings of both local field potential (LFP) and extracellular spiking activity. Especially with the advent of ultrahigh-density probes such as Neuropixels [43, 44] and SiNAPS [45] probes, hundreds of single units may be recorded simultaneously from both cortical and subcortical areas. In addition to single-shank probes, most manufacturers offer multi-shank alternatives for 2- or 3-dimensional recordings [44, 46, 47]. Such arrays or grids of electrodes offer the additional benefit of recording neurons at multiple closely spaced locations, which may offer insights into the mechanisms of intra-area neural processing.

Extracellular electrophysiology allows for an almost direct measurement of neural activity; the method does not rely on a secondary measure such as the blood-oxygen-level-dependent (BOLD) signal as is the case in, e.g., fMRI or functional ultrasound (fUS) imaging or calcium transients for calcium imaging. Transmembrane currents are directly measured as a potential difference, and thus temporal delays between neural events and recording are negligible. This is a unique benefit of electrophysiology, and no current methods offer comparable spatiotemporal performance [48].

The laminar structure of the mammalian cortex has striking anatomical and functional features, which can readily be explored with silicon probes. Electrode sites are generally spaced spanning vertical distance of microns to millimeters, enabling the concurrent positioning of electrodes in the cortex and underlying subcortical structures. Layer identities may be estimated post hoc using current-source density (CSD) and multiunit activity (MUA) analyses [49], where peak MUA activity is associated with cortical layer 5, and a range of CSD profiles predict several cortical layers. Besides physiological landmarks, exact probe trajectories can be determined postmortem by applying a fluorescent dye on probes during the final day of recording. One example of such a dye is DiI, a fluorescent lipophilic dye that diffuses into the lipid bilayers of the cell where it stably remains fluorescent [14]. These combined methods allow researchers to precisely determine the position of electrodes (and consequently single neurons) within the cortical column and study the intracortical mechanisms of multisensory processing.

One limitation of probe electrophysiology is its invasiveness. Probes need to be implanted inside of the cortex through small craniotomies. As with every surgical procedure, extra care is required after surgery. During the experimental period, animal well-being should be intensely monitored. Probes can be implanted chronically [50] or acutely inserted and removed in subsequent recording sessions [13, 27]. In this case, each recording session will slightly damage the insertion area, especially with thicker probes. Multiple acute probe recordings can be done on different days, but inflammation and tissue damage will eventually forbid the continuation of experiments. These factors generally limit data collection to a period of approximately 5–10 days. Additionally, each acute recording session will contain spikes from a different set of neurons, as the probe is reinserted into at slightly different location. Some of the issues mentioned for acute probe recordings can be avoided by chronic implantation. Next generation probes (i.e., Neuropixels) have been successfully implanted for prolonged periods of time, in both rats and mice [50–52]. Chronic implantation eliminates the need for animals to be head-restricted during experiments, opening the door to experiments more closely resembling naturalistic behavior, tracking neurons over prolonged periods, whereas acute recording may only last for the duration of head restriction. Chronic implants have been shown to track the same neurons of days, allowing longitudinal studies to be performed at the single-neuron level [53].

Another drawback of using silicon probes in vivo is the limited available information about the recorded cells. There are currently no methods to univocally determine the identity of recorded neurons, such as in patch-clamp methods, where cells can be morphologically reconstructed. Most anatomical and physiological features of cells cannot readily be identified based purely on electrophysiological data, with the exception of their excitatory or inhibitory nature. A relatively common practice is the division of neurons based on their peak-to-trough delay, as inhibitory neurons have shorter waveforms than excitatory neurons [54–56]. Optogenetics can be used to elicit spiking activity from genetically identifiable neuronal populations, but this approach can be mostly applied to one neuronal subpopulation at a time [57].

Overall, the abovementioned procedures will enable to record ensemble activity (i.e., neuronal activity from multiple cells with single-neuron resolution) across the brain regions that, jointly, process activity coming from different sensory modalities.

3 Analysis of Neuronal Activity

After the previous description of experimental procedures, we now transition to methods for analyzing the neurophysiological data collected in the context of experiments meant to study multisensory processing. Specifically, we will focus on techniques enabling the study of topics such as the integration of visual and auditory cues and see how sound influences vision and vice versa [23], the influence of visual-tactile stimuli in decision-making tasks [58], and how multisensory stimuli are affected by animal behavior [22, 23]. Having collected a rich dataset that captures neural responses across different trials, modalities and time points, the next task is to distill this complexity into comprehensible and actionable insights. We will start with the preprocessing stage, specifically spike sorting. Next, we will discuss analysis methods at the level of single neurons. We will assess various methods such as raster plots, peri-stimulus time histograms (PSTH), tuning curves, receiver operating characteristic (ROC) curves, and generalized linear models (GLM). We will then conclude with population-level analyses, discussing the comparative advantages and disadvantages of two dimensionality reduction techniques and the use of population decoding techniques.

3.1 Preprocessing: Spike Sorting

Our initial focus in preprocessing the data is on spiking sorting, a set of procedures meant to isolate the activity of putative individual neuronal units, and thus enable the investigation of their responses to sensory stimuli [59]. This level of resolution is indispensable for elucidating the neurophysiological mechanisms that govern sensory encoding and multisensory integration. Today, various spike sorting algorithms are at our disposal, including but not limited to Kilosort [59, 60], Klusta [61], SpyKING [62], and IronClust [63]. All these methods can be used and compared via the use of tools such as SpikeInterface [64], a Python package designed to streamline the process of spike sorting and electrophysiological data analysis.

Before being able to apply spike sorting, the raw data needs to be preprocessed to make the detection of action potentials easier. Two common preprocessing steps are common median reference and band-pass filtering [17]. Common median (or mean) reference involves subtracting the median (mean) voltage of all active channels from each individual channel's raw voltage data, which helps reduce common noise sources and improves signal-to-noise ratio. Next, band-pass filtering is used to remove noise and other unwanted signal components (e.g., LFP). A typical frequency range is between 600 and 6000 Hz, although variations exist.

Delving into spike sorting, there are two main approaches that are currently used: waveform-based and template-based methods.

In waveform-based methods, the first step typically involves identifying spike events, often determined by voltage thresholds. Next, their waveforms are aligned to ensure consistent representation. These waveforms are then transformed into a shared low-dimensional space, often through techniques like principal component analysis (PCA) [65]. This feature extraction step simplifies the data while preserving essential features, setting the stage for clustering. Algorithms in this category then usually employ clustering techniques such as Gaussian mixtures or peak-density detection to group similar waveforms into putative single neurons [66].

In contrast, template-based methods such as Kilosort [60] take a different approach. These methods focus on initially generating a set of templates, which serve as typical spike shapes against which detected spikes can be compared. After an initial round of spike detection, the detected spikes are matched against these templates. This template matching is particularly effective in resolving overlapping spikes, where multiple spikes occur almost simultaneously. Finally, the spikes are grouped into distinct clusters, each representing an individual neuron, based on their similarity to the templates.

Regardless of the chosen approach, automated spike sorting has its limitations. It can mistakenly detect noise as spikes, get confused by overlapping spikes from multiple neurons, and misclassify signals due to small shifts in the electrode's position during the experiment, known as "probe drift." For these reasons a manual curation step is necessary after the initial automated phase. This serves three key purposes: first, it combines spike groups that the automated process incorrectly separated, assuming they originated from different neurons; second, it separates spike groups that the automated process wrongly identified as originating from just one neuron; finally, it filters out artefactual and mixed signals, often referred to as "multiunit activity," which are actually undistinguishable action potentials from multiple neurons (Fig. 4). Software tools—such as Phy [67]—have been developed to facilitate this process. During manual curation, operators initiate the process by scrutinizing neurons' spiking autocorrelogram, specifically focusing on a 2-ms window within the refractory period—the time following an action potential when the neuron cannot fire again—to ensure it is free of contamination. If there are no spurious spikes in the refractory period, one can then select the most similar cluster for comparison. Next, waveform shapes, amplitudes, and interspike interval distributions are examined and compared to spot cluster similarities. If two clusters appear very similar and originate from neighboring (or overlapping) recording channels, the clusters are merged. Finally, if the merged cluster shows consistent correlogram patterns, waveforms, amplitudes, and interspike intervals, it is classified as a "good" single unit; otherwise, it is labeled as a "multiunit" activity. Manual curation becomes particularly important

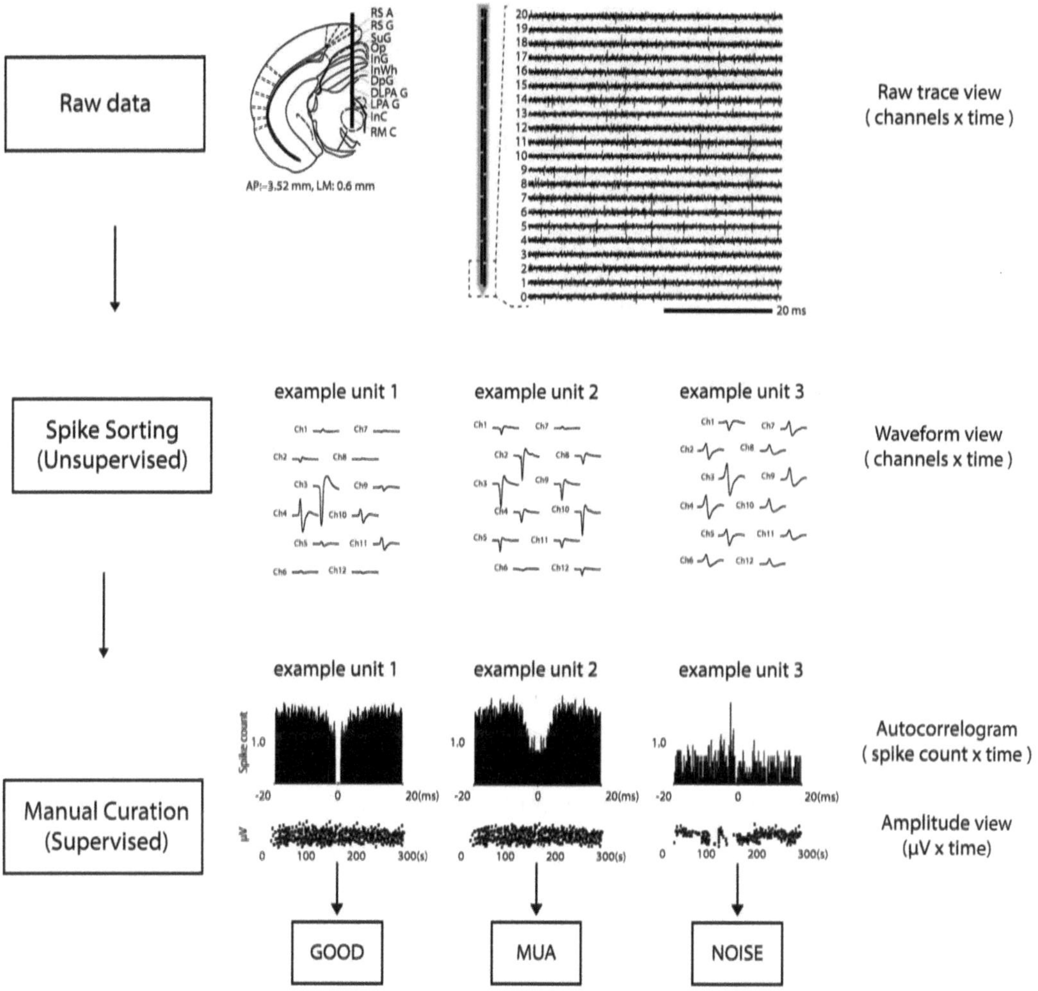

Fig. 4 Spike sorting pipeline: from data collection to manual curation. *Raw data*: This panel provides a schematic representation of the probe track reconstruction within the mouse brain. The targeted areas were situated at −3.52 mm anterior–posterior (AP) and 0.5 mm lateral–medial (LM) axis relative to bregma. Displayed on the right side are the filtered voltage traces, which have undergone a 300 Hz high-pass and 9000 Hz low-pass filter, spanning 50 ms from 20 sample channels located at the probe's tip. *Unsupervised spike sorting*: Here, the output for three example units is showcased. This allows for the visualization of unit activity at each channel by plotting the recorded waveforms. *Manual curation:* This final step involves a detailed examination of the three example neurons from the previous section. Using patterns observed in the autocorrelogram and amplitude, these neurons can be categorized as follows: "good" indicating single unit activity, "MUA" representing multiunit activity, or "noise." (Adapted with permission from Ref. [52])

when dealing with recordings presenting electrode drift. In these scenarios, the waveforms produced by a specific neuron change over time and could be incorrectly allocated to multiple clusters. While new preprocessing software, like Kilosort 3, performs drift corrections, older software lacks this operation [60]. Thus, the curation step helps to ensure these waveforms are correctly associated with their originating neuron.

While spike sorting algorithms and manual curation prepare the data for analysis, it is crucial to understand that these techniques have their own set of limitations.

Data originating from outdated laminar probes (with a low density of recording sites) or tetrodes are more susceptible to issues related to noise or drifting of signals over time. While modern software like SpikeInterface helps select the most appropriate algorithm for each probe, spike sorting algorithms still need to improve in fully addressing these challenges. Moreover, while often effective, manual curation can be both subjective and time-consuming, and its feasibility may be questioned when dealing with vast datasets, for instance, with data coming from neuropixels probes.

In conclusion, spike sorting is a pivotal step in the data analysis pipeline for neural recordings. Carefully processing and curating the data can provide a more precise and accurate picture of the neural activity underlying behavior and cognition. Although limitations exist, ongoing developments in algorithms and analysis methods are helping to address these challenges and further improve the quality and reliability of neural data.

3.2 Analysis of Single Neuron Evoked Responses

The brain's process of combining information from various sensory modalities to fully understand our environment relies on neural encoding and decoding mechanisms.

Encoding allows neurons to transform real-world sensory information from multiple sources into a form the brain can comprehend and use. Decoding, instead, is the process of interpreting these signals, enabling the brain to perceive a coherent, multisensory world representation based solely on the activity of neurons [68, 69].

To understand these processes, analyzing single-neuron activity is often the first required. This is in line with the Sherringtonian view that individual neurons serve as the foundational units for neural function, each contributing uniquely to the brain's overall ability to process and integrate information from multiple sensory modalities—but see [70].

To this end, several techniques for the visualization and quantification of neuronal responses come into play, facilitating both analysis and visualization. Among these techniques are raster plots, which display the temporal patterns of spikes in response to stimuli. Peri-stimulus time histograms (PSTHs) offer a quantified representation of neural responses over time intervals. Tuning curves map a neuron's firing rate in response to varying stimulus parameters. Receiver operating characteristic (ROC) curves are used to assess the discrimination ability of the neurons for different stimuli. Finally, generalized linear models (GLMs) allow for the statistical analysis of relationships between neural activity and external factors like stimuli or behavior. All these methods collectively provide a comprehensive framework for investigating the single-neuron underpinnings of information encoding and processing.

3.2.1 Raster Plots

A raster plot is arguably the simplest and most intuitive method to examine a single neuron's firing rate over time, allowing for visual inspection of the trial-by-trial variability of the time-locked responses to sensory stimuli.

Constructing a raster plot begins by capturing when spikes, or action potentials, occur. Next, these spike times are aligned to the times of particular events of interest, such as a sensory stimulus being presented. Once this alignment process is complete, time is plotted on the x-axis, and the trial number is on the y-axis. Each action potential is represented with a small vertical line or a dot at the corresponding time it occurred (Fig. 5a). As a result, the raster plot provides a clear, intuitive depiction of a neuron's activity across multiple trials relative to a given event, such as the presentation of a sensory stimulus. This visual representation allows researchers to easily identify patterns, response latencies, and variabilities in neuronal responses.

3.2.2 Peristimulus Time Histograms (PSTHs)

PSTHs are commonly used to quantify how a neuron's activity changes over time in response to a specific event or stimulus. This involves counting spikes in defined time intervals before, during, and after the stimulus (or event) onset. The PSTH's foundation is the raster plot, from which a PSTH is basically obtained by aggregating spikes within specific time bins across trial. This provides an average temporal profile of neuronal activity around the stimulus.

Drawing from the raster plot to construct a PSTH, the first step involves selecting an appropriate bin size. This decision is often based on the balance between resolution and statistical reliability. Large bins, like 50–100 ms, blur finer temporal patterns, while excessively narrow bins, like 1–5 ms, are intrinsically noisy. The choice also hinges on the distinction between rate and temporal codes. Broader bins might be sufficient for capturing overall firing rates without detailing precise spike timing. However, if discerning the exact timing of spikes is paramount, smaller bins in the range of 20–50 ms are preferable.

The spike data is then binned. Nonoverlapping binning divides the time range into equally sized sections. Instead, the sliding window method shifts a window incrementally across the time range, frequently overlapping with the previous window to capture subtler temporal nuances.

Once the data has been binned, the next step is smoothing. This process helps reduce noise and accentuate underlying patterns. Common smoothing techniques include Gaussian smoothing and moving average filters. However, it's important to balance smoothing with the preservation of key temporal details, like precise spike timing, that are critical for accurate data interpretation. Afterwards, normalization to spikes per second is always needed so that the PSTH can be used for comparisons across neurons, conditions, and experiments. Finally, z-scoring is often applied as it provides a more

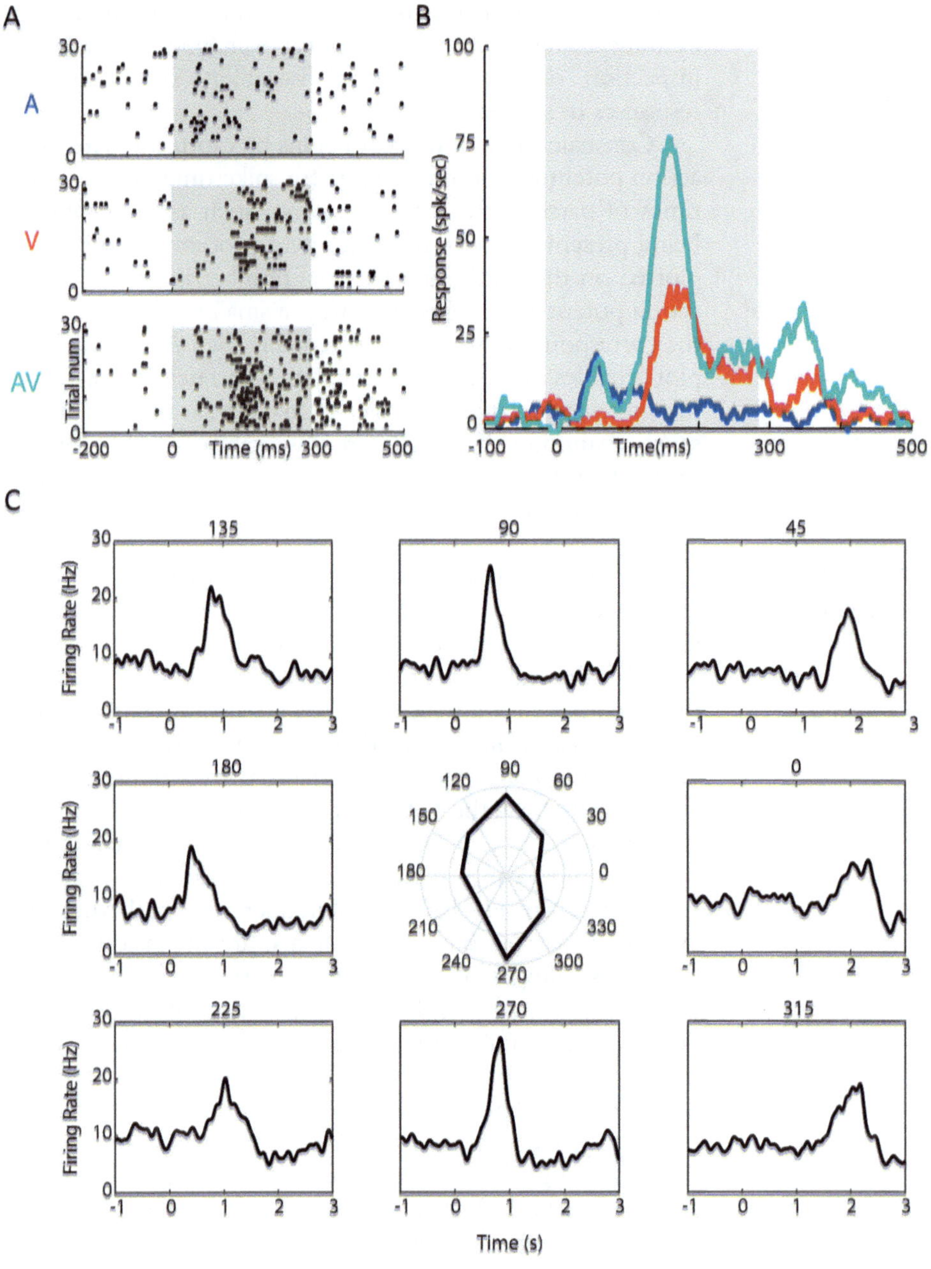

Fig. 5 Raster plot, PSTH, and orientation tuning after audiovisual stimulus presentation: (**a**): Raster plots of auditory (A, blue) visual (V, red) and audiovisual (AV, green) trials of an example neuron exhibiting multisensory enhancement. (**b**): PSTH of the same neuron shown in panel A demonstrating enhanced firing in response to audiovisual stimuli compared with unimodal stimuli (same colors as in panel A). (**c**): PSTH and polar plot demonstrating the orientation tuning and magnitude of response (Hz) of example neuron in V1 of mouse. (Adapted with permission from Refs. [71, 72])

standardized comparison of activity levels throughout the analyzed time window and neurons (Fig. 5b).

In multisensory analysis, PSTHs are useful tools not only to visualize but also to quantify specific features of neuronal activity. Metrics such as peak firing rate and latency to peak can be used to understand how neurons integrate inputs from multiple sensory modalities. For instance, a higher peak firing rate when auditory and visual stimuli are presented together compared to unisensory presentations indicates multisensory enhancement. On the other hand, a decreased peak firing rate indicates response depression. Besides, calculating the area under the PSTH curve provides the neuron's total activity in response to multisensory stimuli, further aiding in identifying enhancement or depression effects [73]. By thoroughly interpreting and quantifying these patterns, PSTHs evolve from mere visual representations to tools that enable preliminary measurements. These insights shed light on neuronal responses' timing, consistency, and magnitude, showing how neurons encode and differentiate information.

3.2.3 Tuning Curves

While PSTHs provide insights into a neuron's response to a single stimulus, tuning curves offer a broader perspective by comparing neuronal responses across multiple stimuli. In constructing a tuning curve, researchers present a series of stimuli during the experiment, such as gratings at different orientations (e.g., 0 to 180 degrees) or pure tones at varying frequencies (e.g., 5–80 kHz) [74]. A neuron is then considered "tuned" when the average response against varying feature values of the same stimulus class is significantly higher than others.

To analyze and construct the tuning curve analysis, a PSTH is first computed for each stimulus. Metrics such as mean firing rate or peak response are then calculated, separately for each stimulus/condition: the former by averaging spikes per bin for each stimulus and the latter by identifying the highest spike count usually during the stimulus presentation window. Polar plots or similar visualizations are often used to display how response magnitudes vary as a function of stimulus features (Fig. 5c). Successively, the stimulus value that induces the peak response from this plot is deemed the neuron's preferred stimulus, indicating the characteristic the neuron is most receptive to. Researchers also often fit a mathematical function, e.g., a Gaussian, cosine, or von Mises curve, to the observed data points. Such curve fitting enhances the estimation of a neuron's tuning but also equips scientists with the ability to predict neuronal responses to untested stimulus values within the tested range. Finally, statistical analyses are indispensable to validate that the identified tuning traits aren't merely random variances. Given the multiple stimuli conditions, an ANOVA is often employed to compare the neuron's response across different orientations or frequencies against its baseline firing rate. Circular statistics are often employed in the visual domain.

Even though PSTHs and tuning curves provide valuable insights into neural activity, they come with limitations. For instance, they fail to capture the variability in neural responses across different trials or within a single trial. This limitation is particularly important in multisensory research, as it can mask interactions between different sensory inputs that may only become apparent in individual trials. Furthermore, these methods can miss the mixed effects in experimental data. For instance, movement artifacts have been observed to alter neural recordings when an animal moves during sound presentation [22, 23]. Physiological changes such as variations in heart rate or pupil size can also introduce noise into neural activity. Because PSTHs and tuning curves rely on averaged measures, these factors can be masked in the aggregation across trials, leading to potentially inaccurate or incomplete interpretations of neural behavior. For these reasons, techniques such as receiver operating characteristics (ROC) curves and generalized linear models (GLM) are often used to supplement raster plots, PSTHs, and tuning curves.

3.2.4 Receiver Operating Characteristic (ROC)

Receiver operating characteristic (ROC) analysis is a powerful tool for studying neuronal responses, particularly during multisensory integration experiments. ROC analysis can quantify a neuron's discriminative capacity, specifically its ability to classify or distinguish between sensory inputs correctly [13]. For instance, considering a neuron responding to both visual and auditory stimuli, while a PSTH might show the neuron's average firing rate in response to each stimulus, ROC analysis would assess at which point in time the neuron differentiates between the two stimuli [13].

ROC analysis requires several stages. After data collection, researchers gather raw data, focusing on the neurons' responses during each trial. Such spike trains are converted into continuous traces by computing single-trial PSTHs. After, setting the decision threshold is crucial. The "threshold" is a cutoff value used for classifying instances—based on their predicted probabilities or scores—as either positive or negative. Initially, this threshold is set at the maximum predicted score, meaning all instances are classified as negative. ROC analysis then proceeds by varying this threshold. By doing so, one can assess a model's performance at different levels of sensitivity and specificity. Sensitivity measures the rate of correctly identified positive instances while specificity measures the rate of correctly identified negative instances. Adjusting the threshold helps understand the trade-offs between these two metrics, optimizing a model's overall performance. After setting the various decision thresholds, the next step is to calculate key metrics at each threshold level: the true positive rate (TPR) and the false positive rate (FPR). The TPR, also known as sensitivity, is the proportion of actual positive cases that are correctly identified by the model. On the other hand, the FPR is the proportion of actual negative cases that are incorrectly classified as positive by the model. With these rates, the visual representation of the ROC analysis, the ROC curve, takes form. In the graph, the FPR is on the x-axis, and the TPR is on the y-axis. Each coordinate on this curve represents a pair of FPR and TPR values associated with distinct thresholds. The ideal ROC curve tends to lean toward the upper-left corner of the graph, symbolizing a high degree of sensitivity and a minuscule rate of false positives (Fig. 6a).

The ROC curve is summarized into a singular metric: the area under the curve (AUC). An AUC of 1 represents perfect discrimination, while 0.5 indicates random guessing. To assess the significance of the AUC against random guessing, corresponding p-values are calculated [13]. A low p-value (typically below 0.05) would imply that the observed AUC is statistically significant and not due to random chance (Fig. 6a).

To visualize the time-resolved discriminative ability of neurons, one can plot the fraction of neurons significantly discriminating (encoding) between two conditions over time. For example, we can observe the time at which neurons in the V1 area encode

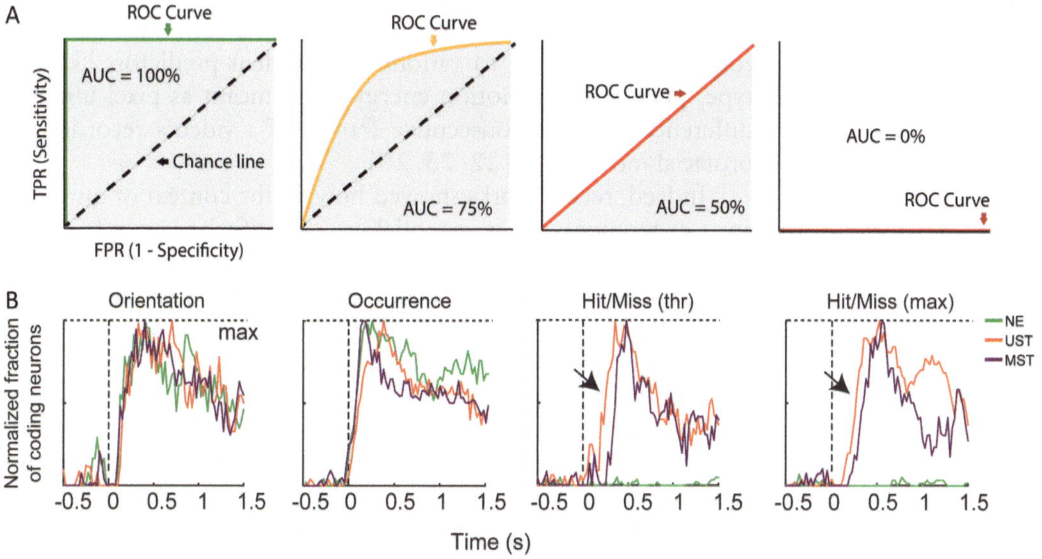

Fig. 6 ROC curve schematic and results: (**a**): ROC curves and area under the curve (AUC) for different discrimination levels. (**b**): We used a single-neuron decoding analysis (ROC) to compute the fraction of neurons (summed over all recordings) coding for task-relevant variables over time. The coding of change occurrence and orientation (visual features) showed similar dynamics across cohorts, whereas hit/miss coding (visual hits vs misses) was only present in UST and MST mice and started earlier in UST than MST mice (highlighted with black arrows). (Adapted with permission from Ref. [13])

different stimulus and task features during an audiovisual change detection task (Fig. 6b). Through ROC analysis, we can pinpoint the specific time points where neurons differentiate between "hit" and "miss" trials, thereby revealing when they are encoding task-relevant feature (Fig. 6b).

ROC analysis thus provides a quantifiable metric for the efficiency of neural encoding across different sensory inputs or conditions. However, ROC analysis has some limitations.

First, its design primarily suits binary classifiers, evaluating two distinct classes such as "response" and "no response." In the multisensory integration context, however, stimuli responses can span more than two categories [14].

Additionally, ROC assumes an equal distribution between the two classes it evaluates. Since neural datasets sometimes conform to this assumption, the AUC might paint an overly positive picture in cases lacking this balance. A skewed dataset might lead to biased classification, falsely elevating perceived accuracy [75]. However, this limitation can be addressed by subsampling one of the two classes to match the size of the other, thereby reducing the potential for biased outcomes.

3.2.5 Generalized Linear Models (GLMs)

While ROC analysis is used for decoding, primarily to assess how well a neuron can distinguish between two conditions, GLMs are encoding models that explore how sensory input, task conditions, and behavioral responses shape neuronal spiking activity [13, 23].

Essentially, a GLM links a dependent variable, such as a neuron's spiking activity, to various independent predictors like trial type, pupil size, or motion energy (here meant as pixel intensity difference between consecutive frames of a video's recording of orofacial movements [22, 23, 26].

Indeed, recent works showed how, in the context of multisensory experiments, it is essential to quantify how much of the observed neural variability is accounted for by the animal's movement, thereby isolating the neural responses that are truly attributable to the sensory inputs [22, 23]. This makes GLMs very useful for multifactorial neuronal activity analysis.

GLMs rely on probability distributions to capture the relationships in neural data. A probability distribution helps to describe how the outcome variable, such as spike count, is distributed across different conditions or time points. By selecting appropriate probability distributions, GLMs can be tailored to various research objectives and data types. These distributions can be Gaussian for continuous data, binomial for binary data, or Poisson for count data like spike-train data. Specifically, when working with spike-train data the Poisson GLM is often preferred, as it accurately models expected spike counts within short time intervals.

After choosing the model type, it is fundamental to choose an appropriate link function. This transformative element bridges nonlinear associations between the dependent variable and predictors into a linearizable format. The specific choice of a link function is informed by the nature of the data and the relationship under study. In the context of Poisson GLMs for spike-train data, the log link function is typically used, as it models the relationship between the expected log count (or rate) and the predictors:

$$\log \lambda_i = \beta_0 + \beta x_i$$

Here:

- $\log\lambda_i$ is the natural logarithm of the expected count or rate for the *i*th observation.
- β_0 is the intercept, representing the expected log count or rate when the predictor is zero.
- β is the coefficient quantifying the change in the expected log count or rate for a one-unit increase in the predictor.
- x_i is the value of the predictor for the *i*th observation.

After selecting the appropriate link function, the next crucial step is to choose the right predictors, which are variables believed to influence neural responses. These predictors can vary widely depending on the research context. For example, in audiovisual experiments aimed at studying multisensory integration, potential predictors could include the timing and intensity of a visual stimulus, the frequency of an auditory stimulus, and behavioral metrics such as eye movements. The model includes these predictors to help account for the variability in neural firing rates and better understand the underlying neural mechanisms (Fig. 7b). Next, single-trial PSTHs for each unit across all trials are generated as this data serves as the foundation for fitting the full model.

The model is then fitted to the data through a process that includes cross-validation. Specifically, cross-validation is carried out over individual trials to fine-tune model parameters and ensure the model's predictive accuracy and robustness. Once the optimal model is identified, a shuffling process is applied to the predictors across various categories. This is done to account for any spurious relationships in the data. The validity of the shuffling is estimated by comparing the model's predictive performance before and after shuffling, helping to distinguish the relationships from those that may have arisen by chance.

Upon completing the analysis, it is important to visually represent the results to discern how our predictors elucidate the variances in firing rates. Typically, plotting how firing rates are explained by each predictor gives a precise view of neural activity patterns and their temporal relationship with the predictors (Fig. 7d). Bar plots can be used for visualization of the contribution

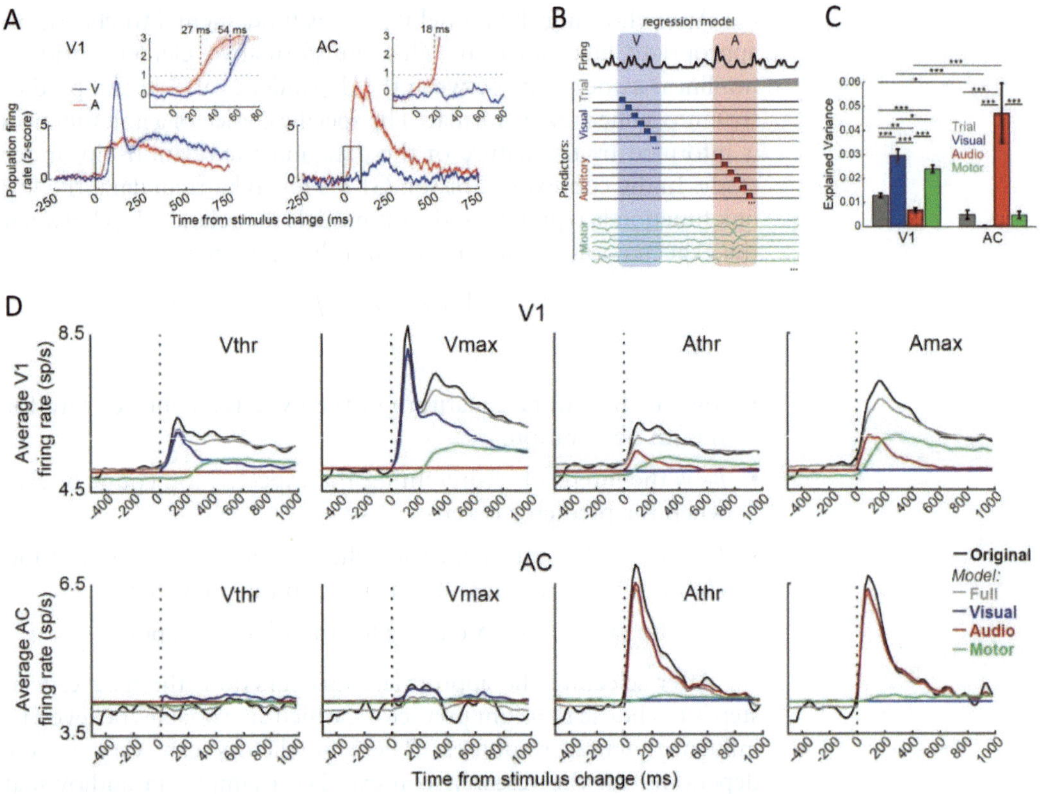

Fig. 7 GLM schematic and output from silicon probe recording in V1 and AC mouse during audiovisual stimulation: (**a**): Population firing rate after auditory and visual stimulus changes in V1 (left) and AC (right). (**b**): Schematic of the model. Single-neuron firing rate was predicted as a linear combination of four sets of predictors. A "trial" predictor captured rate fluctuation throughout the session. For visual (V) and auditory (A) stimuli, a set of separate predictors spanned 1 s after stimulus change per saliency level and feature (orientation and auditory frequency; shown only for one trial type) to capture neuron-specific response patterns. Finally, motor predictors were also included. (**c**): Explained variance during 0–200 ms of all trials for each category of predictors. Mean ± SEM across neurons. (**d**): The firing rate of single neurons was predicted using only subsets of predictors (visual, blue; auditory, red; motor, green) or all predictors (gray) and compared to the original firing rate (black). Shown is the firing rate during different trial types averaged across V1 neurons (upper row) or AC neurons (bottom row). (Adapted with permission from Ref. [23])

of individual predictors (Fig. 7c). Such visual aids not only facilitate a comprehensive understanding of the model's outputs but also provide a rigorous validation of the findings in a format that's both accessible and academically rigorous.

However, GLMs also have limitations. One key limitation is the assumption of a linear relationship between predictors, e.g., the intensity of sensory stimuli, and the response variable, usually neuronal activity. This assumption may not hold in multisensory settings where stimuli from different senses interact complexly to influence neuronal activity. If an incorrect link function is chosen, the model might offer misleading insights into how sensory inputs affect neuronal responses, potentially resulting in flawed conclusions or misdirected future research.

Furthermore, creating the right set of predictors can be challenging. In a multisensory context, these predictors might include factors such as individual trials, variations in stimuli, and related movements. Constructing an appropriate design matrix is crucial: missing important predictors can hide significant effects and reduce sensitivity, while adding unnecessary ones can introduce noise [76].

Lastly, GLMs are sensitive to outliers. Specifically, rare but extreme neuronal responses can disproportionately influence the model. If not addressed, this can make the model less generalizable, resulting in erroneous interpretations of what constitutes "typical" neuronal behavior in response to multisensory stimuli. While robust regression techniques can mitigate this issue, they also introduce additional complexities to the model's architecture, which might complicate its interpretation and application [76].

To conclude this section on the analysis of single-neuron activity, studying the firing properties of individual neurons is crucial for analyzing neural activity, particularly in multisensory experiments. While traditional methods like raster plots, PSTHs, and tuning curves offer basic understanding—they help elucidate the general patterns and primary characteristics of neural activity—they often simplify what neural responses actually reflect. These methods might not fully capture between trial dynamics and the interaction effects between stimuli from different sensory modalities. ROC analysis adds valuable insight by quantifying a neuron's ability to distinguish between different sensory stimuli, excelling in measuring the sensitivity of neuronal responses. However, in multisensory experiments, ROC may face challenges incorporating behavioral aspects like animal movement or licking. For instance, neural activity recorded during a multisensory trial may not solely be influenced by the sensory stimuli presented but also by the animal's behavior at that time [77].

On the other hand, GLMs can elucidate the relationship between a neuron's activity and multiple factors, such as trial type and behavior, making them well-suited for multisensory experiments where multiple inputs interact to influence neuronal responses. However, GLMs operate under certain assumptions, such as a linear relationship between predictors and response and independently distributed errors, which might only sometimes hold since they can also be sensitive to outliers, leading to poorly generalized models.

3.3 Analysis of Population Evoked Responses

After having discussed single-neuron analysis methods, we will now examine population-level dynamics. A philosophical divide distinguishes single-neuron analysis and population-level analysis. The former, rooted in reductionism, focuses on single neurons as the basic computational units of the brain. In contrast, population-level analysis embraces holism, focusing on the collective dynamics of interconnected neural networks and emergent properties [30, 70, 78]. While single-neuron research sheds light on cell-specific activity, population studies offer a broader perspective on how groups of neurons collectively process information. As we categorize the techniques employed in the latter analysis, two primary methodologies emerge, based on, respectively, unsupervised and supervised learning.

Unsupervised strategies, such as principal component analysis (PCA) and tensor component analysis (TCA), are typically dimensionality reduction techniques, and they explore and reveal latent variables within multisensory data, free from the boundaries of predetermined categories. In contrast, supervised techniques, particularly neural decoding, quantify how much information about sensory or behavioral events is embedded within neural firing patterns [14].

In an experiment in which animals must choose between two choices to obtain a reward, neurons may encode multiple aspects of the task, such as the choices presented, the animal's decision, and its confidence in that decision [79]. This results in a "mixed" representation of different task elements by the same set of neurons, making the data challenging to analyze. Dimensionality reduction techniques are instrumental here, systematically capturing most of the data's variance and elucidating how neural representations depend on various task parameters like stimuli, decisions, or rewards. Decoding techniques instead infer variables such as stimuli or behavioral states from observed neural activity. By training models on specific neural patterns aligned with known stimuli or behaviors, decoding can predict these factors from neural datasets. For instance, decoding neural activity in the visual cortex allows to quantify how well neuronal activity in this area can be used to predict the type of visual stimulus presented to an organism [71]. Similarly, decoding techniques have revolutionized brain-machine interface applications when applied to the motor cortex [80], facilitating direct communication between neural circuits and external machinery based on interpreted neural representations of intended actions.

3.3.1 Dimensionality Reduction Techniques

Principal component analysis (PCA) and tensor component analysis (TCA) stand out as techniques employed in neuroscience for dimensionality reduction and extraction of latent variables. PCA is probably the most commonly used tool for reducing the dimensionality of neural data, transforming it into a set of orthogonal axes that capture the most significant patterns. For instance, in a decision-making task with audiovisual stimuli, PCA not only de-noises the data but also isolates dimensions that capture the variance specific to movement and decisions [81]. TCA can be seen as a multidimensional extension of PCA [82]. For example,

TCA can uncover latent dimensions corresponding to multiple task-relevant components (e.g., sensory and motor components), operating simultaneously across neurons, trials, and multiple trial types [14].

3.3.2 Principal Component Analysis (PCA)

Principal component analysis (PCA) is a statistical method that identifies and generates the primary axes or dimensions along which the data varies the most. By doing so, it creates new variables, called principal components, which are linear combinations of the original features. These principal components effectively simplify the dataset's complexity while retaining as much of the original variance as possible. Besides, the two components are orthogonal, denoting their lack of correlation. PCA can, for example, be used to reveal subpopulations of neurons characterized by similar activity patterns [83].

Delving into the PCA process, single-trial PSTHs of individual neurons are collected for each trial and are standardized to have a mean of zero and a standard deviation of one. This ensures that each variable contributes equally to the PCA, regardless of its original scale. Next, we compute the covariance matrix for the standardized single-trials PSTH data. Each row and column in this matrix represent different neurons (or channels), and entries within the matrix illustrate the degree of covariance—or shared variability—between the spike counts of two different neurons across all trials. A high covariance value would suggest that two neurons often fire in a correlated manner when subjected—for instance—to specific sensory stimuli. PCA is achieved by computing the eigenvectors and eigenvalues of the covariance matrix. Eigenvectors of this covariance matrix represent the directions along which the neural data varies the most, and the eigenvalues are proportional to the magnitude of that variance. Together, they form the "principal components" that serve as the data's new vector basis. Sorting and selecting the principal components are the next steps. These eigenvectors are sorted by the magnitude of their eigenvalues in descending order. The eigenvector with the highest eigenvalue becomes the first principal component, capturing the most significant variance within the data. This sorted set of eigenvectors forms a new orthogonal basis set for our data. Transitioning to the application phase, we employ these sorted eigenvectors to perform an orthogonal transformation of the original, standardized data. This transformation projects the high-dimensional neural data into a new orthogonal set of variables, the principal components (Fig. 8). Each principal component explains a unique level of variance, representing different patterns and structures in neural encoding. The final step in PCA is to evaluate how much variance each principal component explains. This helps us decide which principal components are most relevant for our analysis. This

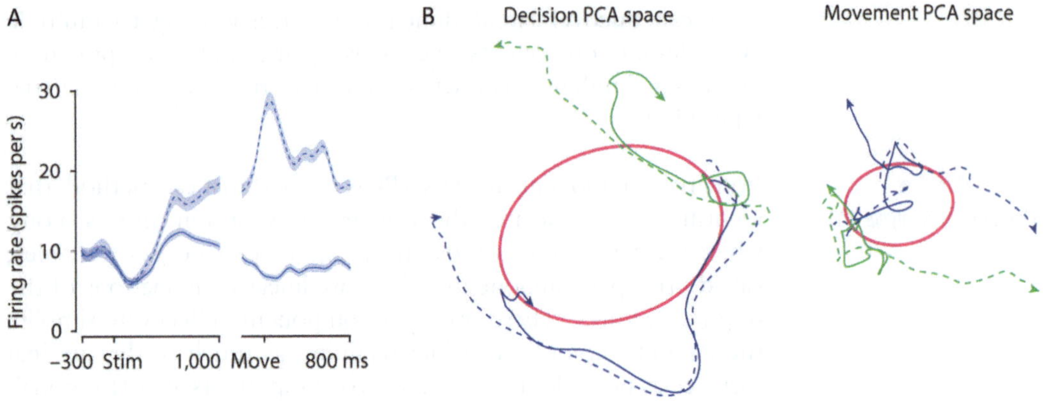

Fig. 8 Standard principal component analyses operations. (**a**): Time-varying firing rates of two example neurons during a stimulus-decision task. The first neuron exhibits a sustained preference for high-rate stimuli (represented by dashed lines) over low-rate ones (solid lines). The responses are aligned to stimulus onset on the left and movement onset on the right. This alignment is necessary due to the slight variation in time between the stimulus end and the animal's movement from trial to trial. Each subplot represents one neuron, with each line indicating one condition. The color codes the stimulus, while the solid/dashed line codes the binary decision. (**b**): Two-dimensional projections of decision-epoch data, consistent across all panels. These projections illustrate the variance in neuronal activity during the decision-making phase. Magenta ellipses indicate 1 standard deviation of the data when projected into the specified space. The spaces for these projections are chosen based on the principal components of the decision-epoch data. (Adapted with permission from Ref. [81])

selection process is based on the ration between the sum of the eigenvalues of the retained components and the sum of all eigenvalues. By focusing on the retained principal components, researchers can isolate the neuronal patterns associated with specific sensory inputs or behavioral outcomes, gaining a deeper understanding of the neural mechanisms at play in multisensory experiments.

PCA, while commonly employed in extracting dominant patterns from high-dimensional data, has certain inherent limitations. One of its primary constraints is its linearity. The method operates under the assumption that relationships among variables are linear. This can challenge capturing complex, nonlinear interactions often present in neural datasets [84]. Alternatives like Kernel PCA or t-distributed stochastic neighbor embedding (t-SNE) offer ways to capture nonlinear relationships and can be more suitable for such complex datasets.

Second, PCA's ranking of components by variance might only sometimes reflect the most biologically relevant features [85]. The method's sensitivity to data scale renders it vulnerable to outliers and noise, a notable concern in electrophysiology. Interpreting the biological significance of principal components, especially when they encapsulate a mix of neural sources, is often complex. Lastly, the orthogonality of principal components, a core assumption of

PCA, may sometimes align with the realities of biological processes. In multisensory research, essential cross-modal interactions could, for instance, be obscured if attention is limited to the dominant components.

3.3.3 Tensor Component Analysis (TCA)

Tensor component analysis (TCA) offers an alternative to traditional PCA for analyzing multidimensional neural data. Unlike PCA, which typically handles two-dimensional data matrices, TCA can manage higher-dimensional structures called tensors [82]. This enables TCA to capture patterns across dimensions like time, trials, and stimuli without requiring data preprocessing steps like concatenation. This feature makes TCA particularly useful in multisensory research. For example, in a study involving auditory and visual cues, TCA could analyze a four-way tensor encompassing time, neurons, trials, and additional factors such as pupil size, providing a more comprehensive view than PCA [14].

After collecting neural data across visual, auditory, and behavioral dimensions, we catalog each trial and time point to construct a foundational tensor (Fig. 9a). A tensor is a multidimensional generalization of a matrix, capable of representing complex data structures beyond just rows and columns. We then apply TCA to this tensor. The TCA method decomposes it into a core tensor and several factor matrices (Fig. 9b). The core tensor is a condensed representation capturing the original data's most significant relationships and variations. In contrast, the factor matrices offer granular insights into individual dimensions such as trials, sensory modalities, and time points. These matrices indicate the contribution of individual components identified by the TCA.

TCA allows us to identify specific latent dynamics associated, for example, with different sensory inputs. Unlike PCA, TCA retains the multidimensional nature of the data, making it easier to detect interactions among various sensory modalities, time points, and trials. For example, TCA can reveal how a group of neurons responds to visual stimuli and auditory cues and how these responses change over time. Visualization can map these complex interactions, providing a more holistic view of multisensory processing (Fig. 9c).

TCA provides a method for unsupervised analysis of latent dimensions underlying neural dynamics that is particularly useful in the context of multisensory integration. However, its application is not without limitations. A significant concern is the computational complexity associated with TCA. As the size and complexity of datasets increase, there is a corresponding rise in computational demands. This is inherent to TCA's capability to process multidimensional data; larger dimensions necessitate increased computational resources and time for effective analysis [86].

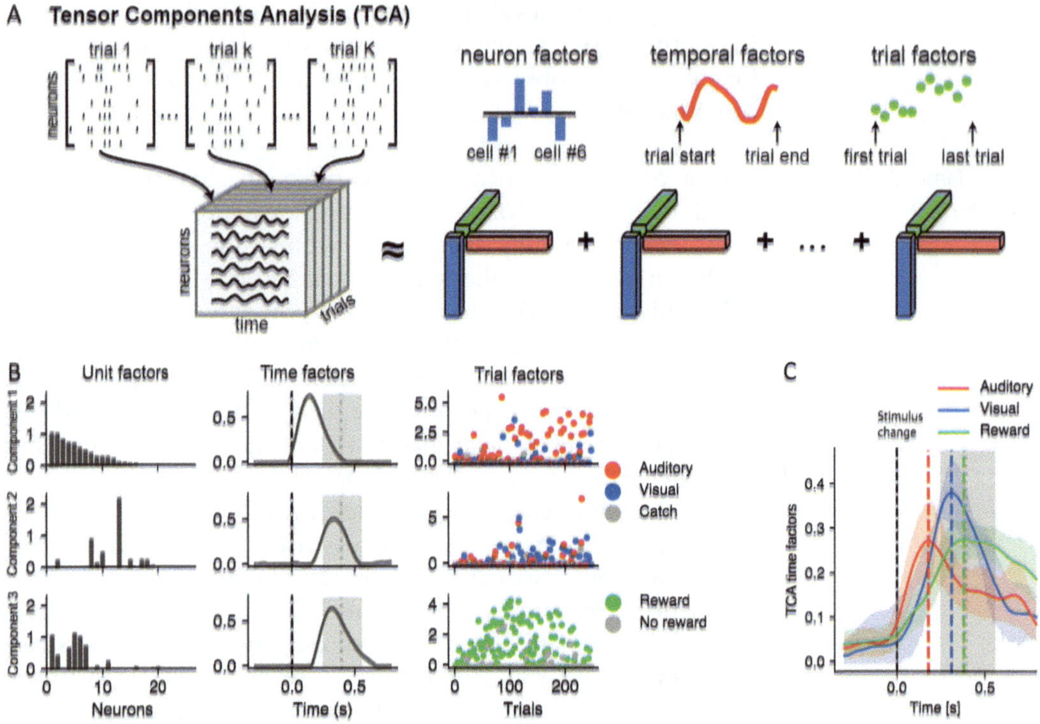

Fig. 9 TCA architecture and example of latent dimensions of parietal population activity correspond to task-relevant components. (**a**): TCA architecture developed by Williams et al. [82]. The dataset is structured into a third-order tensor with dimensions N × T × K. Utilizing TCA, the dataset is approximated by the summation of outer products derived from three orthogonal vectors. This methodology results in an ancillary set of low-dimensional factors, denoted as trial factors (represented in green), which systematically elucidate the temporal variations in neuronal activity across sequential trials. (**b**): Three selected components from a TCA analysis of an example recording session. Each row corresponds to a component (#1–3) that captures a subpopulation of neurons (unit factors, left column) that share a temporal response profile within the trial (temporal factors, central column), which is in turn expressed differently across trials (trial factors, right column). All factors are unitless. Neurons are ranked based on their contribution to the first plotted component, then the second. TCA is performed without information about trial types, colored a posteriori. Nonetheless, the identified latent components show selective expression in auditory (#1), visual (#2), and hit trials (#3). (**c**): Averaging the time factors of all selective components from different sessions showed a distinct temporal profile with a progression of first auditory, then visual, and last, decision components. (Adapted with permission from Refs. [14, 82])

Additionally, the interpretation of results derived from TCA presents challenges. The patterns identified by TCA, despite being multidimensional, may not always correspond directly to established or anticipated neural processes. This necessitates careful consideration when attempting to relate TCA-derived patterns to specific neural mechanisms [82].

3.3.4 Decoding

PCA and TCA are unsupervised methods, focusing on condensing and extracting dominant patterns from high-dimensional datasets without necessarily referring to specific outcomes. Decoding, on the other hand, adopts a supervised approach, aiming to predict specific experimental conditions or behavioral outcomes from patterns of neural activity. This distinction underscores a fundamental difference in goals: while unsupervised techniques aim for dimensionality reduction and feature extraction, supervised decoding is more outcome-oriented, making inferences about external variables based on neural activity.

Decoding methods come in various types, each with its own features. For example, support vector machines are used for classification and regression problems, excelling at finding the best dividing line or "hyperplane" that separates data into different categories [87]. Artificial neural networks are often applied to capturing complex, nonlinear dependencies across data sequences [88]. Finally, K-nearest neighbors is a supervised learning algorithm that classifies new data points by looking at their "k" closest neighbors in the training set and making a prediction based on the most common class or average value among these neighbors [87]. In this section, we will focus on another method: random forest decoding. One primary reason for this choice is its widespread use in neuroscience research since this approach offers high interpretability, can handle nonlinear interactions effectively, and is relatively robust to overfitting [89]. This technique is based on the use of multiple decision trees to make better predictions. A tree is a simple decision-making model that contributes its own "vote" on the outcome and, when combined with other trees, helps make the overall prediction more accurate and robust.

Delving into the process, initial steps such as data preprocessing and normalization are pivotal. Next, the random forest classifier is configured by constructing multiple decision trees. These trees work collectively, enhancing the predictive power and accuracy of the model. The complexity of this model is adjustable by tuning hyperparameters, such as the number of trees and their maximum depth. Finding an equilibrium is essential: while having more trees generally augments robustness, it simultaneously escalates computational requirements. Moreover, controlling the depth of the trees is vital to ensure our model neither underfits nor overfits the data.

In contrast with other decoding approaches, there is an ongoing debate in the field about the need to use cross-validation with random forest classifiers. Some researchers believe this is not needed, because the design of random forests naturally prevents overfitting, thanks to how it uses multiple decision trees. However, using cross-validation adds an extra check to ensure the model works well and is free from biases [90].

In the context of multisensory research, training the random forest involves mapping activity features, such as neural responses to varying sensory stimuli, to a designated target variable. Once the model is trained, it employs a "majority vote" system among the individual trees for classification tasks or aggregates them for regression tasks. Essentially, each tree offers its independent prediction, with the model's final decision stemming from aggregating these numerous insights. This method is especially adept at managing high-dimensional data (Fig. 10a).

This type of decoder is often used to determine how well a set of experimental conditions (e.g., sensory stimulus) can be decoded over time. In this case, independent decoders may be trained at each time bin after—for instance—stimulus onset, to indicate how well neuronal activity in a given area tracks experimental variables (Fig. 10b).

The random forest algorithm, like all techniques, has some drawbacks. For instance, while seminal work on the algorithm highlighted its robustness and versatility, a subtle reading reveals concerns about interpretability [90]. Another aspect to consider is the sensitivity to hyperparameter settings. In machine learning, the careful choice of hyperparameters can differentiate between a successful and an ineffective model [92]. Finally, some studies provide insights into the performance of decoding algorithms, including random forests, across varying data landscapes [93].

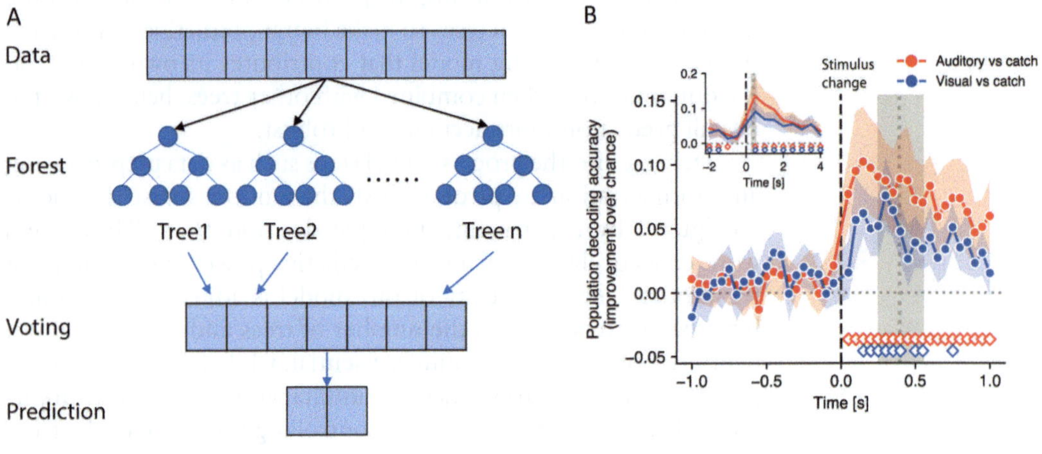

Fig. 10 Random forest architecture and example result: (**a**): Random forest architecture overview: the process begins with the dataset input, characterized by its features and observations. This leads to the construction of the decision trees. Each tree within this ensemble is crafted utilizing distinct subsets of data and features. As these trees process the data, they each arrive at their individual conclusions, culminating with the voting stage. After, this collective decision-making process ultimately conclude in the final prediction. (**b**): Accuracy of population decoding relative to chance plotted over time for random forest classifiers trained to discriminate visual versus catch trials (blue) and auditory versus catch trials (red). Trials were aligned to the time of stimulus change. Accuracy above chance was obtained by subtracting from the average decoding accuracy after repeatedly shuffling the trial labels. (Adapted with permission from Refs. [14, 91])

4 Conclusions

In this chapter, we presented an overview of the major experimental and analytical steps required to study multisensory processing in head-fixed behaving rodents, with a primary focus on silicon probe recordings. We especially paid attention to analysis methods aiming to understand how single neurons and neuronal populations (ensembles) process (multi)sensory information. Here we mainly focused on methods to analyze activity within brain regions. Several approaches are however emerging to analyze interactions between brain regions [94–96]. These will prove very valuable in the context of multisensory processing but are less established compared to the methodologies we surveyed here.

This focus stems from recent development of methods capable of simultaneously recording neuronal activity from hundreds of neurons located across cortical and subcortical regions, with action potential resolution [17, 44]. This remains currently unfeasible with other techniques, such as imaging [16], and is especially relevant for the multisensory field. In fact, multisensory processing has been observed in almost any brain region [6, 10]. However, recent studies have questioned whether all these supposedly multisensory signals actually have a different source (e.g., in terms of correlates of stereotypical motor responses to sensory stimuli) [22, 23]. Having the high-temporal resolution that only electrophysiology enables is essential to disambiguate true multisensory signals from other sources, using methods such as GLM regression models [23]. Nevertheless, all recording techniques can only yield intrinsically correlative insights about the link between brain activity and external factors. For this reason, integrating recording techniques with perturbational approaches such as optogenetics is essential to obtain mechanistic insights. This aspect goes beyond the scope of this chapter, but it must be pointed out that electrophysiology can be more easily integrated with optogenetics or other similar approaches, compared, for example, to two-photon imaging.

Compared to other, more established fields, multisensory processing is still in its infancy. While the basic mechanisms governing multisensory integration in subcortical structures such as the superior colliculus are well-known [7], much less is known about the cortex. In fact, not only the circuit-level mechanisms of cortical multisensory processing are still poorly understood [6, 9, 24], but there is even no consensus on which cortical areas actually process more than one modality [6, 10, 14, 22, 23, 97]. This chapter is meant as a guide for researchers willing to address this fascinating yet still relatively unexplored topic.

References

1. Spence C (2015) Multisensory flavor perception. Cell 161:24–35
2. Parr WV, Geoffrey White K, Heatherbell DA (2003) The nose knows: influence of colour on perception of wine aroma. J Wine Res 14:79–101
3. Gleiss S, Kayser C (2014) Acoustic noise improves visual perception and modulates occipital oscillatory states. J Cogn Neurosci 26:699–711
4. Iurilli G, Ghezzi D, Olcese U, Lassi G, Nazzaro C, Tonini R, Tucci V, Benfenati F, Medini P (2012) Sound-driven synaptic inhibition in primary visual cortex. Neuron 73:814–828
5. Chang S, Xu J, Zheng M, Keniston L, Zhou X, Zhang J, Yu L (2022) Integrating visual information into the auditory cortex promotes sound discrimination through choice-related multisensory integration. J Neurosci 42:8556–8568
6. Meijer GT, Mertens PEC, Pennartz CMA, Olcese U, Lansink CS (2019) The circuit architecture of cortical multisensory processing: distinct functions jointly operating within a common anatomical network. Prog Neurobiol 174:1–15
7. Stein BE, Stanford TR (2008) Multisensory integration: current issues from the perspective of the single neuron. Nat Rev Neurosci 9:255–266
8. Angelaki DE, Gu Y, DeAngelis GC (2009) Multisensory integration: psychophysics, neurophysiology, and computation. Curr Opin Neurobiol 19:452–458
9. van Atteveldt N, Murray MM, Thut G, Schroeder CE (2014) Multisensory integration: flexible use of general operations. Neuron 81:1240–1253
10. Ghazanfar AA, Schroeder CE (2006) Is neocortex essentially multisensory? Trends Cogn Sci 10:278–285
11. Wimmer RD, Schmitt LI, Davidson TJ, Nakajima M, Deisseroth K, Halassa MM (2015) Thalamic control of sensory selection in divided attention. Nature 526:705–709
12. Lee CCY, Diamond ME, Arabzadeh E (2016) Sensory prioritization in rats: behavioral performance and neuronal correlates. J Neurosci 36:3243–3253
13. Oude Lohuis MN, Pie JL, Marchesi P, Montijn JS, de Kock CPJ, Pennartz CMA, Olcese U (2022) Multisensory task demands temporally extend the causal requirement for visual cortex in perception. Nat Commun 13:2864
14. Oude Lohuis MN, Marchesi P, Pennartz CMA, Olcese U (2022) Functional (ir)relevance of posterior parietal cortex during audiovisual change detection. J Neurosci 42:5229
15. Carandini M, Churchland AK (2013) Probing perceptual decisions in rodents. Nat Neurosci 16:824–831
16. Machado TA, Kauvar IV, Deisseroth K (2022) Multiregion neuronal activity: the forest and the trees. Nat Rev Neurosci 23:683–704
17. Jun JJ, Steinmetz NA, Siegle JH et al (2017) Fully integrated silicon probes for high-density recording of neural activity. Nature 551:232–236
18. Olcese U, Iurilli G, Medini P (2013) Cellular and synaptic architecture of multisensory integration in the mouse neocortex. Neuron 79:579–593
19. Deneux T, Harrell ER, Kempf A, Ceballo S, Filipchuk A, Bathellier B (2019) Context-dependent signaling of coincident auditory and visual events in primary visual cortex. elife 8. https://doi.org/10.7554/eLife.44006
20. Lippert MT, Takagaki K, Kayser C, Ohl FW (2013) Asymmetric multisensory interactions of visual and somatosensory responses in a region of the rat parietal cortex. PLoS One 8:e63631
21. Shams L, Kamitani Y, Shimojo S (2000) Illusions. What you see is what you hear. Nature 408:788
22. Bimbard C, Sit TPH, Lebedeva A, Reddy CB, Harris KD, Carandini M (2023) Behavioral origin of sound-evoked activity in mouse visual cortex. Nat Neurosci 26:251–258
23. Oude Lohuis MN, Marchesi P, Olcese U, Pennartz CM (2024) Triple dissociation of visual, auditory and motor processing in mouse primary visual cortex. Nat Neurosci 27(4):758–771
24. Fetsch CR, DeAngelis GC, Angelaki DE (2013) Bridging the gap between theories of sensory cue integration and the physiology of multisensory neurons. Nat Rev Neurosci 14:429–442
25. Pouget A, Deneve S, Duhamel J-R (2002) A computational perspective on the neural basis of multisensory spatial representations. Nat Rev Neurosci 3:741–747
26. Stringer C, Pachitariu M, Steinmetz N, Reddy CB, Carandini M, Harris KD (2019) Spontaneous behaviors drive multidimensional, brain-wide activity. Science 364:255
27. Steinmetz NA, Zatka-Haas P, Carandini M, Harris KD (2019) Distributed coding of

choice, action and engagement across the mouse brain. Nature 576:266–273

28. Zatka-Haas P, Steinmetz NA, Carandini M, Harris KD (2021) Sensory coding and the causal impact of mouse cortex in a visual decision. elife 10:e63163
29. Cunningham JP, Yu BM (2014) Dimensionality reduction for large-scale neural recordings. Nat Neurosci 17:1500–1509
30. Saxena S, Cunningham JP (2019) Towards the neural population doctrine. Curr Opin Neurobiol 55:103–111
31. Glasenapp A, Bankstahl JP, Bähre H, Glage S, Bankstahl M (2023) Subcutaneous and orally self-administered high-dose carprofen in male and female mice: pharmacokinetics, tolerability and impact on cage-side pain indicators. bioRxiv, 2023–06
32. Guo ZV, Hires SA, Li N et al (2014) Procedures for behavioral experiments in head-fixed mice. PLoS One 9. https://doi.org/10.1371/journal.pone.0088678
33. Tucci V, Hardy A, Nolan PM (2006) A comparison of physiological and behavioural parameters in C57BL/6J mice undergoing food or water restriction regimes. Behav Brain Res 173:22–29
34. Urai AE, Aguillon-Rodriguez V, Laranjeira IC, Cazettes F, Laboratory TIB, Mainen ZF, Churchland AK (2021) Citric acid water as an alternative to water restriction for high-yield mouse behavior. eNeuro 8:ENEURO.0230. https://doi.org/10.1523/ENEURO.0230-20.2020
35. Pinto L, Koay SA, Engelhard B, Yoon AM, Deverett B, Thiberge SY, Witten IB, Tank DW, Brody CD (2018) An accumulation-of-evidence task using visual pulses for mice navigating in virtual reality. Front Behav Neurosci 12:36
36. Dombeck DA, Harvey CD, Tian L, Looger LL, Tank DW (2010) Functional imaging of hippocampal place cells at cellular resolution during virtual navigation. Nat Neurosci 13:1433–1440
37. Keller GB, Bonhoeffer T, Hübener M (2012) Sensorimotor mismatch signals in primary visual cortex of the behaving mouse. Neuron 74:809–815
38. Green DM, Swets JA (1966) Signal detection theory and psychophysics. Wiley, New York
39. Sridharan D, Steinmetz NA, Moore T, Knudsen EI (2014) Distinguishing bias from sensitivity effects in multialternative detection tasks. J Vis 14:1–32
40. Nicolai EN, Michelson NJ, Settell ML, Hara SA, Trevathan JK, Asp AJ, Stocking KC, Lujan JL, Kozai TDY, Ludwig KA (2018) Design choices for next-generation Neurotechnology can impact motion artifact in electrophysiological and fast-scan cyclic voltammetry measurements. Micromachines (Basel) 9:494
41. Zhuang J, Ng L, Williams D, Valley M, Li Y, Garrett M, Waters J (2017) An extended retinotopic map of mouse cortex. elife 6:e18372
42. Juavinett AL, Nauhaus I, Garrett ME, Zhuang J, Callaway EM (2017) Automated identification of mouse visual areas with intrinsic signal imaging. Nat Protoc 12:32–43
43. Steinmetz NA, Koch C, Harris KD, Carandini M (2018) Challenges and opportunities for large-scale electrophysiology with Neuropixels probes. Curr Opin Neurobiol 50:92–100
44. Steinmetz NA, Aydin C, Lebedeva A et al (2021) Neuropixels 2.0: a miniaturized high-density probe for stable, long-term brain recordings. Science 372:eabf4588
45. Angotzi GN, Boi F, Lecomte A, Miele E, Malerba M, Zucca S, Casile A, Berdondini L (2019) SiNAPS: an implantable active pixel sensor CMOS-probe for simultaneous large-scale neural recordings. Biosens Bioelectron 126:355–364
46. Neural Probes | Cambridge NeuroTech (2023). https://www.cambridgeneurotech.com/neural-probes. Accessed 25 Jul 2023
47. Electrode Arrays – NeuroNexus (2023). https://neuronexus.com/products/electrode-arrays/. Accessed 25 Jul 2023
48. Hong G, Lieber CM (2019) Novel electrode technologies for neural recordings. Nat Rev Neurosci 20:330–345
49. Senzai Y, Fernandez-Ruiz A, Buzsáki G (2019) Layer-specific physiological features and Interlaminar interactions in the primary visual cortex of the mouse. Neuron 101:500–513.e5
50. Juavinett AL, Bekheet G, Churchland AK (2019) Chronically implanted Neuropixels probes enable high-yield recordings in freely moving mice. elife 8:e47188
51. Luo TZ, Bondy AG, Gupta D, Elliott VA, Kopec CD, Brody CD (2020) An approach for long-term, multi-probe Neuropixels recordings in unrestrained rats. elife 9:e59716
52. van Daal RJJ, Aydin Ç, Michon F, Aarts AAA, Kraft M, Kloosterman F, Haesler S (2021) Implantation of Neuropixels probes for chronic recording of neuronal activity in freely behaving mice and rats. Nat Protoc 16:3322–3347
53. Okun M, Lak A, Carandini M, Harris KD (2016) Long term recordings with immobile silicon probes in the mouse cortex. PLoS One 11:e0151180

54. Connors BW, Gutnick MJ (1990) Intrinsic firing patterns of diverse neocortical neurons. Trends Neurosci 13:99–104
55. McCormick DA, Connors BW, Lighthall JW, Prince DA (1985) Comparative electrophysiology of pyramidal and sparsely spiny stellate neurons of the neocortex. J Neurophysiol 54: 782–806
56. Nowak LG, Azouz R, Sanchez-Vives MV, Gray CM, McCormick DA (2003) Electrophysiological classes of cat primary visual cortical neurons in vivo as revealed by quantitative analyses. J Neurophysiol 89:1541–1566
57. Kvitsiani D, Ranade S, Hangya B, Taniguchi H, Huang JZ, Kepecs A (2013) Distinct behavioural and network correlates of two interneuron types in prefrontal cortex. Nature 498: 363–366
58. Fiorilli J, Marchesi P, Ruikes T, Huis in 't Veld G, Buckton R, Duque Quintero M, Reiten I, Bjaalie J, Pennartz CM (2023) Neural correlates of object identity and reward outcome in the corticohippocampal hierarchy: double dissociation between perirhinal and secondary visual cortex. bioRxiv (2023): 2023-05
59. Pachitariu M, Steinmetz N, Kadir S, Carandini M, Kenneth DH (2016) Kilosort: realtime spike-sorting for extracellular electrophysiology with hundreds of channels. BioRxiv061481
60. Pachitariu M, Sridhar S, Stringer C (2023) Solving the spike sorting problem with Kilosort. BioRxiv2023.01.07.523036
61. Rossant C, Kadir SN, Goodman DFM et al (2016) Spike sorting for large, dense electrode arrays. Nat Neurosci 19:634–641
62. Yger P, Spampinato GL, Esposito E et al (2018) A spike sorting toolbox for up to thousands of electrodes validated with ground truth recordings in vitro and in vivo. elife 7: e34518
63. IronClust (RRID:SCR_018903), https://github.com/flatironinstitute/ironclust
64. Buccino AP, Hurwitz CL, Garcia S, Magland J, Siegle JH, Hurwitz R, Hennig MH (2020) SpikeInterface, a unified framework for spike sorting. elife 9:e61834
65. Lewicki MS (1998) A review of methods for spike sorting: the detection and classification of neural action potentials. Network 9:R53–R78
66. Tolias AS, Ecker AS, Siapas AG, Hoenselaar A, Keliris GA, Logothetis NK (2007) Recording chronically from the same neurons in awake, behaving primates. J Neurophysiol 98:3780–3790
67. Introduction – phy (2023). https://phy.readthedocs.io/en/latest/. Accessed 2 Sep 2023
68. Wen H, Shi J, Zhang Y, Lu K-H, Cao J, Liu Z (2018) Neural encoding and decoding with deep learning for dynamic natural vision. Cereb Cortex 28:4136–4160
69. Xu L, Xu M, Jung T-P, Ming D (2021) Review of brain encoding and decoding mechanisms for EEG-based brain–computer interface. Cogn Neurodyn 15:569–584
70. Barack DL, Krakauer JW (2021) Two views on the cognitive brain. Nat Rev Neurosci 22:359–371
71. Oude Lohuis MN, Canton AC, Pennartz CMA, Olcese U (2021) Higher order visual areas enhance stimulus responsiveness in mouse primary visual cortex. Cereb Cortex 32(15):3269–3288
72. Reches A, Gutfreund Y (2009) Auditory and multisensory responses in the tectofugal pathway of the barn owl. J Neurosci 29:9602–9613
73. Stevenson RA, Ghose D, Fister JK, Sarko DK, Altieri NA, Nidiffer AR, Kurela LR, Siemann JK, James TW, Wallace MT (2014) Identifying and quantifying multisensory integration: a tutorial review. Brain Topogr 27:707–730
74. Knöpfel T, Sweeney Y, Radulescu CI, Zabouri N, Doostdar N, Clopath C, Barnes SJ (2019) Audio-visual experience strengthens multisensory assemblies in adult mouse visual cortex. Nat Commun 10(1):5684
75. Movahedi F, Padman R, Antaki J (2021) Limitations of receiver operating characteristic curve on imbalanced data: assist device mortality risk scores. J Thorac Cardiovasc Surg 165: 1433. https://doi.org/10.1016/j.jtcvs.2021.07.041
76. Poline J-B, Brett M (2012) The general linear model and fMRI: does love last forever? NeuroImage 62:871–880
77. Meijer GT, Marchesi P, Mejias JF, Montijn JS, Lansink CS, Pennartz CMA (2020) Neural correlates of multisensory detection behavior: comparison of primary and higher-order visual cortex. Cell Rep 31:107636
78. Vaccari FE, Diomedi S, Filippini M, Hadjidimitrakis K, Fattori P (2022) New insights on single-neuron selectivity in the era of population-level approaches. Front Integr Neurosci 16:929052
79. Kobak D, Brendel W, Constantinidis C, Feierstein CE, Kepecs A, Mainen ZF, Qi X-L, Romo R, Uchida N, Machens CK (2016) Demixed principal component analysis of neural population data. elife 5:e10989
80. Gallego JA, Makin TR, McDougle SD (2022) Going beyond primary motor cortex to improve brain–computer interfaces. Trends Neurosci 45:176–183

81. Raposo D, Kaufman MT, Churchland AK (2014) A category-free neural population supports evolving demands during decision-making. Nat Neurosci 17:1784–1792
82. Williams AH, Kim TH, Wang F, Vyas S, Ryu SI, Shenoy KV, Schnitzer M, Kolda TG, Ganguli S (2018) Unsupervised discovery of demixed, low-dimensional neural dynamics across multiple timescales through tensor component analysis. Neuron 98:1099–1115.e8
83. Ringnér M (2008) What is principal component analysis? Nat Biotechnol 26:303–304
84. Vaccari FE, Diomedi S, Filippini M, Galletti C, Fattori P (2021) A Poisson generalized linear model application to disentangle the effects of various parameters on neurophysiological discharges. STAR Protocols 2:100413
85. Björklund M (2019) Be careful with your principal components. Evolution 73:2151–2158
86. Kolda TG, Bader BW (2009) Tensor decompositions and applications. SIAM Rev 51(3): 455–500
87. Fawzy H, Rady EHA, Abdel Fattah AM (2020) Comparison between support vector machines and k-nearest neighbor for time series forecasting. J Math Comput Sci 10(6):2342–2359
88. Zhang H, Li L, Li X (2022) Exponential synchronization of coupled neural networks under stochastic deception attacks. Neural Netw 145: 189–198
89. Schooten SV, Harel R, Ercan S, Groot ED (2014) Applying feature selection methods on fMRI data
90. Breiman L (2001) Random forests. Mach Learn 45:5–32
91. Kong Y, Yu T (2018) A deep neural network model using random forest to extract feature representation for gene expression data classification. Sci Rep 8:16477
92. Probst P, Wright MN, Boulesteix AL (2019) Hyperparameters and tuning strategies for random forest. Wires Data Min Knowl J 9(3): e1301
93. Hastie T, Tibshirani R, Friedman JH, Friedman JH (2009) The elements of statistical learning: data mining, inference, and prediction, vol 2. springer, New York, pp 1–758
94. Gokcen E, Jasper AI, Semedo JD, Zandvakili A, Kohn A, Machens CK, Yu BM (2022) Disentangling the flow of signals between populations of neurons. Nat Comput Sci 2:512–525
95. Semedo JD, Jasper AI, Zandvakili A, Krishna A, Aschner A, Machens CK, Kohn A, Yu BM (2022) Feedforward and feedback interactions between visual cortical areas use different population activity patterns. Nat Commun 13:1099
96. Semedo JD, Gokcen E, Machens CK, Kohn A, Yu BM (2020) Statistical methods for dissecting interactions between brain areas. Curr Opin Neurobiol 65:59–69
97. Pennartz CMA, Oude Lohuis MN, Olcese U (2023) How "visual" is the visual cortex? The interactions between the visual cortex and other sensory, motivational and motor systems as enabling factors for visual perception. Philos Trans R Soc Lond Ser B Biol Sci 378: 20220336

Chapter 2

Detection of Cell Assemblies in High-Density Extracellular Electrophysiological Recordings

Gabriel Makdah, Sidney I. Wiener, and Marco N. Pompili

Abstract

Cell assemblies, i.e., concurrently active groups of neurons, likely underlie neural processing for higher brain functions. Recent technological progress has enabled large-scale recording of neuronal activity, permitting the exploration and analysis of cell assembly dynamics. This review aims to provide both conceptual insights and practical knowledge pertaining to principal methodologies used for detecting cell assemblies in the last 13 years. The goal is to assist readers in comparing and selecting various protocols to optimize their data processing and analysis pipeline. Each algorithm is explained with its fundamental principles, their application in neuroscience for cell assembly detection, and illustrated with published studies. Recognizing the similarities, advantages, and drawbacks of diverse methodologies may pave the way for developing new procedures for cell assembly identification to facilitate future endeavors in the understanding of brain activity.

Key words Electrophysiological recordings, Cell assemblies, Dimensionality reduction, Network analysis, Agglomerative algorithms

1 Introduction

In 1949, Hebb [1] introduced the term "cell assembly" to define a group of neurons that are frequently coactive, thereby establishing strong interconnections and acting together as a functional unit, which could underlie cognitive function [2]. Cell assemblies are often regarded as representational entities, combining the behavioral correlates of individual neurons, perhaps binding together diverse features into an element for cognitive processing, e.g., a percept [3]. Some experimental questions also examine the occurrence of activations of assembly neurons in specific sequences, which could represent events in a temporal order.

Initial draft: GM and MNP. Simulations and testing of methods: GM. Graphics and visualization: GM and MNP. Writing: GM, SIW, and MNP.

Luis Carrillo-Reid (ed.), *Identification, Characterization, and Manipulation of Neuronal Ensembles*, Neuromethods, vol. 215, https://doi.org/10.1007/978-1-0716-4208-5_2,

Complementarily, assemblies can be studied in terms of how their activations are processed by downstream targets [4, 5], where a cell assembly is defined by the activity it elicits. As in the case of binding, cell assemblies can span multiple brain areas [6]. Note also that assemblies may be found at different time scales depending upon the brain system studied [7].

Currently, two principal experimental approaches provide multiple single-neuron activity data, albeit at different time scales. In vivo and in vitro recordings with multiple microelectrodes typically resolve single unit activity on the order of milliseconds. Alternatively, calcium imaging allows the recording of large groups of neurons, and hence cell assemblies with known detailed anatomical distributions, over long periods of time [8, 9]. Moreover, optical tools also provide the ability to stimulate cell assemblies [10]. However, the duration of calcium transients is on the order of hundreds of milliseconds [11, 12], resulting in lower temporal resolution compared to microelectrode recordings. Additionally, with current methods, only 10–20% of individual action potentials can be detected [13], with calcium transients mostly signaling when a given neuron is bursting. This limits the possibility to detect cell assembly activations when individual neurons fire single-action potentials. As such, cell assemblies detected from calcium imaging data mostly consist of groups of neurons that burst together on timescales of around 1 second [14, 15].

Various methods have been developed to detect cell assemblies from microelectrode recordings and calcium imaging data, based on a range of theoretical and methodological frameworks. However, several existing techniques have only been tested on simulated recordings or have not been utilized outside of the laboratory where they were developed, and we hope that this review will encourage future studies to remedy this. Here, our focus is on methods that have been used to detect cell assemblies in in vivo extracellular electrophysiological recordings from multiple microelectrodes and that have been employed by researchers other than their developers. For recent reviews of methods used for detecting cell assemblies in calcium imaging, see [14, 16].

Early assembly detection methods focused on event-based activity, identifying groups of neurons with firing correlated with specific behavioral variables, such as reward delivery, sensory stimuli, or the animal's position in space (e.g., [17, 18]), potentially inducing bias. More recently, there has been increased interest in methods capable of detecting cell assemblies independently of environmental events. These can be particularly useful for examining brain activity without clear behavioral correlates. None of the approaches described here require information about the behavioral correlates of the neurons.

This review will provide an overview of several methodologies. Some techniques have the explicit ability to identify specific groups of recorded neurons as participating in cell assemblies [7, 19, 20], while other methods identify components or patterns for which all recorded neurons participate to some extent [21–24]. Furthermore, while some techniques are limited to detecting the co-activation of multiple cells within a given time window, others can detect cell assemblies as patterns of sequential activations of cells [7, 23, 24]. The subject of detection of sequential activations is beyond the scope of this review and is only presented when the method detects assemblies as well.

Most techniques begin with the detection of co-activation of single pairs of neurons. This can then be expressed as co-activation matrices (such as correlation, similarity, or functional connectivity matrices), which are then used to extract more complex, multicellular, functional relationships. Three principal approaches will be discussed in separate sections below: dimensionality reduction (principal component analysis, independent component analysis, non-negative matrix factorization), network analysis (community detection, Markov processes), and agglomerative construction. We discuss the applications of each method as well as their strengths and weaknesses. This review aims to help investigators compare and choose among different protocols to find the best pipeline to detect and characterize cell assemblies in their data. Example publications are cited to provide more details useful for applying these methods, as well as examples of some of the types of results that they can yield.

2 Detecting Cell Assemblies with Dimensionality Reduction

Dimensionality reduction maps high-dimensional data into a lower-dimensional space and is commonly employed for analyzing large-scale datasets [25]. In neuroscience, an example of a high-dimensional space is one formed by the activity of a neural population. Each neuron represents one dimension within this space. As such, a recording of 100 neurons over time would represent a 100-dimensional space and can be reduced to a smaller space of, for instance, 10 dimensions (or factors). The resulting ten factors aim to preserve most of the information about neuronal coactivity in the original recording. In essence, each dimension in the new reduced space represents a recurrent pattern of population activity in the original (high-dimensional) spike matrix.

In linear dimensionality reduction techniques, each dimension in the reduced space corresponds to a linear combination of all neurons, with each neuron having a weight quantifying its contribution to that factor. The interpretation of the obtained population pattern depends on the dimensionality reduction technique employed. For instance, if a technique is applied to a correlation matrix computed with Pearson's pairwise correlations [26], multiple cells sharing high weights within a given factor would indicate that they have a high Pearson's correlation score. The following section outlines the most popular dimensionality reduction techniques applied in the context of cell assembly detection: principal component analysis, independent component analysis, and non-negative matrix factorization.

2.1 Principal Component Analysis (PCA)

2.1.1 Basic Principles

PCA is likely the most widely used linear technique for dimensionality reduction in neuroscience. Beyond neurophysiology, it is also useful to reduce high-dimensional behavioral datasets (e.g., [27]). PCA reduces the original dataset into a set of orthogonal dimensions (or principal components, the cell assemblies) defined by the eigenvectors of the covariance matrix, each resulting from the linear combination of the original variables (i.e., activity of individual neurons) [28, 29]. Each component is associated with an eigenvalue that quantifies the proportion of variance in the original dataset explained by that component. PCA is typically applied to a smoothed spike matrix representing binned spike counts (equivalent to firing rates), which are z-transformed. The first principal component explains the largest amount of variance (i.e., in coactivity of neurons), with subsequent components being necessarily orthogonal to the preceding ones and explaining as much of the remaining variance as possible. By definition, there are as many principal components as variables (neurons) in the original dataset; however, in practical applications, only a subset of them is needed to explain most of the variance.

2.1.2 Detection of Cell Assemblies with PCA

PCA is effective in detecting groups of neurons with correlated activity. Initially used in neuroscience to compress and synthesize neuronal responses across multiple trials [30–33], it was later employed by Nicolelis and colleagues [34, 35] as an indirect representation of cell assembly activity. In its most common implementation, as developed by Peyrache et al. [21], the algorithm starts by constructing a pairwise correlation matrix (Fig. 1a, b) from the original spike matrix using Pearson's pairwise correlation, a linear correlation measure that is normalized to the firing rates of the respective cells. The pairwise correlation matrix is then factorized into a set of orthogonal eigenvectors, each associated with an eigenvalue. Peyrache et al. [21] identified those principal components that explain a suprathreshold proportion of variance (termed "significant components") using a threshold computed with the Marčenko-Pastur distribution [36]. Indeed, for a random normalized matrix, it has been demonstrated that eigenvalues follow a well-defined probability function and are limited in magnitude by an upper bound [21]. The eigenvectors whose eigenvalues exceed this upper bound (Fig. 1c) are considered to represent cell assembly

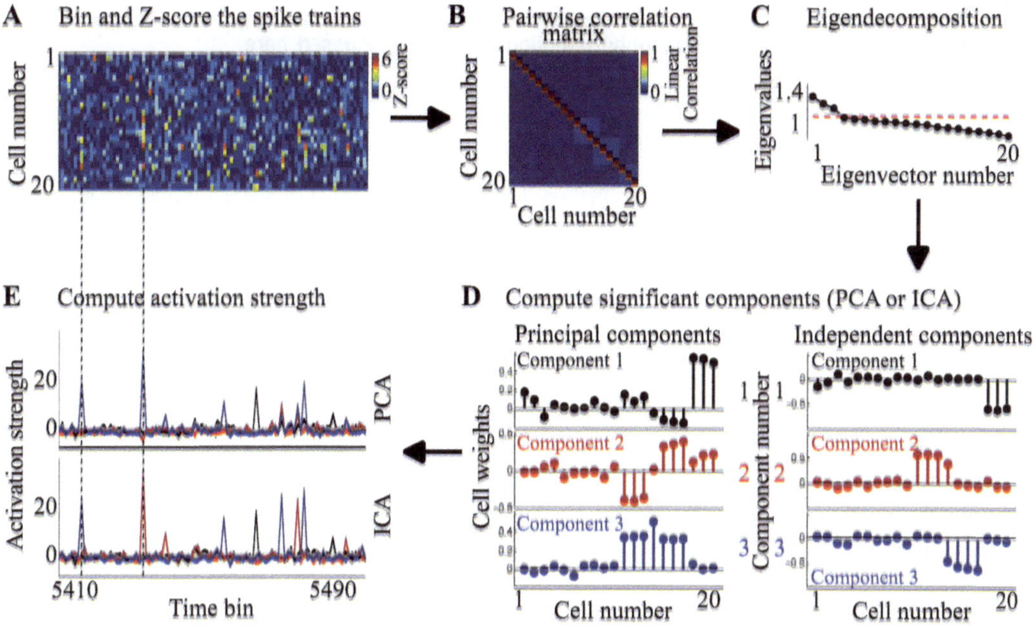

Fig. 1 PCA and ICA implementation. (Example with synthetic data. Figure adapted from Ref. [22]). (**a**). The dataset (a fragment of which is shown here) composed of N (here 20) cells is binned in a window *dt* set by the user, and then firing is Z-scored. (**b**). A pairwise correlation matrix is created, quantifying the coactivity of all pairs of cells in the dataset. (**c**). The eigenvectors and eigenvalues are then extracted. The M (here, three) eigenvectors that have eigenvalues greater than a threshold (determined using the Marčenko-Pastur distribution, red dashed line) are considered as assemblies. (**d**). Principal components and independent components obtained from the correlation matrix in B (see Table 1 for source code link). (**e**). Activation strengths of principal and independent components in this dataset (as described in the Detection of Assembly Activations section). Color codes are as in D

patterns (Fig. 1d). It remains to be proven to what extent this approach's assumptions (such as the Gaussianity of the underlying signal) may bias results across different experimental conditions. Alternative approaches to detect significant components have been proposed [19, 22].

Peyrache et al. [37] used this method to detect cell assemblies in simultaneous recordings of dozens of prefrontal cortex neurons in behaving rats learning rules in a Y-maze. In this, and a subsequent paper [38], the authors identified cell assemblies in the medial prefrontal cortex that emerged upon learning and were activated at the decision point of the maze. These cell assemblies were then replayed during slow wave sleep, synchronously with hippocampal sharp wave ripples. This was interpreted as supporting a role for coordinated hippocampal-prefrontal activity in memory consolidation. Since then, others have used this approach for cell assembly detection [e.g., 39–44]. Sjulson et al. [45] used PCA to detect cell assemblies and the directionality of information flow between the hippocampus and the medial striatum. They found

Table 1
Availability of code for the cell assembly detection methods discussed here

Algorithm	Code language	Repository link
PCA Peyrache et al. [21]	Python, Matlab	Python: https://github.com/PeyracheLab/pynacollada Matlab: https://fmatoolbox.sourceforge.net/
ICA Lopes-dos-Santos et al. [22]	Matlab	https://github.com/tortlab/Cell-Assembly-Detection
seqNMF Mackevicius et al. [24]	Matlab	https://github.com/FeeLab/seqNMF
STC Humphries [19]	Matlab	https://github.com/mdhumphries/SpikeTrainCommunitiesToolBox
MSCD Billeh et al. [20]	Matlab	https://github.com/CellAssembly/Detection Note: Code outputting the final set of assemblies is not provided.
HMM Maboudi et al. [23]	Python	https://github.com/nelpy
CAD Russo and Durstewitz [7]	Matlab	https://github.com/DurstewitzLab/Cell-Assembly-Detection

that medial striatal cell assemblies decode spatial information from the hippocampus and that conditioning to a cocaine stimulus increases the coupling of assemblies between these two structures.

2.1.3 Advantages and Drawbacks of PCA

The main advantages of PCA are its operating speed, ease of use, and scalability up to very large datasets, even for thousands of neurons and tens of hours of recording. Computing the correlation matrix as well as matrix factorization are rapid operations and do not require the user to manually input values for multiple parameters. The only parameter that needs to be set by the user is the width of the binning window for detecting population patterns.

However, PCA shares some drawbacks with other dimensionality reduction algorithms, and these will be discussed in subheading 2.2. A drawback specific to PCA is the orthogonality between principal components. While the first component is optimized based on the direction of the largest variance in the multidimensional spike activity space, subsequent components are constrained to be orthogonal to it. Thus, a single neuron generally does not have a high weight in more than one assembly. This might not reflect the optimal segregation of groups of correlated cells [46], leading to a lower probability of attributing appropriate weights to neurons participating in multiple cell assemblies [22].

2.1.4 Choice of Binning Window

Most methods discussed in this review require a binning window, which corresponds to the timescale at which cell assemblies are active. The choice of the binning window is intrinsically linked to the specific brain region(s) under study and the experimental hypotheses, and different binning windows have been used in different studies ([47], 150 ms; [45], 25 ms; [38], 30 ms and 125 ms bins, synchronized to the theta rhythm). From a theoretical perspective, the time constant across which presynaptic spikes are integrated by neurons in the cerebral cortex during the waking state is 10–30 ms [48]. Consistent with this time window, the timescale of cell assemblies in the hippocampus of a waking rat has been shown to be 10–30 ms [49]. For studies that target specific oscillatory rhythms, time windows are adaptively adjusted (e.g., ~100 ms for theta, or ~ 250 ms for delta).

2.2 Independent Component Analysis ICA

2.2.1 Basic Principles

Independent component analysis (ICA) has gained prominence as a data analysis technique in neuroscience [50]. Initially developed to segregate mixed sources into linearly additive subcomponents [51], ICA addresses challenges like the "cocktail party problem," where individual conversations must be distinguished in a noisy room. Similarly, in neuronal recordings, the activity is comprised of a set of assemblies. The objective then is to reconstruct the identity of individual cell assemblies from the amalgam of activity from multiple cell assemblies and background noise.[1] The central limit theorem is the underlying basis of ICA. It stipulates that the result of combining two independent random variables is more Gaussian than the original variables. In contrast with PCA, ICA improves the independence of components by rotating them and optimizing the non-Gaussianity of their combinations resulting in non-orthogonal components.

2.2.2 Detection of Cell Assemblies with ICA

ICA was first introduced in neuroscience to identify groups of neurons exhibiting correlated firing activity [52]. The algorithm proposed by Lopes-dos-Santos and colleagues is now widely used for detecting cell assemblies [5, 6, 53–68]. Typical implementation of ICA for cell assembly detection involves a two-step procedure [22]. The algorithm first begins by reducing the dataset via PCA to yield significant principal components, as above, and then uses ICA to rotate these components, yielding independent components.

Using this approach, Trouche et al. [63] identified spatially tuned cell assemblies in electrophysiological recordings. These assemblies encoded spatial information, and optogenetic inhibition of these cell assemblies impaired behavior during a spatial memory test. Guan et al. [69] utilized ICA to detect cell assemblies in

---

[1] We refer to background noise as the spikes of neurons not belonging to the assemblies and spikes of assembly neurons outside of assemblies' activations

bilateral recordings in the CA1 region of the hippocampus, identifying some assemblies restricted to either the left or the right hippocampus, while others extended bilaterally.

2.2.3 Advantages and Drawbacks of ICA

Since the components in ICA are independent and not orthogonal, ICA is more suitable than PCA for including the same high-weight neuron in more than one cell assembly. Moreover, ICA inherits PCA's advantageous traits: it is fast, scalable to large datasets, and user-friendly with only one parameter to set (the width of the binning window).

However, ICA comes with its own set of drawbacks. First, it assumes that the distribution of underlying variables is non-Gaussian, as components are retained when their non-Gaussianity is maximized. This assumption may limit the applicability of ICA in cases where the data distribution is Gaussian. Furthermore, due to the iterative nature of the optimization process, ICA is sensitive to the initialization of the weight vectors and is susceptible to converge to a local minimum [70], i.e., the best result within a certain region of the solution space, but not necessarily the overall best solution. In the FastICA algorithm utilized in cell assembly detection [22, 71], the weight vectors are initialized randomly. As such, repeated applications of ICA can return different results, and thus, different sets of cell assemblies. The effects of convergence to local minima can be reduced by always initializing the weight vectors to the same value (e.g., initial weight values are set to 0) to yield the same set of cell assemblies between runs. However, this will not ensure that ICA converges to its best solution. More complex algorithms may offer a higher likelihood of finding the globally optimal solution [72]. Additionally, different ICA algorithms can yield significantly different results based on their implementations [73], introducing variability that researchers must consider when interpreting findings.

A limitation shared by ICA and PCA is the arbitrary sign of highly weighted cells within a component. Highly positive or negative weighted cells within a component can both indicate that cells are highly correlated [21]. It is crucial to note that the presence of both highly positive and negative weighted cells within a single component doesn't necessarily represent two groups of anticorrelated neurons, but rather, this may result from poor separation of two assemblies [5, 6].

2.3 Non-negative Matrix Factorization (NMF)

2.3.1 Basic Principles

Non-negative matrix factorization (NMF) is a method designed to extract sparse features from high-dimensional datasets [74, 75], wherein each component represents groups of variables that are correlated. NMF decomposes the initial dataset into two lower-rank matrices, such that multiplying them together yields a result approximately equal to the initial dataset (Fig. 2a). To achieve this approximation, values in the lower-rank matrices are iteratively

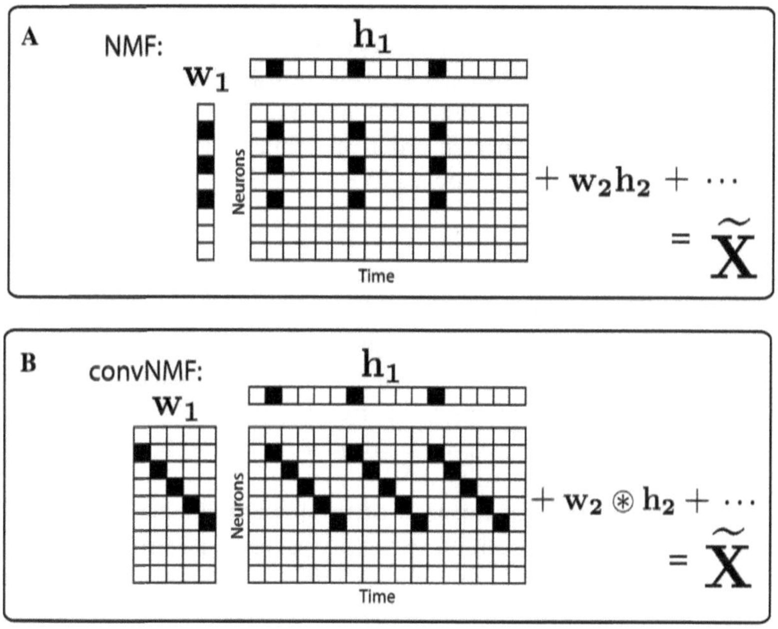

Fig. 2 NMF and convNMF algorithms. (Adapted from Ref. [24]). (**a**). NMF seeks to approximate a data matrix X comprising N neurons by the product of two matrices, W and H. W is a matrix with N rows and K columns, where N corresponds to the number of recorded neurons and K is set by the user and represents the number of cell assemblies. H is a matrix with K rows and T columns, where T is the number of time bins in the data matrix. Matrices W and H can be decomposed into vectors $w_1...w_k$ and $h_1...h_k$. Each vector in W represents an assembly pattern (indicating which cells are highly active in the assembly), and each vector in H represents a temporal pattern (indicating the time bins where the cell assembly is highly active and the relative amplitudes of this activity). (**b**). In convNMF, W is composed of matrices $w_1...w_k$, and each matrix represents a sequential pattern of neuronal firing

updated to minimize the approximation error between their product and the initial dataset. The process terminates when the error converges—changes in the matrices become negligible—or when a user-preset maximum iteration limit is reached.

2.3.2 Detection of Cell Assemblies with NMF

In its application for cell assembly detection, NMF is employed on a binned spike matrix containing only positive values. It produces two lower-rank matrices of size K, where K denotes the number of patterns in the underlying dataset, typically chosen by the user. The first matrix denotes the identity of cell assemblies, with each vector representing a cell assembly (Fig. 2a). The second matrix denotes assembly activations over time, where each vector represents moments when the corresponding cell assembly is active (Fig. 2a).

With convolutional kernels, the algorithm can detect sequences of firing, capturing cell assembly activations with temporal structure across multiple time bins (Fig. 2b, [24, 76]). Tingley and Buzsáki [77] applied this method to identify patterns of sequential cell activations in the hippocampus during sharp-wave ripple events. These assemblies were then used to assess whether neuronal activity in the lateral septum was related to the temporal organization of neuronal activations within sharp-wave ripples and revealed a correlation with the magnitude of neuronal co-activity rather than the specific order of cellular activity. See Terada et al. [108] for another application of NMF to detect sequencing in cell assemblies.

2.3.3 Advantages and Drawbacks of NMF

The convolutional NMF algorithm offers the advantage of detecting assemblies with only positive weights, simplifying the interpretation of results compared to PCA and ICA. Additionally, it identifies sequences of neuron activations, enabling a more detailed analysis of temporal patterns within cell assemblies.

However, this method is not without its limitations. Similar to ICA, the results of NMF may vary between runs and are contingent on the initialization of the lower-ranked matrices. Convolutional NMF can exhibit sensitivity to noise, potentially yielding correlated factors, or representing individual neuron activity through overfitting of the dataset [78]. To mitigate this issue, a regularization algorithm is often implemented [79, 80]. In terms of computational efficiency, convolutional NMF is slower compared to PCA and ICA, requiring more time to extract cell assembly patterns. Moreover, in the implementation by Mackevicius et al. [24], the method involves setting 3 parameters (up to 6 optionally) including the number of cell assemblies, the bin size, and the sequence length (and, optionally, regularization parameters) [24, 76]. The necessity for the user to preset the number of assemblies to find in the recording is a significant limitation, since in most cases it is impossible to know how many assemblies are to be found. It is worth noting that the authors provide tools in their toolboxes for automatic estimation of these parameters through recursive runs of the algorithm on the dataset, choosing those that optimize certain metrics (such as minimizing redundancy between factors). However, this automated approach contributes to slower implementation. Note that convolutional NMF has been extended using point process methods that don't require binning the data [81].

2.4 Summary: Advantages and Disadvantages of Dimensionality Reduction for Cell Assembly Detection

Dimensionality reduction algorithms exhibit notable advantages, first of all, their relatively rapid run times. These approaches are particularly effective in detecting recurrent and stereotypical neuronal activation patterns, as evidenced in applications like bird songs [24, 82]. However, they also share some limitations. One common challenge is the necessity to create a binned spike matrix, introducing edge effects where millisecond-scale changes in firing times may misattribute an action potential to a preceding or succeeding bin. To address this, datasets are often binned in overlapping intervals or smoothed using algorithms to distribute neuronal activity across multiple bins, mitigating edge effects [47, 83, 84]. Nevertheless, this can extend algorithm run time and potentially introduce duplicate patterns between bins, biasing cell assembly detection.

Dimensionality reduction methods factorize the dataset into a smaller number of variables, meaning that the number of components/assemblies that can be obtained is limited to the number of cells in that dataset. This number is then further reduced when only the components explaining significant amounts of variance are retained. However, according to the concept of cell assemblies as fundamental computational blocks of brain function [1], the number of cell assemblies (i.e., combinations of coactive neurons) is not necessarily so limited. Indeed, one can argue that nervous system processing might benefit from more cell assemblies than individual neurons. If the fundamental computational blocks were individual cells, then the brain's capabilities would be limited by their overall number. Instead, if the basic computational units were all the possible combinations of neurons, this would extend computational capabilities multifold. Therefore, even in a given recording of a limited sample of neurons, it is possible that the number of cell assemblies exceeds the number of cells. Dimensionality reduction methods would be unable to detect all these patterns as independent cell assemblies. Additionally, results are presented as continuous weights, representing each cell's contribution to a component. Consequently, all neurons contribute to each detected assembly, and the exact identity of cells participating in a cell assembly is not specified. To overcome this limitation, threshold measures have been proposed to identify those cells most involved in specific cell assemblies [5, 6, 55, 56, 59, 60, 85]. Finally, note that, like other approaches, no distinction is made between co-activations related to functional relations between the neurons or from common driving from another source.

3 Cell Assembly Detection with Network Analysis

Beyond dimensionality reduction, other methodological approaches have been proposed to detect cell assemblies in high density neural recordings using a variety of statistical or analytical approaches. Some of them are based on techniques developed to analyze complex networks.

3.1 Community Detection

3.1.1 Basic Principles

Community detection is a network analysis method designed to examine complex graph structures. A graph consists of nodes interconnected by edges, where edges may carry numerical weights quantifying relationships between nodes. Networks often exhibit communities: subsets of nodes with stronger internal connections and weaker connections to other parts of the graph. Originally devised for social networks and biological ecosystems [86], community detection algorithms have been applied in various fields with graphs containing highly connected networks separated by loosely connected nodes.

Community detection has also been proposed for the detection of cell assemblies in neurophysiological recordings [19, 20]. In this case, the nodes correspond to neurons within the dataset. Between two nodes, the edge weights correspond to the degree of co-activation of the two neurons. Detected communities would represent groups of highly connected neurons in the graph. It is important to note that with standard community detection methods, each node can only belong to a single community. Thus, in the implementation of neuronal data, each neuron can only belong to a single cell assembly.

3.1.2 Detecting Cell Assemblies with Spike Train Communities (STC)

Humphries [19] introduced Spike Train Communities (STC), a community detection-based method for finding cell assemblies. The algorithm begins by creating a pairwise similarity matrix based on cosine similarity of spike trains at a user-defined timescale (Fig. 3a, b), although other similarity metrics could be used, such as the Hamming similarity, Pearson's correlation coefficient, or other metrics specific to spike trains (see [87] for an overview). The cosine similarity is influenced by the pattern of neuronal activations and the relative activity of each cell within the population vector, but not the scale (or absolute value) of that activity. To ensure that all neurons contribute equally to community identification, the weight of similarity between spike trains of each pair of neurons is then adjusted based on each neuron's activity's overall level of similarity with all other neurons. The obtained matrix is then factorized with eigen-decomposition, yielding a set of eigenvectors and their corresponding eigenvalues. Only the M eigenvectors that have positive eigenvalues are retained (Fig. 3c). These eigenvectors are used to define a set of coordinates within an M-dimensional eigenspace. This is then used to cluster the data into groups of highly correlated neurons via k-means (Fig. 3d). On each iteration, k-means detects a different number of clusters, numbered between 2 and M + 1 (Fig. 3e). The optimal clustering is then determined based on a consensus clustering algorithm [88]. In addition to detecting cell assemblies, this method has been used to identify similar spike trains fired by a single neuron across multiple trials [57, 89–91].

McMahon et al. [91] applied this method to recordings of spinal cord interneurons in cats performing an air stepping test. They identified two cell assemblies which included interneurons. This allowed the authors to assess the anatomical location of cells participating in these cell assemblies. They found no correlation between a given cell's location along the rostro-caudal axis of the spinal cord and its participation in one or the other cell assemblies.

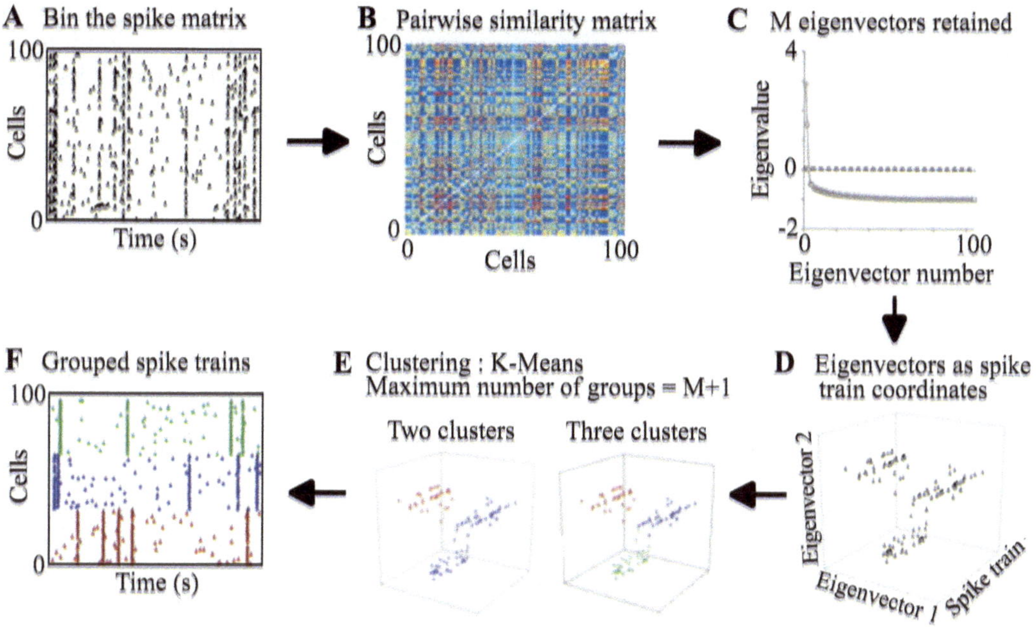

Fig. 3 Implementing STC. (Figure adapted from Ref. [19]). (**a**). The dataset composed of N (here 100) cells is binned into windows *dt*, a value set by the user. (**b**). A pairwise similarity matrix is created describing the coactivity of all pairs of cells in the dataset. Red indicates highly correlated cell pairs, while deep blue is anticorrelated pairs. This matrix is then normalized based on the overall similarity level of each node. (**c**). The eigenvectors and eigenvalues are then computed. The M eigenvectors that have positive eigenvalues are retained. (**d**). Spike trains are represented in the coordinates composed of the M eigenvectors with the highest eigenvalues. The eigenvectors are each N-entries long. Then, each eigenvector defines a set of coordinates in an M-dimensional space for every spike train. Within this eigenspace, the closer two spike trains are, the more similar are their activities. (**e**). Clustering is carried out on the spike trains using k-means, iteratively detecting communities $k = 2, \ldots,$ M. The optimal clustering is then determined based on a consensus clustering algorithm. (**f**). The spike trains from A, grouped according to the communities they belong to. Colored vertical lines correspond to cell assembly activations

They also found that activations of these two cell assemblies were respectively correlated to flexor or extensor muscle bursts during stepping.

3.1.3 Advantages and Disadvantages of STC

Similarly to PCA and ICA, this method does not require the user to set any parameters other than the timescale at which assemblies will be detected. STC can be implemented in a binless version, centered on spikes rather than time bins, yielding superior results but slowing computation by an order of magnitude. Note also that k-means is sensitive to the initialization of parameters [92] and could converge to local minima across different runs, resulting in different cell assemblies at each run. The authors take these limitations of k-means into account and employ methods to minimize these problems [19, 93]. Another disadvantage is that a single neuron cannot participate in more than one assembly.

3.2 Using Markov Chains to Detect Communities

3.2.1 Basic Principle

Markov processes are stochastic models that infer the probabilities of sequences of states in a system [94]. Each state corresponds to an observable variable or event. The future state of the system depends only on the current state and the current connections within the graph, or in the case of a neuronal recording, a pairwise connectivity metric.

In the context of network analysis, the Markov stability framework utilizes Markov processes to identify communities within a given graph [95]. The fundamental idea involves assessing interactions between nodes to pinpoint groups that are densely connected and closely associated. In the case of its application to neuronal recordings, neurons would constitute the nodes within the graph, with the connections between two nodes being proportional to the connectivity (or coactivity) of the two neurons. The detection of communities can be thought of as a diffusion process, a drop of ink propagating in a network of pipes. This drop of ink slowly propagates throughout the graph, prioritizing wider paths (or stronger connections) where diffusion is easier. After a certain period of time (called a Markov time), the nodes that have been reached by the drop of ink are evaluated; these constitute a community (a cell assembly). Each neuron can only belong to a single community. For briefer Markov times, the diffusion process only extends to a limited part of the graph: the neurons that are most strongly connected. This would lead to the detection of small cell assemblies that are comprised of highly coactive neurons. For longer Markov times, the "drop of ink" can propagate further, encompassing more nodes and, as such, detecting larger cell assemblies. The choice of Markov time influences the granularity of identified cell assemblies, offering flexibility to adapt the analysis based on the desired level of detail or comprehensiveness.

3.2.2 Markov Stability Community Detection (MSCD) to Detect Cell Assemblies

To detect cell assemblies, Billeh et al. [20] proposed a method employing Markov Stability for Community Detection (MSCD). First, the algorithm constructs a pairwise coactivity matrix by computing the so-called "functional connectivity" between neurons (note that this does not necessarily correspond to anatomical connectivity). The authors define this metric as the expected contribution of one neuron's activity to predicting that of another. The algorithm proceeds by constructing a template for each action potential fired by a neuron over time. The functional connectivity between two neurons A and B is determined as a function of the likelihood that neuron B fires shortly after neuron A. Excitatory connections are modeled as an activation that decays exponentially with time (Fig. 4a). Inhibitory connections are modeled using an inverted exponential profile (Fig. 4b). This process is repeated for all pairs of neurons in the dataset, resulting in a pairwise functional connectivity matrix. Given the asymmetry of relationships between neurons, the matrix values can be likened to directional connections in a graph.

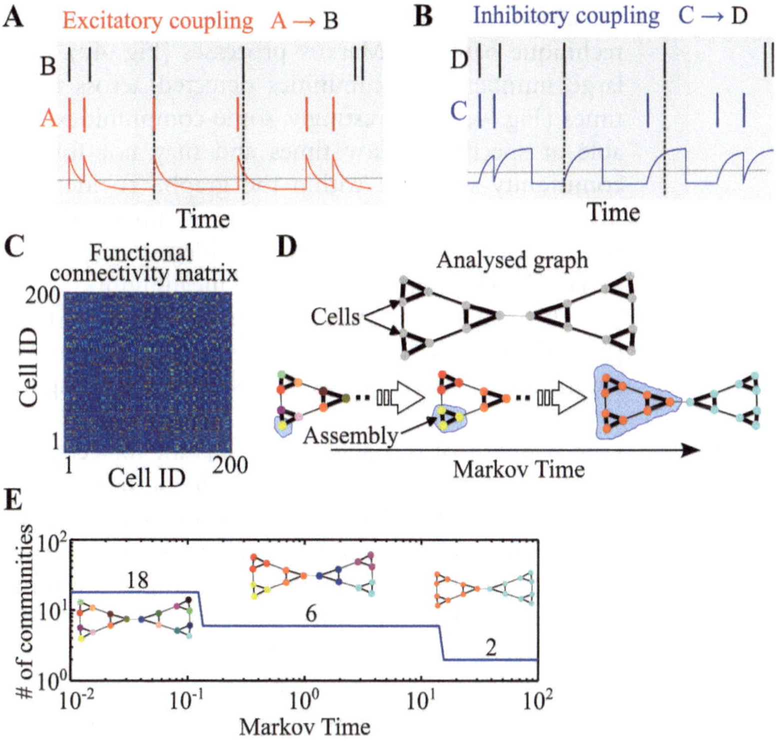

Fig. 4 Schematic representation of the MSCD method. (Adapted from Ref. [20]). (**a**). Computation of functional connectivity between pairs of neurons. Predictability of neuron A firing for neuron B firing is modeled as a positive prediction that decays exponentially with time. The top two lines are spike trains of the neurons. The third row is the association function that governs the degree to which spikes of B are taken to be predicted by A. The association function shown is normalized so that the signal's mean is zero (gray horizontal line). This coupling is highest when neuron B fires immediately after neuron A (left vertical dashed line) but is reduced for longer delays (about 50% for the right dashed vertical line). (**b**). Negative coupling between neuron C and neuron D is modeled using an inverse exponential profile. Note that when neuron D fires immediately after neuron C, the predictive association is nil (left vertical dashed line), but the negative predictive association is partial when neuron D fires at a greater delay after neuron C (right vertical dashed line). (**c**). The functional connectivity is computed for all pairs of neurons. For each pair of neurons, the functional connectivity represents the weight of the connection between the two respective nodes in the graph. (**d**). A process evolving within the associational connectivity graph can be used to detect communities. In this example graph, there are 18 nodes (neurons). Nodes are connected through edges. The width of the edge represents the strength of the functional connectivity between two nodes. At longer Markov times, the process can explore larger parts of the graph, yielding larger communities (cell assemblies; examples are highlighted in light blue). At each Markov Time, the nodes of a given color represent a different community. At the briefest Markov time, each neuron is considered to be a community. (**e**). The number of detected communities is plotted as a function of Markov time. At very brief Markov times, 18 communities are detected. At longer Markov times, six communities are detected, corresponding to the small triangular structures, each of which is composed of three highly connected cells. At even longer Markov times, two communities are detected, each corresponding to nine cells that are more strongly connected

Communities within this graph are then identified using a technique based on Markov processes (Fig. 4c). This can yield a large number of communities detected across multiple Markov times (Fig. 4d). Interestingly, some communities are only detectable at specific Markov times and may not fully represent the community structure within the graph. To identify consistently detected cell assemblies across Markov times, the authors propose two measures of robustness, considering the most consistently detected communities as the most meaningful.

This algorithm was implemented by Miyawaki et al. [96]. Gonzalez et al. [15] also employed a method based on this algorithm to detect cell assemblies in calcium imaging recordings.

3.2.3 Advantages and Disadvantages of MSCD

One strength of this algorithm, relevant for certain experimental questions, stems from the directionality of the computed functional connectivity matrix. Unlike the similarity matrices described earlier, the relationship between neurons here is asymmetrical and dependent on precise spike times. For example, the predictability of neuron A's spike from those of neuron B might not be equal to predictions in the opposite direction. Furthermore, the user can specify which neurons are excitatory and which are inhibitory, and the influence of their activations is modeled differently (Fig. 4a, b, respectively). Also, the exact identities of cells composing a cell assembly are known.

However, this algorithm is computationally more demanding than those described above. Moreover, the results can vary between runs since this community detection algorithm is sensitive to the initialization of parameters. Consequently, the community detection framework is typically executed between 100 and 500 times per Markov time to identify the most consistently detected cell assemblies and reduce dependence on the initialization of parameters. Additionally, at a given Markov time, any given neuron can only belong to a single cell assembly, limiting the number of cell assemblies that can be detected in the dataset.

3.3 Hidden Markov Models (HMM)

3.3.1 Basic Principle

Hidden Markov models (HMM) extend the concept of Markov processes, and early applications included speech recognition and DNA analysis. In the Markov processes discussed above, nodes represent the observable variables of the underlying dataset (neuronal activations). However, the interactions between neurons may be influenced by unmeasured variables in the original dataset, such as global dynamics (UP and DOWN states in the cortex or ripple events in the hippocampus for instance) and cell assemblies. Instead of using a diffusion process to detect highly connected parts of the graph (as described in the Markov stability framework), HMM introduces hidden nodes to represent these unmeasured variables which are inferred from observable measures. Observable variables are the data that can be experimentally measured. In the case of neuronal recordings, these would be the individual neurons and their action potentials. The hidden variables represent the elements that influence and orchestrate neuronal activity. Hidden Markov models attempt to uncover these hidden variables based on the observable data. Observable variables can contribute to several hidden nodes with different weights, meaning that a single neuron can contribute to several different cell assemblies. HMM computes the hidden variables that best explain the interactions between the observable variables. These hidden nodes (cell assemblies) are generally expressed as linear combinations of the observable variables (activity of neurons), each neuron having a weight quantifying its contribution to a given cell assembly.

3.3.2 Detection of Cell Assemblies with HMM

Several authors have employed HMM to detect cell assemblies [23, 97, 98]. Maboudi et al. [23] constructed a hidden Markov model using population burst events, defined as time windows where the number of spikes fired by the neuron population is at least three standard deviations greater than the mean over the entire recording. In all implementations, the user must select, as a parameter, the number of hidden nodes describing the unmeasured variables (here, cell assemblies) that characterize spiking activity relations between neurons in the dataset. A principal goal of the algorithm is to assign to each neuron a weight defining its contribution to each of the assemblies.

This method allowed Maboudi and colleagues [23] to model sequentially active neurons forming assemblies in hippocampal recordings, predicting the positions of behaving rats from the place fields of the neurons. Furthermore, distinct sequences of replay in the hippocampus were identified without knowledge of the behavioral correlates of the individual neurons. The method described by Maboudi and colleagues was also implemented in other studies [61, 99, 100]. Noguchi et al. [99] used this method to detect sequential patterns in hippocampal neuronal activity during sharp-wave ripples. This allowed the authors to study how the activity of interneurons preceding sharp-wave ripples influenced the structure of subsequent neuronal sequences.

3.3.3 Advantages and Disadvantages of HMM

This method allows a single neuron to participate in several cell assemblies (unlike community detection algorithms). Like the convolutional NMF algorithm, it can also detect patterns of sequential activations of cells. Furthermore, the number of cell assemblies that can be detected by this algorithm is theoretically unlimited and can exceed the number of neurons in the dataset.

A main limitation of this method is that the user needs to estimate how many cell assemblies are present in the recording, which is difficult. If the user overestimates the number of assemblies, this leads to the distribution of noise among the assemblies; the algorithm may seek to explain residual spike activity as cell assemblies. On the other hand, a low estimate will lead to missing some assemblies. Due to the higher number of user-preset variables and the need to learn hidden variables from the observable dataset, a main drawback of hidden Markov models relative to other Markov process methods is the long computation time.

4 Agglomerative Construction of Groups

4.1 Basic Principle

Instead of reducing dimensions or clustering the population of cells based on functional relationships between them, agglomerative algorithms start with small templates, such as a pair of coactive cells. The process then progressively enlarges the group by finding another cell that tends to be coactive with the existing group, creating a three-cell group. The algorithm continues this agglomerative construction iteratively, searching for neurons significantly coactive with an n-cell template to create an n + 1 template. This agglomerative construction persists until no new neurons can be added to the template. Various statistical methods can be employed to assess neuronal coactivity and its significance.

4.2 Agglomerative Detection of Cell Assemblies with CAD

Russo and Durstewitz [7] developed the CAD (cell assembly detection) algorithm to identify groups of cells that are active together above chance levels. The method first creates a binned spike matrix and then tests all combinations of pairs of cells for significant co-activation (Fig. 5a, b). For any pair of neurons, the number of spikes fired by the two neurons is counted at several time lags (time lag of 1 bin, 2 bins, etc.), and the time lag with the largest number of coupled spikes is retained. The counts are then tested for significance (Fig. 5c), and cell assemblies' identities are progressively constructed (Fig. 5d, e). A group of cells is considered as a cell assembly when no new cells can be added (Fig. 5f).

In recordings from a real brain, the number of spikes fired by a neuron can vary depending on the behavioral state of the animal, common driving inputs, and neural oscillations, giving rise to non-stationarities. To ensure that these global variations in cell firing rate do not influence the correlation statistics between cells, Russo and Durstewitz devised a novel pairwise test (Fig. 5b). The algorithm measures the asymmetry between the spikes of two neurons (or a neuron and a group of cells), comparing the number of spikes in the spike train ("forward direction") and the spike train replayed in reverse order. Indeed, if two cells A and B emit action potentials independently, it would be expected that the number of spikes fired by cell A after cell B to be equal to the number of spikes fired by cell B after cell A. If this number is skewed in one direction or another, it is inferred that the two cells have a functional relationship. This difference can then be used to compute the significance of the interaction between cells A and B.

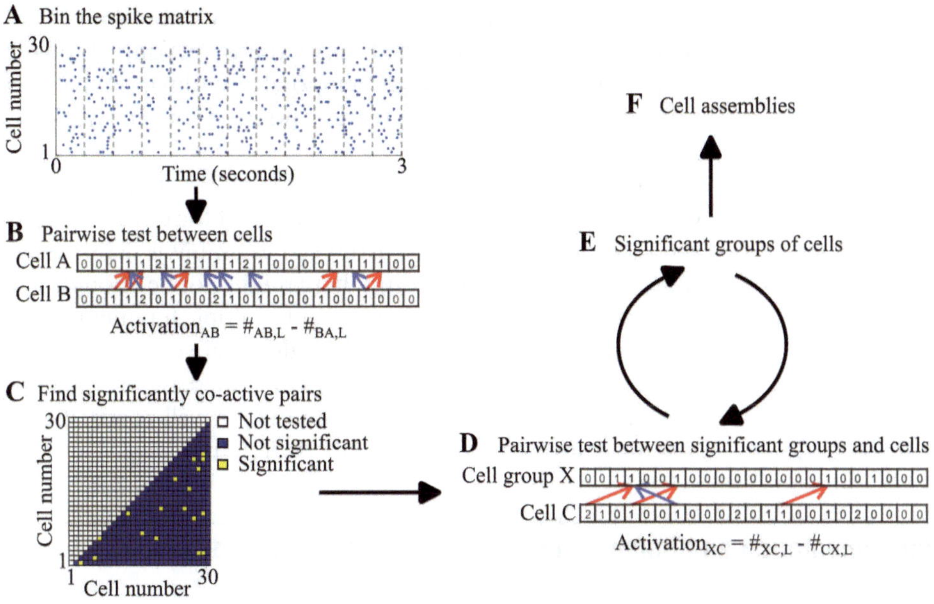

Fig. 5 The agglomerative assembly detection algorithm described by Russo and Durstewitz. (Adapted from Ref. [7]). (**a**). The dataset composed of N cells is binned in a window *dt* set by the user. (**b**). The coactivity of all pairs of cells in the dataset is assessed. $\#_{AB,L}$ is the number of times cell B is active after cell A at the time lag L; $\#_{BA,L}$ is the number of times cell A is active after cell B at the time lag L. Activation$_{AB}$ is the difference between the two. (**c**). The significantly coactive pairs of neurons are determined. (**d**, **e**). Cell groups are progressively expanded. The pairwise test is performed between significant cell group X found in the previous iteration and each remaining cell. Cells are added to the cell group if they are significantly coactive with the cell group. (**f**). The cell assemblies are obtained when no new cells can be added to the cell groups

This method was employed to detect cell assemblies in high-density electrophysiological recordings [101, 102]. Oettl et al. [101] analyzed midbrain dopaminergic and striatal projection neurons with this method and found that striatal projection neuron activity preceded midbrain dopaminergic neuron activity. Furthermore, cell assembly activity changed during the trials and was correlated to the value of a stimulus. They use a faster version of the algorithm, called CADopti, that only returns cell assemblies at their most significant time scale.

4.2.1 Advantages and Disadvantages of CAD

With agglomerative methods, the number of cell assemblies that can be detected is very high and could theoretically extend to the number of combinations among the recorded neurons. Furthermore, as cells are assessed individually for their participation in cell assemblies, the final identity of neurons composing each cell assembly is known. The same neuron can be included in more than one assembly. The method can also detect patterns of sequential activations of cells. Additionally, the statistical tests designed by Russo and colleagues are robust against non-stationarities in the data, thereby reducing false positives when applied to non-stationary time series. This feature also allows independent testing of different time scales in the recorded spike trains, making it possible to determine the timescale of temporal coordination between the cells of an assembly. Thus, there is no need for bin width to be an initialization parameter.

Note that, like many methods, the sensitivity of agglomerative clustering is dependent upon the activity of the cell assembly during the recording. Indeed, if a large cell assembly is active only a few times across a given recording there might not be enough statistical power to detect it as an assembly, but multiple fragments of it could be detected. The feasibility of agglomerative clustering in large datasets is challenged by an exponential increase in the number of statistical tests that must be carried out, as well as statistical power loss for datasets with large numbers of neurons.

5 Implementation and Comparison of the Respective Methods

For all the algorithms described above, code availability and links to the official repositories are presented in Table 1. To help the user select the best method for their analysis, Fig. 6 summarizes their main features.

For comparison, we implemented these methods on a simulated recording of 50 cells forming 50 cell assemblies over 60 min. Each assembly was composed of three cells (with individual cells participating in multiple assemblies). Cells were simulated using a Poisson point process with a mean firing rate of 1 spike per second. Cell assembly activations were added at random time points in the dataset and were active once per minute. All neurons composing the cell assembly fired together within a 30 ms window by adding random jitter to spike timings. Within cell assembly activations, the member neuronal activity was simulated with a Poisson point process with a mean firing rate of two spikes per assembly activation. The different methods were implemented using the toolboxes available online (Table 1). The code from Billeh et al. [20] was modified to run faster by reducing the binning window tenfold, optimizing the thresholding function, and enabling parallel processing on multiple cores (these modifications only impacted algorithm run time without altering the result). Where necessary, a binned spike matrix was constructed using a 30 ms binning window and a 15 ms overlap between two successive windows. In the methods that required setting a time window for cell assembly detection, this was set to 30 ms. For the remaining parameters, the default settings were used for the various methods. The results are presented in Table 2. There are large variations in run times, with some methods completing the cell assembly detection task in only a few seconds (PCA, ICA) and others taking

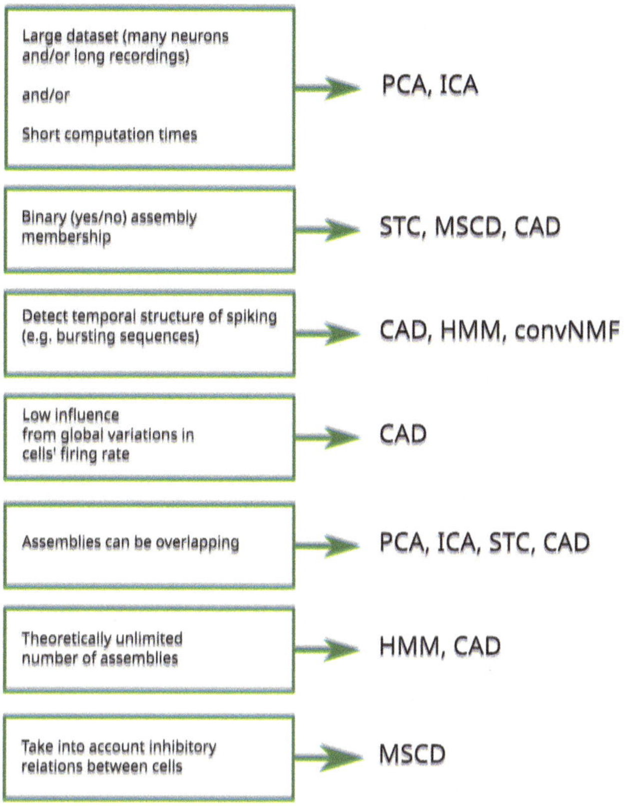

Fig. 6 Decision chart for the selection of cell assembly algorithms based upon experimental aims

minutes (MSCD, HMM, CAD). Note that none of the approaches detected the exact identity of the 50 cell assemblies present in the underlying dataset. Furthermore, the results also differed between methods and no two algorithms returned identical sets of cell assemblies. Of course, performance could be different with different data sets.

6 Detection of Assembly Activations

Some techniques can identify members of cell assemblies [7, 19, 20], while other methods identify components or patterns for which all recorded neurons participate to some extent [21–24] (but see [5, 6, 55] for methods to extract individual members from their components). After identifying cell assemblies, one may wish to detect the exact moments in time when the activations of a cell assembly occurred. This can be used to characterize behavioral correlates associated with these activations or the reactivation of cell assemblies during various behavioral and brain states such as

Table 2
Performance and metrics of the methods used in this review. The dataset was a simulated recording of 50 cells forming 50 cell assemblies over 60 min. Each assembly was composed of 3 cells. See text for the selected parameters for these implementations

Algorithm	Run time (s)	Number of assemblies detected	User-set parameters	Notes
PCA Peyrache et al. [21]	0.4	18	1	Parameter: time bin duration
ICA Lopes-dos-Santos et al. [22]	4.3	18	1	Parameter: time bin duration
seqNMF Mackevicius et al. [24]	12.0	(User set to 50 here)	3–6	Parameters: number of assemblies to detect (among others) Some parameters can be optimized with algorithms in the toolbox
STC Humphries [19]	12.0	10	1–2	Parameters: time bin and possibility to optionally choose the correlation method
MSCD Billeh et al. [20]	120.4	2	1	Parameter: Markov time. List of Markov times must be provided in advance
HMM Maboudi et al. [23]	97.4	(User set to 50 here)	2	Parameters: time bin; number of assemblies
CAD Russo and Durstewitz [7]	2918.7	90	2	Parameters: range of time bins, maximal lag to be tested Possibility to optionally choose the significance threshold

sleep (e.g., [103]). Determining the occurrence and eventually the strength of cell assemblies' activations is a problem distinct from the identification of the assemblies themselves.

As evoked above, certain methods described here return the identity of cell assemblies, but not the times at which these assemblies are active (PCA, ICA, STC, MSCD). To obtain the timing of activations of an assembly, typically, matrix convolution is used. Matrix convolution involves the multiplication of a matrix with a kernel (a pattern) to extract relevant features from the former. The kernel can be represented by either a vector or a matrix, generally of smaller size than the data matrix. These vectors contain either the binary identities of the members of a cell assembly (i.e., whether a cell is an assembly member or not) or continuous weights representing the degree of participation of the neuron to the given assembly. The vector resultant of the convolution represents the activation of the assembly over time. This will either be a discrete variable for assemblies with binary membership or continuous for assemblies with weights for each recorded cell.

In the case of weighted vectors, to overcome the limits of matrix convolution, Peyrache et al. [21] introduced a new metric to compute the similarity of the population vector in the binned spike matrix (Fig. 1e). The method computes the reactivation strength as the square of the weighted vector (with a nulled diagonal) projected onto the binned and z-scored spike matrix. By squaring the weighted vector, this method takes into account the coactivity of cells, rather than the firing rates of individual cells. The resulting reactivation strength can then be z-scored, and epochs surpassing a certain user-set z threshold can be retained as significant reactivations of the assembly. By utilizing a weighted cell vector, this method takes into account population activity as a whole (silent as well as coactive neurons, and positively and negatively correlated neurons). Furthermore, reactivation strength is sensitive to the bursting behavior of cells. Epochs when the cell assembly displays bursting behavior will have higher reactivation strengths than others. However, using this method, it is difficult to control the number of cells whose co-activation is required in order to consider a specific epoch as being a reactivation of the cell assembly. Indeed, bursts of activity of a subset of cells in the assembly can be sufficient to reach significance. As such, it is possible that two nonoverlapping sets of cells could result in detected activations of the same cell assembly.

7 Conclusions and Perspectives

Despite the diversity of the methods designed to detect cell assemblies using various mathematical approaches, they all share some common limitations.

It is important to note that in recordings of large nervous systems, such as in mammals, it is highly unlikely that any method can detect all members of a cell assembly, due to technical limitations in the number of cells that can be recorded simultaneously. Thus, an assembly consisting of only three members may represent the "tip of the iceberg," which contains many other unrecorded members.

The cell assembly is not simply a static representation but rather functions by transmitting a signal that may well be processed by other assemblies spanning the same, overlapping, or different structures. Currently developed methods seek to detect cell assemblies by analyzing the activity of typically a single structure sending the signal (but see [6]).

The techniques presented here are best at detecting assemblies that recur fairly frequently in the dataset. However, some cell assemblies, perhaps involved in processing rare behavioral or cognitive events, may be active only sporadically and thus may not be reliably detected.

At this time, to identify cells within assemblies, no empirical evidence exists to justify the choice between continuous weights or binary membership, and thus it is difficult to compare the accuracy of these respective methods. The choice of the method should pertain to the theoretical basis of the experimental question and the experimental design.

Synchrony between neurons can come from local connectivity, or one group of neurons exciting another, but also from mutual driving by a third structure, a sensory input, or another event such as UP/DOWN state transitions, etc. There is a risk that neuronal correlation measures could be influenced by these events. Non-stationary spike trains arising due to such events can be reduced by detecting cell assemblies during a neutral period such as sleep (taking into account eventual sleep oscillations). The method of Russo and Durstewitz [7] is another approach for dealing with non-stationarities.

Modern tools allow the recording of hundreds or even thousands of neurons simultaneously across different cerebral structures. The development of modern computational tools for the study of cerebral circuits enables the exploration of population activity in novel ways including the detection of cell assemblies, which previously were only theoretically hypothesized. Some methods, such as those based on dimensionality reduction or clustering approaches, also simplify the analysis of the dataset by reducing it into a smaller number of variables while preserving its principal properties. Others can detect numbers of cell assemblies larger than the number of neurons in the original dataset and thus may be able to explore and assess the dynamics and interactions of cell assemblies more thoroughly. Furthermore, most algorithms are based on neuronal activity discretized into a binned spike matrix, and some evidence suggests that binless detection algorithms may return superior results [19]. Future algorithms could address this issue as well as enable a more thorough view of the available cell assemblies through more sensitive algorithms as well as algorithms that are better optimized to run on very large datasets (e.g., [104–106]). Finally, in recent years, population activity has been investigated with the use of machine learning and neural networks (e.g., [107]), and such approaches could be employed for cell assembly detection.

Acknowledgments

We thank Dr. Ralitsa Todorova for constructive discussions and Dr. Mark Humphries, Noé Hamou, Vinicius Lima, Dr. Vitor Lopes dos Santos, Dr. Emily Mackevicius, and Pr. Eleonora Russo for useful comments on the manuscript. This work was funded by a LabEx MemoLife grant to SIW and MNP and by an FFRE grant to MNP. We apologize to the numerous authors inspiring this work, but whose studies we could not cite, and acknowledge that the selection of references has left out many important contributions.

References

1. Hebb DO (1949) The organization of behavior; a neuropsychological theory. Wiley, Oxford
2. Eichenbaum H (2018) Barlow versus Hebb: when is it time to abandon the notion of feature detectors and adopt the cell assembly as the unit of cognition? Neurosci Lett 680: 88–93. https://doi.org/10.1016/j.neulet.2017.04.006
3. Gray CM, Konig P, Engel AK, Singer W (1989) Oscillatory responses in cat visual cortex exhibit inter-columnar synchronization which reflects globalstimulus properties. Nature 338:334–337
4. Buzsáki G (2010) Neural syntax: cell assemblies, synapsembles, and readers. Neuron 68: 362–385. https://doi.org/10.1016/j.neuron.2010.09.023
5. Pompili MN, Boucly CJ, Todorova R, Leroux EM, Wiener SI, Zugaro M (2022) Flexible communication between cell assemblies and "reader" neurons. bioRxiv 2022.09.06.506754
6. Oberto VJ, Boucly CJ, Gao HY, Todorova R, Zugaro MB, Wiener SI (2022) Distributed cell assemblies spanning prefrontal cortex and striatum. Curr Biol 32:1–13. https://doi.org/10.1016/j.cub.2021.10.007
7. Russo E, Durstewitz D (2017) Cell assemblies at multiple time scales with arbitrary lag constellations. elife 6:1–31. https://doi.org/10.7554/eLife.19428
8. Yang W, Yuste R (2017) In vivo imaging of neural activity. Nat Methods 14:349–359. https://doi.org/10.1038/nmeth.4230
9. Carrillo-Reid L, Yuste R (2020) Playing the piano with the cortex: role of neuronal ensembles and pattern completion in perception and behavior. Curr Opin Neurobiol 64: 89–95. https://doi.org/10.1016/j.conb.2020.03.014
10. Carrillo-Reid L, Yang W, Bando Y, Peterka DS, Yuste R (2016) Imprinting cortical ensembles. Science 353:691–694
11. Chen TW, Wardill TJ, Sun Y, Pulver SR, Renninger SL, Baohan A, Schreiter ER, Kerr RA, Orger MB, Jayaraman V, Looger LL, Svoboda K, Kim DS (2013) Ultrasensitive fluorescent proteins for imaging neuronal activity. Nature 499:295–300. https://doi.org/10.1038/nature12354
12. Ali F, Kwan AC (2019) Interpreting in vivo calcium signals from neuronal cell bodies, axons, and dendrites: a review. Neurophotonics 7:1. https://doi.org/10.1117/1.nph.7.1.011402

13. Huang L, Ledochowitsch P, Knoblich U, Lecoq J, Murphy GJ, Reid RC, de Vries SEJ, Koch C, Zeng H, Buice MA, Waters J, Li L (2021) Relationship between simultaneously recorded spiking activity and fluorescence signal in gcamp6 transgenic mice. elife 10:1–19. https://doi.org/10.7554/eLife.51675
14. Mölter J, Avitan L, Goodhill GJ (2018) Detecting neural assemblies in calcium imaging data. BMC Biol 16:1–20. https://doi.org/10.1186/s12915-018-0606-4
15. Gonzalez WG, Zhang H, Harutyunyan A, Lois C (2019) Persistence of neuronal representations through time and damage in the hippocampus. Science 365:821–825. https://doi.org/10.1126/science.aav9199
16. Carrillo-Reid L, Yang W, Kang Miller JE, Peterka DS, Yuste R (2017) Imaging and optically manipulating neuronal ensembles. Annu Rev Biophys 46:271–293. https://doi.org/10.1146/annurev-biophys-070816-033647
17. Wilson MA, McNaughton BL (1994) Reactivation of hippocampal ensemble memories during sleep. Science 265:676–679
18. Kudrimoti HS, Barnes CA, McNaughton BL (1999) Reactivation of hippocampal cell assemblies: effects of behavioral state, experience, and EEG dynamics. J Neurosci 19: 4090–4101. https://doi.org/10.1523/jneurosci.19-10-04090.1999
19. Humphries MD (2011) Spike-train communities: finding groups of similar spike trains. J Neurosci 31:2321–2336
20. Billeh YN, Schaub MT, Anastassiou CA, Barahona M, Koch C (2014) Revealing cell assemblies at multiple levels of granularity. J Neurosci Methods 236:92–106. https://doi.org/10.1016/j.jneumeth.2014.08.011
21. Peyrache A, Benchenane K, Khamassi M, Wiener SI, Battaglia FP (2010) Principal component analysis of ensemble recordings reveals cell assemblies at high temporal resolution. J Comput Neurosci 29:309–325
22. Lopes-dos-Santos V, Ribeiro S, Tort ABL (2013) Detecting cell assemblies in large neuronal populations. J Neurosci Methods 220: 149–166. https://doi.org/10.1016/j.jneumeth.2013.04.010
23. Maboudi K, Ackermann E, de Jong LW, Pfeiffer BE, Foster D, Diba K, Kemere C (2018) Uncovering temporal structure in hippocampal output patterns. elife 7:e34467
24. Mackevicius EL, Bahle AH, Williams AH, Gu S, Denisenko NI, Goldman MS, Fee MS (2019) Unsupervised discovery of temporal sequences in high-dimensional datasets, with applications to neuroscience. elife 8:e38471
25. Cunningham JP, Yu BM (2014) Dimensionality reduction for large-scale neural recordings. Nat Neurosci 17:1500–1509. https://doi.org/10.1038/nn.3776
26. Freedman DA (2009) Statistical models: theory and practice. Cambridge University Press
27. Demars F, Todorova R, Makdah G, Forestier A, Krebs MO, Godsil BP, Jay TM, Wiener SI, Pompili MN (2022) Post-trauma behavioral phenotype predicts the degree of vulnerability to fear relapse after extinction in male rats. Curr Biol 32:3180–3188. https://doi.org/10.1016/j.cub.2022.05.050
28. Pearson K (1901) On lines and planes of closest fit to systems of points in space. Philos Mag 2:559–572
29. Jolliffe IT (2002) Principal component analysis, second edition. Springer
30. Richmond BJ, Optican LM, Podell M, Spitzer H (1987) Temporal encoding of two-dimensional patterns by single units in primate inferior temporal cortex. I. Response characteristics. J Neurophysiol 57:132–146
31. Richmond BJ, Optican LM, Spitzer H (1990) Temporal encoding of two-dimensional patterns by single units in primate primary visual cortex. I. Stimulus-response relations. J Neurophysiol 64:351–369
32. Mcclurkin JW, Optican LM, Richmond BJ, Gawne TJ (1991) Concurrent processing and complexity of temporally encoded neuronal messages in visual perception. Science 253:675–677
33. Kjaer TW, Hertz JA, Richmond BJ (1994) Decoding cortical neuronal signals: network models, information estimation and spatial tuning. J Comput Neurosci 1:109–139. https://doi.org/10.1007/BF00962721
34. Nicolelis MAL, Baccala LA, Lin RCS, Chapin JK (1995) Sensorimotor encoding by synchronous neural ensemble activity at multiple levels of the somatosensory system. Science 268:1353–1358. https://doi.org/10.1126/science.7761855
35. Chapin JK, Nicolelis MAL (1999) Principal component analysis of neuronal ensemble activity reveals multidimensional somatosensory representations. J Neurosci Methods 94: 121–140. https://doi.org/10.1016/S0165-0270(99)00130-2
36. Marchenko VA, Pastur LA (1967) Distribution of eigenvalues for some sets of random matrices. Mat Sb 114:507–536
37. Peyrache A, Khamassi M, Benchenane K, Wiener SI, Battaglia FP (2009) Replay of rule-

learning related neural patterns in the prefrontal cortex during sleep. Nat Neurosci 12: 919–926. https://doi.org/10.1038/nn.2337

38. Benchenane K, Peyrache A, Khamassi M, Tierney PL, Gioanni Y, Battaglia FP, Wiener SI (2010) Coherent theta oscillations and reorganization of spike timing in the hippocampal- prefrontal network upon learning. Neuron 66:921–936. https://doi.org/10.1016/j.neuron.2010.05.013
39. Gulati T, Ramanathan DS, Wong CC, Ganguly K (2014) Reactivation of emergent task-related ensembles during slow-wave sleep after neuroprosthetic learning. Nat Neurosci 17:1107–1113. https://doi.org/10.1038/nn.3759
40. Ramanathan DS, Gulati T, Ganguly K (2015) Sleep-dependent reactivation of ensembles in motor cortex promotes skill consolidation. PLoS Biol 13:e1002263. https://doi.org/10.1371/journal.pbio.1002263
41. Gulati T, Won SJ, Ramanathan DS, Wong CC, Bodepudi A, Swanson RA, Ganguly K (2015) Robust neuroprosthetic control from the stroke perilesional cortex. J Neurosci 35: 8653–8661. https://doi.org/10.1523/JNEUROSCI.5007-14.2015
42. Tang W, Shin JD, Frank LM, Jadhav SP (2017) Hippocampal-prefrontal reactivation during learning is stronger in awake compared with sleep states. J Neurosci 37:11789–11805. https://doi.org/10.1523/JNEUROSCI.2291-17.2017
43. Rothschild G, Eban E, Frank LM (2017) A cortical–hippocampal–cortical loop of information processing during memory consolidation. Nat Neurosci 20:251–259. https://doi.org/10.1038/nn.4457
44. Chenani A, Sabariego M, Schlesiger MI, Leutgeb JK, Leutgeb S, Leibold C (2019) Hippocampal CA1 replay becomes less prominent but more rigid without inputs from medial entorhinal cortex. Nat Commun 10: 1–13. https://doi.org/10.1038/s41467-019-09280-0
45. Sjulson L, Peyrache A, Cumpelik A, Cassataro D, Buzsáki G (2018) Cocaine place conditioning strengthens location-specific hippocampal coupling to the nucleus accumbens. Neuron 98:926–934.e5. https://doi.org/10.1016/j.neuron.2018.04.015
46. Deolindo CS, Kunicki ACB, Brasil FL, Moioli RC (2014) Limitations of principal component analysis as a method to detect neuronal assemblies. In: 2014 IEEE 16th international conference on E-health networking, applications and services (Healthcom). IEEE, pp 24–30
47. Dejean C, Courtin J, Karalis N, Chaudun F, Wurtz H, Bienvenu TCM, Herry C (2016) Prefrontal neuronal assemblies temporally control fear behaviour. Nature 535:420–424. https://doi.org/10.1038/nature18630
48. Koch C, Rapp M, Segev I (1996) A brief history of time (constants). Cereb Cortex 6: 93–101. https://doi.org/10.1093/cercor/6.2.93
49. Harris KD (2005) Neural signatures of cell assembly organization. Nat Rev Neurosci 6: 399–407. https://doi.org/10.1038/nrn1669
50. Brown GD, Yamada S, Sejnowski TJ (2001) Independent component analysis at the neural cocktail party. Trends Neurosci 24:54–63
51. Comon P (1994) Independent component analysis, a new concept? Signal Process 36: 287–314
52. Laubach M, Shuler M, Nicolelis MAL (1999) Independent component analyses for quantifying neuronal ensemble interactions. J Neurosci Methods 94:141–154
53. Almeida-Filho DG, Lopes-dos-Santos V, Vasconcelos NAP, Miranda JGV, Tort ABL, Ribeiro S (2014) An investigation of Hebbian phase sequences as assembly graphs. Front Neural Circuits 8:1–13. https://doi.org/10.3389/fncir.2014.00034
54. Bower MR, Stead M, Bower RS, Kucewicz MT, Sulc V, Cimbalnik J, Brinkmann BH, Vasoli VM, St Louis EK, Meyer FB, Marsh WR, Worrell GA (2015) Evidence for consolidation of neuronal assemblies after seizures in humans. J Neurosci 35:999–1010. https://doi.org/10.1523/JNEUROSCI.3019-14.2015
55. van de Ven GM, Trouche S, McNamara CG, Allen K, Dupret D (2016) Hippocampal offline reactivation consolidates recently formed cell assembly patterns during sharp wave-ripples. Neuron 92:968–974. https://doi.org/10.1016/j.neuron.2016.10.020
56. Trouche S, Perestenko PV, van de Ven GM, Bratley CT, McNamara CG, Campo-Urriza-N, Black SL, Reijmers LG, Dupret D (2016) Recoding a cocaine-place memory engram to a neutral engram in the hippocampus. Nat Neurosci 19:564–567. https://doi.org/10.1038/nn.4250
57. Conde-Ocazionez S, Altavini TS, Wunderle T, Schmidt KE (2018) Motion contrast in primary visual cortex: a direct comparison of single neuron and population encoding. Eur J Neurosci 47:358–369

58. Middleton SJ, Kneller EM, Chen S, Ogiwara I, Montal M, Yamakawa K, McHugh TJ (2018) Altered hippocampal replay is associated with memory impairment in mice heterozygous for the scn2a gene. Nat Neurosci 21:996–1003. https://doi.org/10.1038/s41593-018-0163-8
59. Deolindo CS, Kunicki ACB, da Silva MI, Lima Brasil F, Moioli RC (2018) Neuronal assemblies evidence distributed interactions within a tactile discrimination task in rats. Front Neural Circuits 11:114
60. See JZ, Atencio CA, Sohal VS, Schreiner CE (2018) Coordinated neuronal ensembles in primary auditory cortical columns. elife 7: e35587
61. Giri B, Miyawaki H, Mizuseki K, Cheng S, Diba K (2019) Hippocampal reactivation extends for several hours following novel experience. J Neurosci 39:866–875. https://doi.org/10.1523/JNEUROSCI.1950-18.2018
62. Todorova R, Zugaro MB (2019) Isolated cortical computations during delta waves. Science 366:377–381
63. Trouche S, Koren V, Doig NM, Ellender TJ, El-Gaby M, Lopes-dos-Santos V, Reeve HM, Perestenko PV, Garas FN, Magill PJ, Sharott A, Dupret D (2019) A hippocampus-accumbens tripartite neuronal motif guides appetitive memory in space. Cell 176:1393–1406.e16. https://doi.org/10.1016/j.cell.2018.12.037
64. Oliva A, Fernández-Ruiz A, Leroy F, Siegelbaum SA (2020) Hippocampal CA2 sharp-wave ripples reactivate and promote social memory. Nature 587:264–269. https://doi.org/10.1038/s41586-020-2758-y
65. Fernández-Ruiz A, Oliva A, Soula M, Rocha-Almeida F, Nagy GA, Martin-Vazquez G, Buzsáki G (2021) Gamma rhythm communication between entorhinal cortex and dentate gyrus neuronal assemblies. Science 372: eabf3119. https://doi.org/10.1126/science.abf3119
66. McKenzie S, Huszár R, English DF, Kim K, Christensen F, Yoon E, Buzsáki G (2021) Preexisting hippocampal network dynamics constrain optogenetically induced place fields. Neuron 109:1040–1054.e7. https://doi.org/10.1016/j.neuron.2021.01.011
67. El-Gaby M, Reeve HM, Lopes-dos-Santos V, Campo-Urriza N, Perestenko PV, Morley A, Strickland LAM, Lukács IP, Paulsen O, Dupret D (2021) An emergent neural coactivity code for dynamic memory. Nat Neurosci 24:694–704. https://doi.org/10.1038/s41593-021-00820-w
68. Pompili MN, Hamou N, Wiener SI (2024) Differential encoding of fear learning and fear expression in the ventral and dorsal hippocampus. BioRxiv preprint. https://doi.org/10.1101/2024.01.26.577384
69. Guan H, Middleton SJ, Inoue T, McHugh TJ (2021) Lateralization of CA1 assemblies in the absence of CA3 input. Nat Commun 12: 6114
70. Himberg J, Hyvärinen A, Esposito F (2004) Validating the independent components of neuroimaging time series via clustering and visualization. NeuroImage 22:1214–1222
71. Hyvärinen A, Oja E (2000) Independent component analysis: algorithms and applications. Neural Netw 13:411–430. https://doi.org/10.1016/S0893-6080(00)00026-5
72. Hyvärinen A, Ramkumar P (2013) Testing independent component patterns by inter-subject or inter-session consistency. Front Hum Neurosci 7:94
73. Klemm M, Haueisen J, Ivanova G (2009) Independent component analysis: comparison of algorithms for the investigation of surface electrical brain activity. Med Biol Eng Comput 47:413–423
74. Paatero P, Tapper U (1994) Positive matrix factorization: a non-negative factor model with optimal utilization of error estimates of data values. Environmetrics 5:111–126
75. Lee DD, Seung HS (1999) Learning the parts of objects by non-negative matrix factorization. Nature 401:788–791
76. Peter S, Kirschbaum E, Both M, Campbell L, Harvey B, Heins C, Durstewitz D, Diego F, Hamprecht FA (2017) Sparse convolutional coding for neuronal assembly detection. Adv Neural Inf Process Syst 30
77. Tingley D, Buzsáki G (2020) Routing of hippocampal ripples to subcortical structures via the lateral septum. Neuron 105:138–149.e5. https://doi.org/10.1016/j.neuron.2019.10.012
78. Chen Z, Cichocki A (2005) Nonnegative matrix factorization with temporal smoothness and/or spatial decorrelation constraints. Lab Adv Brain Signal Process RIKEN, Tech Rep 68
79. Choi S (2008) Algorithms for orthogonal nonnegative matrix factorization. Neural Netw:1828–1832
80. Hastie T, Tibshirani R, Friedman JH (2009) The elements of statistical learning: data mining, inference, and prediction. Springer
81. Williams AH, Degleris A, Wang Y, Linderman SW (2020) Point process models for sequence detection in high-dimensional neural spike

trains. Adv Neural Inf Process Syst 33:14350–14361

82. Robotka H, Thomas L, Yu K, Wood W, Elie JE, Gahr M, Theunissen FE (2023) Sparse ensemble neural code for a complete vocal repertoire. Cell Rep 42:112034
83. Ghandour K, Ohkawa N, Fung CCA, Asai H, Saitoh Y, Takekawa T, Okubo-Suzuki R, Soya S, Nishizono H, Matsuo M, Osanai M, Sato M, Ohkura M, Nakai J, Hayashi Y, Sakurai T, Kitamura T, Fukai T, Inokuchi K (2019) Orchestrated ensemble activities constitute a hippocampal memory engram. Nat Commun 10:1–14. https://doi.org/10.1038/s41467-019-10683-2
84. Grosmark AD, Sparks FT, Davis MJ, Losonczy A (2021) Reactivation predicts the consolidation of unbiased long-term cognitive maps. Nat Neurosci 24:1574–1585
85. Pietri T, Romano SA, Pérez-Schuster V, Boulanger-Weill J, Candat V, Sumbre G (2017) The emergence of the spatial structure of tectal spontaneous activity is independent of visual inputs. Cell Rep 19:939–948
86. Girvan M, Newman MEJ (2002) Community structure in social and biological networks. Proc Natl Acad Sci USA 99:7821–7826. https://doi.org/10.1073/pnas.122653799
87. Satuvuori E, Kreuz T (2018) Which spike train distance is most suitable for distinguishing rate and temporal coding? J Neurosci Methods 299:22–33. https://doi.org/10.1016/j.jneumeth.2018.02.009
88. Lancichinetti A, Fortunato S (2012) Consensus clustering in complex networks. Sci Rep 2: 336. https://doi.org/10.1038/srep00336
89. Shimono M, Beggs JM (2015) Functional clusters, hubs, and communities in the cortical microconnectome. Cereb Cortex 25: 3743–3757. https://doi.org/10.1093/cercor/bhu252
90. Pinotsis DA, Miller EK (2023) In vivo ephaptic coupling allows memory network formation. Cereb Cortex 33:9877–9895. https://doi.org/10.1093/cercor/bhad251
91. McMahon C, Kowalski DP, Krupka AJ, Lemay MA (2022) Single-cell and ensemble activity of lumbar intermediate and ventral horn interneurons in the spinal air-stepping cat. J Neurophysiol 127:99–115. https://doi.org/10.1152/jn.00202.2021
92. Arthur D, Vassilvitskii S (2007) K-means++: the advantages of careful seeding. Stanford
93. Bruno AM, Frost WN, Humphries MD (2015) Modular deconstruction reveals the dynamical and physical building blocks of a locomotion motor program. Proceedings of the eighteenth annual ACM-SIAM symposium on Discrete algorithms 86:304–318
94. Levin D, Peres Y (2017) Markov chains and mixing times. American Mathematical Society
95. Delvenne JC, Yaliraki SN, Barahona M (2010) Stability of graph communities across time scales. Proc Natl Acad Sci USA 107: 12755–12760. https://doi.org/10.1073/pnas.0903215107
96. Miyawaki H, Billeh YN, Diba K (2017) Low activity microstates during sleep. Sleep 40. https://doi.org/10.1093/sleep/zsx066
97. Ponce-Alvarez A, Nácher V, Luna R, Riehle A, Romo R (2012) Dynamics of cortical neuronal ensembles transit from decision making to storage for later report. J Neurosci 32:11956–11969. https://doi.org/10.1523/JNEUROSCI.6176-11.2012
98. Mazzucato L, Fontanini A, La Camera G (2015) Dynamics of multistable states during ongoing and evoked cortical activity. J Neurosci 35:8214–8231. https://doi.org/10.1523/JNEUROSCI.4819-14.2015
99. Noguchi A, Huszár R, Morikawa S, Buzsáki G, Ikegaya Y (2022) Inhibition allocates spikes during hippocampal ripples. Nat Commun 13:1280. https://doi.org/10.1038/s41467-022-28890-9
100. Bollmann L, Baracskay P, Stella F, Csicsvari J (2023) Sleep stages antagonistically modulate reactivation drift. BioRxiv preprint. https://doi.org/10.1101/2023.10.13.562165
101. Oettl LL, Scheller M, Filosa C, Wieland S, Haag F, Loeb C, Durstewitz D, Shusterman R, Russo E, Kelsch W (2020) Phasic dopamine reinforces distinct striatal stimulus encoding in the olfactory tubercle driving dopaminergic reward prediction. Nat Commun 11:3460. https://doi.org/10.1038/s41467-020-17257-7
102. Londei F, Arena G, Ferrucci L, Russo E, Ceccarelli F, Genovesio A (2024) Connecting the dots in the zona incerta: a study of neural assemblies and motifs of inter-area coordination in mice. Iscience 27:108761
103. Pompili MN, Todorova R (2022) Discriminating sleep from freezing with cortical spindle oscillations. Front Neural Circuits 16:783768. https://doi.org/10.3389/fncir.2022.783768
104. Jun JJ, Steinmetz NA, Siegle JH, Denman DJ, Bauza M, Barbarits B, Lee AK, Anastassiou CA, Andrei A, Aydın Ç (2017) Fully integrated silicon probes for high-density recording of neural activity. Nature 551:232

105. Steinmetz NA, Aydin C, Lebedeva A, Okun M, Pachitariu M, Bauza M, Beau M, Bhagat J, Böhm C, Broux M, Chen S, Colonell J, Gardner RJ, Karsh B, Kloosterman F, Kostadinov D, Mora-Lopez-C, O'Callaghan J, Park J, Putzeys J, Sauerbrei B, van Daal RJJ, Vollan AZ, Wang S, Welkenhuysen M, Ye Z, Dudman JT, Dutta B, Hantman AW, Harris KD, Lee AK, Moser EI, O'Keefe J, Renart A, Svoboda K, Häusser M, Haesler S, Carandini M, Harris TD (2021) Neuropixels 2.0: a miniaturized high-density probe for stable, long-term brain recordings. Science 80:372. https://doi.org/10.1126/science.abf4588

106. Ghestem A, Pompili MN, Dipper-Wawra M, Quilichini PP, Bernard C, Ferraris M (2023) Long-term near-continuous recording with Neuropixels probes in healthy and epileptic rats. bioRxiv 1–19. https://doi.org/10.1101/2023.02.16.528689

107. Pandarinath C, O'Shea DJ, Collins J, Jozefowicz R, Stavisky SD, Kao JC, Trautmann EM, Kaufman MT, Ryu SI, Hochberg LR, Henderson JM, Shenoy KV, Abbott LF, Sussillo D (2018) Inferring single-trial neural population dynamics using sequential auto-encoders. Nat Methods 15:805–815. https://doi.org/10.1038/s41592-018-0109-9

108. Terada et al (2022) Adaptive stimulus selection for consolidation in the hippocampus. Springer Nature. https://doi.org/10.1038/s41586-021-04118-6

Chapter 3

A General and Accurate Method for Neuronal Ensemble Detection in Spiking Neural Networks

Rubén Herzog-Amunátegui, Soraya Mora, Garance Prada, Jean-Gabriel Minonzio, Maria Jose Escobar, Rodrigo Cofre, and Adrián G. Palacios

Abstract

We present a novel and scalable approach to accurately identify neuronal ensembles in spiking neuron populations. This method, which is part of a previously published methodology, uses minimal parameter tuning requirements and improved computational efficiency, making it a valuable tool for researchers studying complex ensemble activity in neural circuits. Using clustering of synchronous activities, our methodology allows neurons to be part of multiple ensembles, demonstrating its effectiveness across a wide range of simulation parameters. We also demonstrate the versatility of this technique, applying it to both artificially generated data and spike trains obtained from retinal ganglion cells via multielectrode array recordings. A comparison of the results reveals the superior performance of the method and its wider applicability compared to other prevalent techniques in the field. Our investigations uncover a consistent pattern of stimuli-induced activity, in addition to spontaneously active and sporadic ensembles. These findings suggest the potential compartmentalization of the early visual system into specific functional ensembles. To make this innovative method more accessible and easier to use, we provide a user-friendly graphical interface in this chapter. Our goal is to provide readers with a comprehensive understanding and the resources needed to implement a reliable technique, thus advancing our collective knowledge of the intricate functional networks of the central nervous system.

Key words Neural ensembles, Spiking neural networks, Clustering

1 Introduction

The scientific community has long been fascinated by the intricate dynamics of neuronal networks. Donald Hebb's concept of "cell assemblies" proposed that neurons' networks can act together to encode certain pieces of information [1]. Subsequent research has shown the importance of neuronal ensembles in a variety of brain functions, from sensory processing to memory formation [2]. It is now widely accepted that neurons do not act independently, and

Luis Carrillo-Reid (ed.), *Identification, Characterization, and Manipulation of Neuronal Ensembles*, Neuromethods, vol. 215,
https://doi.org/10.1007/978-1-0716-4208-5_3,

their coordinated firing has been extensively studied to uncover the coding mechanisms of the brain [3]. More recently, spiking neural networks (SNNs) have been developed to more accurately simulate discrete firing events of biological neurons, providing a bridge between biological observation and computational modeling [4].

The utilization of high-density electrode matrix recording techniques has caused a dramatic rise in the quantity of neural data, which necessitates the creation of methods for analysis that are both efficient and precise. As datasets become larger, the capacity of analysis methods to scale is essential, but also critical, to guarantee that the detected ensembles are genuine representations and not artifacts. Furthermore, investigating and comprehending neuronal ensembles can provide insight into visual information processing and have practical applications, such as the development of brain-machine interfaces, neural prosthetics, and neuromorphic computing systems. In recent years, the ability to record and analyze the activity of hundreds of neurons simultaneously has been made possible [5]. This has enabled researchers to uncover synchronous activity between neurons on different spatial and temporal scales [6–8]. These groups of neurons that fire together, known as neuronal ensembles (also called cell assemblies), are believed to be a basic unit of neural processes and form the basis of organized collective brain dynamics [1, 9–11].

It has been suggested that functional units of neural processes are not individual neurons, but rather groups of neurons that fire together, which has been a major shift in the field of computational neuroscience [9]. This concept of neuronal ensembles has been proposed as a key factor in the dynamics of the entire brain related to cognitive functions, such as integration and segregation [6]. Recent advances in technology, such as multi-electrode arrays (MEA) and calcium imaging, have allowed the recording of hundreds and even thousands of neurons simultaneously [8, 12–16] determining the analysis of neuronal ensembles and the exploration of how collective neuronal activity is generated in the brain [17]. These studies have revealed that a common feature of population activity is the organization of neurons into groups, providing new insights into the neural code [12, 18–20]. In addition, activation of certain ensembles has been associated with spontaneous and stimulus-evoked brain function [21]. Neurological and mental impairments can disrupt population activity and thus affect neuronal ensembles, as demonstrated by in vivo two-photon calcium imaging data in mice with epilepsy, schizophrenia, and autism [22–25].

Identifying and extracting features from high-dimensional spiking data of neuronal ensembles is a difficult task. Ensembles can vary in size and activity rate, with some neurons participating in multiple ensembles and not all neurons firing when the ensemble is active. Furthermore, ensembles can be temporally extended, overlapping, or organized hierarchically or modularly, making it difficult to differentiate between them [5]. Various techniques and ideas have been used to detect and interpret neuronal ensembles (*see* [3, 18, 26] for reviews). Examples include principal component analysis (PCA) [27–30], correlation between neurons [30–32], correlation between population vectors [21], statistical evaluation of patterns [33–36], and non-negative matrix factorization [29, 37]. However, most of these methods are computationally expensive and require tuning of multiple parameters, making them difficult to use by the scientific community.

Carrillo and Yuste have elucidated the inconsistencies in the conceptualization of neural ensembles [38]. They quantitatively characterize neural ensembles as population vectors, with the dimensionality dictated by the population size. In this chapter, we adopt their definition, interpreting the binary response of the population within each time-bin interval as the population vector. To further elucidate, we specify that a neural ensemble should encompass a core group of neurons whose temporal activity exhibits significant correlation with the overall temporal activity of the ensemble. Conversely, noncore cells may or may not be part of the ensemble, but their correlation with the ensemble's activation times does not achieve statistical significance. The core of an ensemble comprises the subgroup of neurons that most accurately represents the ensemble's activity, indicating functional coupling among these neurons. For the detection of neural ensembles in simultaneous neuronal recordings, we have developed a method that integrates dimensionality reduction, clustering, and nonparametric statistical tests. Our approach utilizes principal component analysis (PCA) to project population vectors into a low-dimensional space, thereby facilitating reliable clustering. Following clustering, we apply a nonparametric statistical test to identify the core cells. This method ensures that our definition of ensembles as multidimensional population vectors with a core is adhered to. This approach not only provides a robust framework for defining neural ensembles but also offers a precise method for their detection, which is crucial for advancing our understanding of neural dynamics and functional connectivity.

We have tested and validated our method using the first synthetic data, demonstrating its accuracy and wide applicability to different scenarios created by synthetic data with various characteristics [17]. Our tests on simulated data show excellent detection performance for the number of components, the activation times of the components, and the detection of core cells across a wide range of parameters. Furthermore, we demonstrate the functionality of our method in spiking neuronal data recorded using MEA from mouse in vitro retinal ganglion cells (RGCs) obtained during simple ON-OFF stimuli, i.e., consecutive changes between light and darkness [17]. The retina, our biological model here, is part of the central nervous system composed of thousands of neurons of various types [39–41], organized into stratified nuclear and plexiform layers [42]. The neural network of the retina can process various features of the visual scene [43], whose "computational" results are transmitted to the brain through the optic nerve, a neural tract composed mainly of RGC axons [41, 42, 44]. One of the most remarkable and straightforward examples of retinal processing is the separation of ON, OFF, or ON-OFF responses of RGCs, an increase or decrease in the firing rate responding to changes in light intensity. In this case, connectivity between RGCs and bipolar cells in the inner plexiform layer plays an important role, determining the tendency of RGCs to fire when light increases, decreases, or both [42, 45] and represents the broadest functional classification of RGCs into ON, OFF, and ON-OFF cell types.

We hypothesize that the retinal ensembles possess properties analogous to those observed in certain parts of the brain. However, there are also notable distinctions between the retinal and brain circuits in terms of their structural connectivity. The retina, different from the brain, does not have clear internal feedback circuitry and is heavily reliant on the temporal synchrony imposed by the visual stimulus for neural coactivation. Therefore, the neural assemblies in the retina have both a structural and a functional component; the method described in this chapter focuses on the latter as we previously demonstrated [17].

2 Materials and Methods

To make our technique available, we have created a graphical user interface (GUI) and the respective Matlab and Python code that implement our algorithm, designed to provide a fast, efficient, and precise solution to the problem of recognizing neuronal ensembles in multiunit recordings.

We consider binary spike trains arranged in a matrix $\mathbf{S}_{N \times T}$ where N corresponds to the number of neurons and T the number of time bins. The entries in the matrix denoted by $\mathbf{S}_{n,t}$ are equal to one if the n-th neuron is active in the t-th time bin and zero otherwise. At each time bin t, there is a binary population vector of active and silent neurons $(\mathbf{S}_{1,t}, \ldots, \mathbf{S}_{N,t})$.

2.1 Ethics Statement

Regarding retinal data, animal manipulation and breeding and the corresponding experiments were approved by the Bioethics Committee of the University of Valparaíso, according to the bioethics regulation of the National Research Agency (ANID) and international protocols.

2.2 Ensemble Detection

We use PCA to reduce the dimensionality of population vectors. This method extracts a set of orthogonal directions that capture the most significant variability in the population vectors. We excluded population vectors with fewer than three spikes (a parameter that can be adjusted in the GUI) and projected the remaining population vectors onto the first six principal components (PCs). However, the optimal number of PCs may differ depending on the data. The selection of this parameter is based on the cumulative explained variance of the PCs and the cutoff point of the eigenvalue spectrum of the PC decomposition, which is available in the GUI. Finally, we calculate the Euclidean distance between the population vectors projected onto the PC space to obtain a distance matrix. This matrix describes the dissimilarity between population vectors and is essential for the clustering step.

We employ a hard clustering algorithm based on a systematic examination of the distance matrix. Each data point (i.e., a population vector) was characterized by two parameters: the density ρ and the minimum distance to a point with higher density δ. The density of a point reflects how close (in terms of the distance matrix) its closest neighbors are and was calculated as $\rho_i = 1/\overline{d_{i,v}}$, where $\overline{d_{i,v}}$ was the average distance from point i to its closest neighbors (generally 2% were considered). Population vectors with relatively high values of ρ and δ were considered potential centroids, as they have the most points in their vicinity and are far from other points with high density. To detect centroids of the clusters, we used the method described in [46], which is a modified version of the technique developed in [47]. We fit a power law to the δ vs. ρ curve, using the 99.9% upper confidence boundary as a threshold. All points that fell outside this boundary were considered centroids, and the rest of the points were assigned to their closest centroids, forming the clusters. Since each population vector is associated with a time bin, the clustering step yields the cluster activation times.

Core cells are the most telling examples of a neural ensemble, and their existence, which can be established through a statistical test, implies a link between the level of individual neurons and the level of the ensemble, that is, a statistical dependence on multiple scales. After identifying potential groups and centroids, the Pearson correlation coefficient between the activation times of the neuron n and the group e, indicated by corr (n, e), was calculated for each pair (n, e). This correlation is 1 if the neuron n and the group e are always active in the same time intervals and it is -1 if they are never active in the same time intervals. The values between signify combinations of neurons and clusters that tend to be active in the same or different time intervals.

To differentiate between a core and a noncore cell, we established a threshold based on a null hypothesis generated from shuffled versions of the (n, e) pair. For each cluster $e \in E$, the number of times it was active was kept constant, but the temporal activation was randomly rearranged, resulting in a new temporal sequence of activation, denoted e_r. This process was repeated 5000 times for each group, producing a distribution of corr (n, e_r) for the neuron n in the group e, which represented the null hypothesis distribution associated with the correlation between a neuron and a group given by chance. We set a significance threshold of $p < 0.001$ as the 99.9th percentile of this distribution (free parameter). Therefore, if corr $(n, e) > \theta_e$, we conclude that n was a core cell of the group e, as its correlation was above the significance level. We evaluated the internal composition of the cluster by computing the mean correlation between all the core cells of the cluster. In addition, we calculated the average pairwise correlation of the entire population to set a standard for ensemble selection.

By [48], two criteria were employed to identify neuronal ensemble clusters: a minimum cluster size and a within-cluster average correlation. This was done to ensure that the detected group met the definition of a neural group, which is a cluster of population vectors with a core. Three cells were chosen as the minimum number of core cells, but this number is adjustable in the GUI and the codes. The second criterion compared the average correlation within the group with the average pairwise correlation of the population. If the latter exceeded the former, the cluster was discarded. This threshold is also adjustable in the GUI and is expressed in terms of the standard deviation of the pairwise correlations of the entire population. This was done to ensure that only clusters with a minimum size (in terms of core cells) and relatively high (compared to the population) average correlation within the cluster were kept. If a cluster did not meet either of these two criteria, it was excluded from the analysis. Otherwise, it was considered an ensemble.

2.3 Results

2.3.1 Method Overview

Before a systematic evaluation of the performance of our method, we summarize its core elements using the example shown in Fig. 1 [17].

First, we discard any population vector (Fig. 1a) with fewer than three active neurons (free parameters in the GUI). In this way, we focus only on high-order interactions, leaving pairwise interaction out of the analysis. Then, we performed a PCA using the selected population vectors as observations (Fig. 1b). We calculated the Euclidean distance between all the vectors projected onto the first six principal components (*see* Methods (Subheading 2) for more information on the selection of the number of PCs). After the pairwise distances between the population vectors were calculated, the local density, ρ, and the distance to the closest denser point, δ, were calculated for each population vector. Based on these two measures and a power law fit to the curve ρ vs. δ, we automatically identified the cluster centroids (Fig. 1c) and assigned the remaining vectors to their closest centroid, forming the clusters

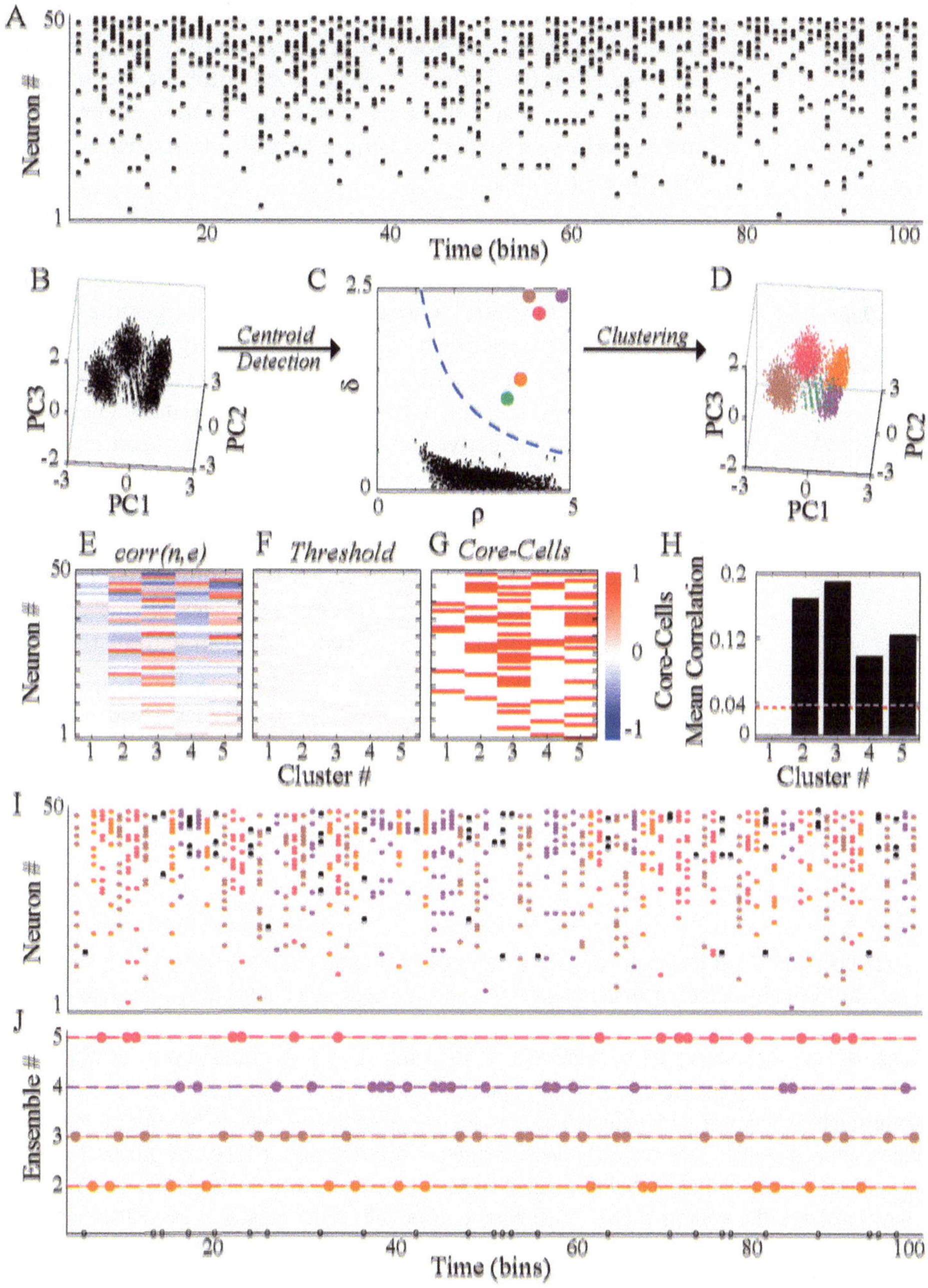

Fig. 1 Methodological scheme. (**a**) We created synthetic spike trains of 50 neurons over a 100-time bins. We then generated four ensembles, as described in the Methods section (Subheading 2). (**b**) PCA is performed on all population vectors of more than two active neurons, and in this case, we used three. For visualization purposes, we plot each of them as a point in the 3D space spanned by the first three PC. Points tend to group in clusters. (**c**) The Euclidean distance is computed between all points in the PC coordinates. The density of each data point ρ and its distance from the next data point with a higher density of δ are obtained. Points above the threshold (blue dashed line; *see* Methods (Subheading 2) for details) are considered cluster centroids. (**d**) All points are assigned to their closest centroid, creating clusters of population vectors. (**e**) The correlation matrix between neurons and clusters, corr (*n*, *e*) is represented by the symmetric red-blue color map. (**f**) A threshold matrix is calculated to define the significance of corr (*n*, *e*) values. (**g**) Core cells are

(Fig. 1d). Since the core of our algorithm is the density-based clustering procedure, we call it *density-based*. To find the core cells, we computed corr (n, e), the correlation between the neuronal spike train n and the activation times of the cluster e (Fig. 1e), and tested its significance using a threshold associated with a null hypothesis obtained.

2.4 Detection of Ensembles on Synthetic Data

2.4.1 Synthetic Spike Trains

We created a set of ensembles E characterized by their *core cells* and an *activation sequence*. Each ensemble consisted of a fixed number of core cells randomly chosen from the entire population of neurons, allowing repetition of core cells among the ensembles. We generate, for each ensemble, a binary column vector c_e of dimension N, with $c_e(n) = 1$ if the neuron n is a core cell of the ensemble e and 0 otherwise. The activation sequence of each ensemble $e \in E$ followed a time-homogeneous Bernoulli process with parameter p_e, denoted $B(p_e)$. We created a binary vector row denoted by a_e of dimension T with $a_e(t) = 1$ if the ensemble e was active in the time bin t and 0 otherwise. We allowed at most one active ensemble per bin, so each population vector could be part of one ensemble or none. To implement this idea, we first randomly draw the activation times of the first ensemble and then remove these times from the list of possible time points available for the second ensemble. We continued this process until we reached P_E.

With the defined core cells and the activation sequence of each ensemble, a spike train was generated by the product $c_e a_e$ (matrix) of dimension $N \times T$.

Fig. 1 (continued) detected based on corr (n, e) values that exceed the threshold (red if the core is white; otherwise). (**h**) The inner cluster mean correlation is compared against a threshold of noncore cell correlation, discarding any cluster below this threshold (cluster 1, in this case). This final filtering yields a set of ensembles. (**i**) Each population vector shown in A is colored according to the cluster to which it belong. Black vectors were discarded by the minimum number of spikes or the threshold rejection criteria. (**j**). The detected ensemble sequence is represented as a vector integer sequence, where the color corresponds to the population vector color in (**i**). Note that the rejection criteria discarded cluster 1, so only groups 2–5 are shown, and black circles denote non-clustered population vectors from shuffled versions of data (**f**). If corr (n, e) was above the threshold, the neuron n was considered a core cell of the cluster e (**g**). Finally, to obtain the ensembles, the pairwise correlation between all core cells of a given cluster was calculated and averaged, representing the average correlation within the cluster. To define a cluster as an ensemble, we compared the average correlation within the cluster with the average pairwise correlation of the entire network (**h**). If the within-cluster was significantly higher (based on a threshold), we considered an *ensemble* to be present; otherwise, the cluster was discarded due to the lack of internal correlation. With this procedure, we were able to separate the population vectors into ensemble and nonensemble vectors (**i**) and to obtain the activation times of different ensembles in time (**j**) with their corresponding core cells. We clarify that our method does not provide any analytical tool to evaluate the statistical significance of the ensemble temporal sequence, rather, it generates it as an integer sequence which we refer to as a global sequence (**j**). (This figure is from a previous publication in PlosOne [17])

$$S^E(n,t) = \sum_{e \in E} C_e(n) a_e(t). \quad (1)$$

To maintain the probabilistic nature of the spiking neurons, the firing rates of each neuron were taken from a corrected Gaussian distribution (only positive values) with a variable standard deviation (s.d.). The larger the s.d., the higher the firing rates in the spike train. This parameter enabled control of the density of the spike train (the total number of spikes with respect to the duration of the spike train). We randomly added or removed spikes to/from each neuron's spike train until it matched the target firing rate $P(n)$. If spikes were removed, the population vector located in the time bins where a given ensemble was active ended up having fewer active neurons than defined; on the other hand, if spikes were added, the opposite effect occurred. We used three different spike train densities: low (s.d. = 0.05), medium (s.d. = 0.1), and high (s.d. = 0.2). These densities reproduce a wide range of spiking responses observed experimentally; however, the density in experimental data will depend on the bin size. Our retinal data correspond to the medium case. For the low-density case, the match of the target firing rates usually required the removal of spikes.

This method distorted the vectors related to the activity of the group by underrepresenting their core cells. In the medium case, we had a balance between adding and removing spikes, and, in the last case, most neurons required the addition of spikes to match the desired firing rate, making neurons participate in ensembles they did not belong to by design.

In conclusion, we created a spike train by following these steps: (1) Generate a set of ensembles, each with its own population vector made up of core cells. (2) Fill the spike train with the activation times of the ensembles, using a time-homogeneous Bernoulli process for each ensemble, with a proportion of P_E of the total population vectors. The remaining population vectors are filled with a population vector of only silent neurons. (3) Draw the firing rates of each neuron from a rectified positive Gaussian distribution. (4) Randomly add or remove spikes from each neuron's spike train until they match the target firing rate of each neuron. This process produces a random spike train based on the activity of the ensembles.

2.4.2 Matching Detected Ensembles with Ground Truth Ensembles

To evaluate our method, we compared GT ensembles with detected ones by computing the correlation between the activation sequence of GT and the detected ensembles. We sought the detected ensemble that maximized the correlation with the GT ensembles. To do this, we generated synthetic data where neuronal ensemble activity can be parametrically controlled (*see* Methods (Subheading 2) for details). As an example, we created seven ensembles with a network of 100 neurons and 5000 bins (Fig. 2). The temporal activity of the

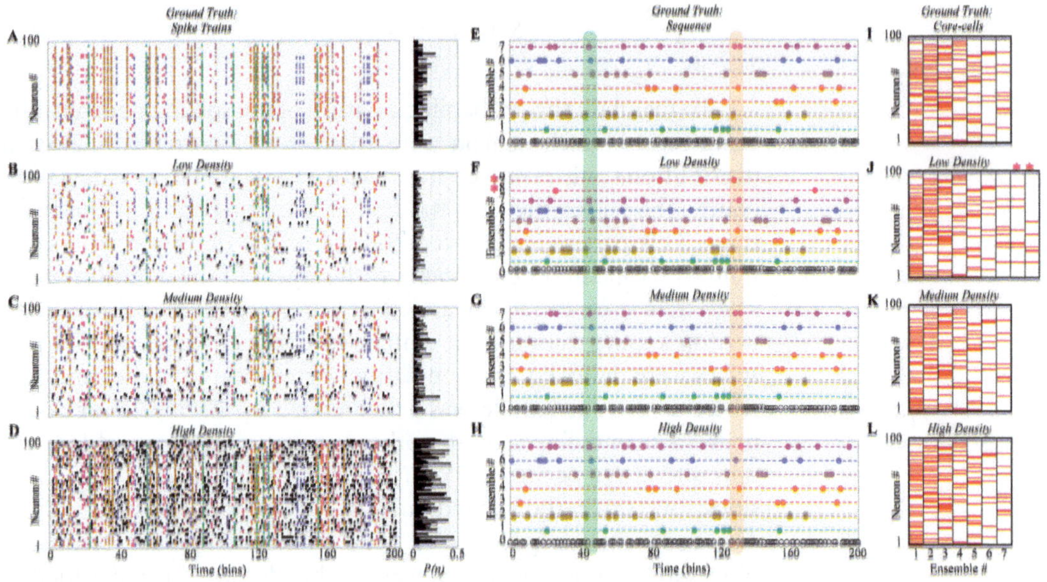

Fig. 2 Robust ensemble detection on synthetic data under different density regimes. (**a**) The spatial (core cells) and temporal (activation times) properties of seven ensembles are synthetically generated on a spike train of 100 neurons. The right panel shows the firing probability of each neuron in each time bin denoted by *P*(*n*). Each ensemble is colored differently. (**b**–**d**) Based on the ground truth (GT) spike trains, three spike trains with different densities were generated: low (**b**), medium (**c**), and high-density (**d**). The firing probability of each neuron is shown on the right as in (**a**). Population vectors are colored as in (**a**). The black dots are population vectors that do not belong to any ensemble. (**e**) Activation times (i.e., sequence) of the GT ensembles. The color corresponds to (**a**). (**f**–**h**) The sequence of the detected ensemble for the three spike trains with different densities, classified according to the GT indexing and their colors. At low density, the method detects two additional ensembles, denoted by a colored asterisk. The green vertical region shows an example of an ensemble that was correctly detected for the three spike trains. The red vertical region shows an example where the method correctly detects the GT vector for medium density, but partially fails in the other two cases. (**i**) GT core cells are sorted in descending order according to the number of core cells. Each column represents an ensemble, where red indicates a core cell. (**j**–**l**) Detected core cells for the three cases. Despite the two additional ensembles found in (**j**) (8 and 9, asterisks), the core cells agree well with GT. (This figure is from a previous publication in PlosOne [17])

ensembles comprised 80% of the sample, and the core cells comprised 20–40 neurons. We then randomly added or removed spikes to each neuron to satisfy a given firing probability for each neuron, $P(n)$, which controls the spike train density (Fig. 2b–d). To evaluate our method, we compared the GT ensembles with the detected ones by computing the correlation between the GT's activation sequence and the detected ensembles. We searched for the detected ensemble that maximized the correlation with the GT ensembles. To do this, we designed an algorithm to generate synthetic sample activity of neuronal ensembles that can be parametrically controlled (*see* Methods (Subheading 2) for details).

We found two ensembles more than expected (seven) for the low-density spike trains (Fig. 2f, red asterisks), while for the medium and high-density ones, we found the expected number of ensembles (Fig. 2g, h, respectively).

We then compared the ensemble sequence of GT (Fig. 2e) with the ensemble sequence detected in each density scenario (Fig. 2f–h). We found almost perfect agreement between both, except for a few false positives in the low- and high-density cases. Furthermore, we compared the detected core cells with the GT (Fig. 2j–l) and found a good agreement. Despite the over detection in the low-density regime (red asterisk), the other ensembles were in good agreement with the GT core cells. This example demonstrates that our method can accurately detect the number, temporal sequence, and core cells of neuronal ensembles in noisy spike trains with different densities. In the next section, we evaluate the detection of these three features for different sample sizes and networks.

2.4.3 Scaling Performance and Comparison with an Alternative Method

To quantify and evaluate the performance of our density-based method, we generated synthetic data with varying network and sample sizes. We then compared our algorithm with the current state-of-the-art ensemble detection method published by Carrillo et al. (SVD-based) [21]. The implementation of the latter can be found in https://github.com/hanshuting/SVDEnsemble. We use the default parameters that accompany these codes. Due to the computational cost of the SVD-based method, we only compared the scaling with the sample size for both methods. For the density-based method, we also calculated the scaling of the performance with the network size. The description of the parameters used in our method can be found in the supplemental information from [17].

However, they are easily settable in the provided GUI. We generate a synthetic spike train with a fixed network size ($N = 300$), number of ensembles ($E = 12$), number of core cells equal to 35, ensemble probability ($P_E = 0.8$), and medium density. We then varied the length of the recording from $T = 500$ to $T = 10^4$. We found that both methods increased the computational time with the sample size, as expected, but the SVD-based method scaled exponentially, while the density-based method is two orders of magnitude faster for $T = 10^4$ (Fig. 3a).

Regarding the activity of the ensemble, our method can accurately detect the number of ensembles for samples as small as 1000, while the SVD-based method tends to underestimate the number of ensembles (Fig. 3b). When the detected ensembles were compared to their closest ground truth ensemble (*see* Methods (Subheading 2) for details), we measured the correlation between both

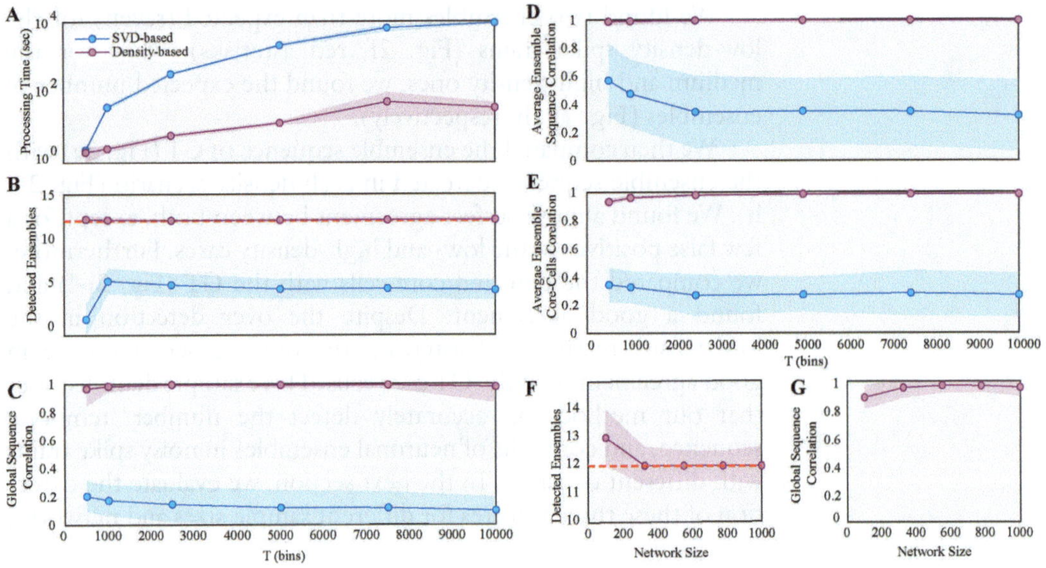

Fig. 3 Performance scaling with sample size and network size. (**a**) The purple line in the graph shows the computational processing time of our method on the log scale, compared to the blue line of Carrillo et al. [21], as a function of sample size (bins). Synthetic data was generated using 300 neurons, medium density, and 35 neurons per ensemble. The dots represent the average, and the shaded regions are ±1 standard deviation of 100 repetitions. Graph (**b**) shows the number of detected ensembles, and the dashed red line represents the real number of ensembles. Graph (**c**) displays the correlation between the detected sequence and the true global ensemble. Graph (**d**) shows the average correlation between the detected and true individual ensemble sequence, and Graph (**e**) between the detected and true ensemble core cells. Graph (**f**) shows the number of ensembles detected as a function of the size of the network, with the number of core cells per ensemble corresponding to 35% of the size of the network. Graph (**g**) is the same as graph (**c**) but as a function of network size. Due to computational cost, only our method was evaluated as a function of the size of the network with $T = 5000$. (This figure is from a previous publication in PlosOne [17])

sequences, finding that our method detects ground truth with excellent performance across the entire range of sample sizes. However, the SVD-based method consistently fails to detect the global sequence (Fig. 3c).

Furthermore, we measured the average correlation between the detected and GT individual sequences, and between the detected and GT core cells, again finding that our method achieved excellent performance for both features even for small sample sizes (Fig. 3d, e, respectively).

Finally, to evaluate the performance for different network sizes, we fixed the sample size while also keeping the other parameters fixed (number of ensembles, number of core cells, and P_E), and varied the network size from $N = 50$ to $N = 1000$, finding that our method slightly overestimated the number of ensembles for small N, but produced accurate results for larger N (Fig. 3f, g).

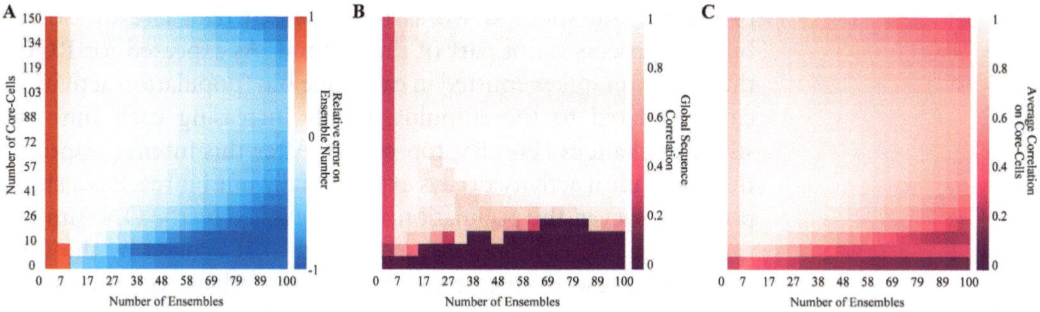

Fig. 4 Detection performance over a wide range of spike-train parameters. (**a**) The relative error of detecting the ensemble number for spike trains with different ensembles and the different number of core cells. Synthetic data was generated using medium density, $N = 300$, and $T = 5000$. Heat maps represent the average over 100 repetitions using the same spike-train parameters. Red/blue represent over/under detection, respectively. (**b**) The correlation between the detected sequence and the GT ensemble. (**c**) Average correlation between the detected cells and the GT core. (This figure is from a previous publication in PlosOne [17])

We conclude that our method reliably performs on a wide range of sample and network sizes, giving a more scalable and accurate solution to the ensemble detection problem than the alternative SVD-based method.

2.4.4 Reliable Performance Over a Wide Range of Spike Train Parameters

Here, we show the performance of our method when the number of ensembles and the number of core cells vary independently.

We generated synthetic spike-trains with fixed network size ($N = 300$), sample size ($T = 5000$), ensemble probability ($PE = 0.8$), and medium density, while the number of ensembles and core cells varied, as shown in Fig. 4. We found a wide combination of these parameters where the method detects the number of ensembles with a small relative error $(V_O - V_T)/V_T$ where V_O stands for the observed value and V_T for theoretical value (Fig. 4a), accurately detects the global ensemble sequence (Fig. 4b), and the corresponding core cells (Fig. 4c). Further explorations of other parameters and combinations of the same are considered work to be developed. To this end, we provide the computational codes and a GUI at https://github.com/brincolab/NeuralEnsembles for Matlab and for Python. This GUI allows one to perform all the analyses presented here on multivariate recordings of single events (e.g., spiking data, calcium events, arrival times in a sensor).

2.5 Detection of Ensembles on RGC Under a Simple ON-OFF Stimulus

We demonstrate the effectiveness of our method on a single simultaneous recording of a mouse retinal patch in vitro using MEA (*see* Methods (Subheading 2) for experimental and preprocessing details). Our aim is to illustrate how to interpret the basic results of our method in a particular physiological context rather than to uncover the functional properties of the collective activity of the retina. We binarized the data using a standard bin size of 20 ms

[49, 50]. Our method works on binarized data already, so the binning process is not part of the method. As expected for RGCs, the sum of all spikes emitted in each time bin (population activity) is closely related to the stimulus, briefly increasing each time the stimulus changes (Fig. 5a, top panel). After this intense response, the population activity decays exponentially until it reaches a stable point. However, the population activity evoked by the ON stimulus is different in amplitude and shape from that evoked by the OFF stimulus. These distinct evoked responses led us to anticipate at least four types of ensembles: two ensembles related to transitions (one for the ON-OFF and one for the OFF-ON) and two ensembles related to the decaying-stable activity after the ON-OFF and OFF-ON transitions, respectively. To test this hypothesis, we applied our ensemble detection algorithm (*see* Methods (Subheading 2) for details and parameters) to the spiking activity of 319 RGCs during 120 seconds of MEA recording. We found 10 ensembles that comprise approximately 68% of the population vectors in the analyzed recording. Their activity was strongly associated with the stimulus (Fig. 5a middle and bottom, and d). We discovered two transiently active ensembles (one for the ON-OFF and one for the OFF-ON transition), whose activity was only evoked by the stimulus transition, showing no activity either before the stimulus start (black arrow in Fig. 5a) or during the decaying or stable response. The other eight ensembles were active before the stimulus presentation, but at a lower rate, and during the stimulus presentation, the decaying or stable response. Four of them (Ens. 7–10) are preferentially active in the OFF stimulus and the other four during the ON stimulus (Ens. 2–5). These results demonstrate that the detected retinal ensembles are preferentially tuned to the features of the stimulus, indicating that without any stimulus-related information, our method can obtain ensembles whose activity is functionally coupled with the stimulus. We demonstrate the effectiveness of our approach on a single simultaneous recording of a mouse retinal patch in vitro using MEA (*see* Methods (Subheading 2) for experimental and preprocessing details). Our goal is to illustrate how to interpret the basic results of our method in a particular physiological context rather than to uncover the functional properties of the collective activity of the retina. Retina. We binarized the data using a standard bin size of 20 ms [49, 50]. Our method works on binarized data already, so the binning process is not part of the method. As expected for RGCs, the sum of all spikes emitted in each time bin (population activity) is closely related to the stimulus, briefly increasing each time the stimulus changes (Fig. 5a, top panel). After this intense response, the population activity decays exponentially until it reaches a stable point. However, the population activity evoked by the ON stimulus is different in amplitude and shape from that evoked by the OFF stimulus. These distinct evoked responses led us to anticipate at least four

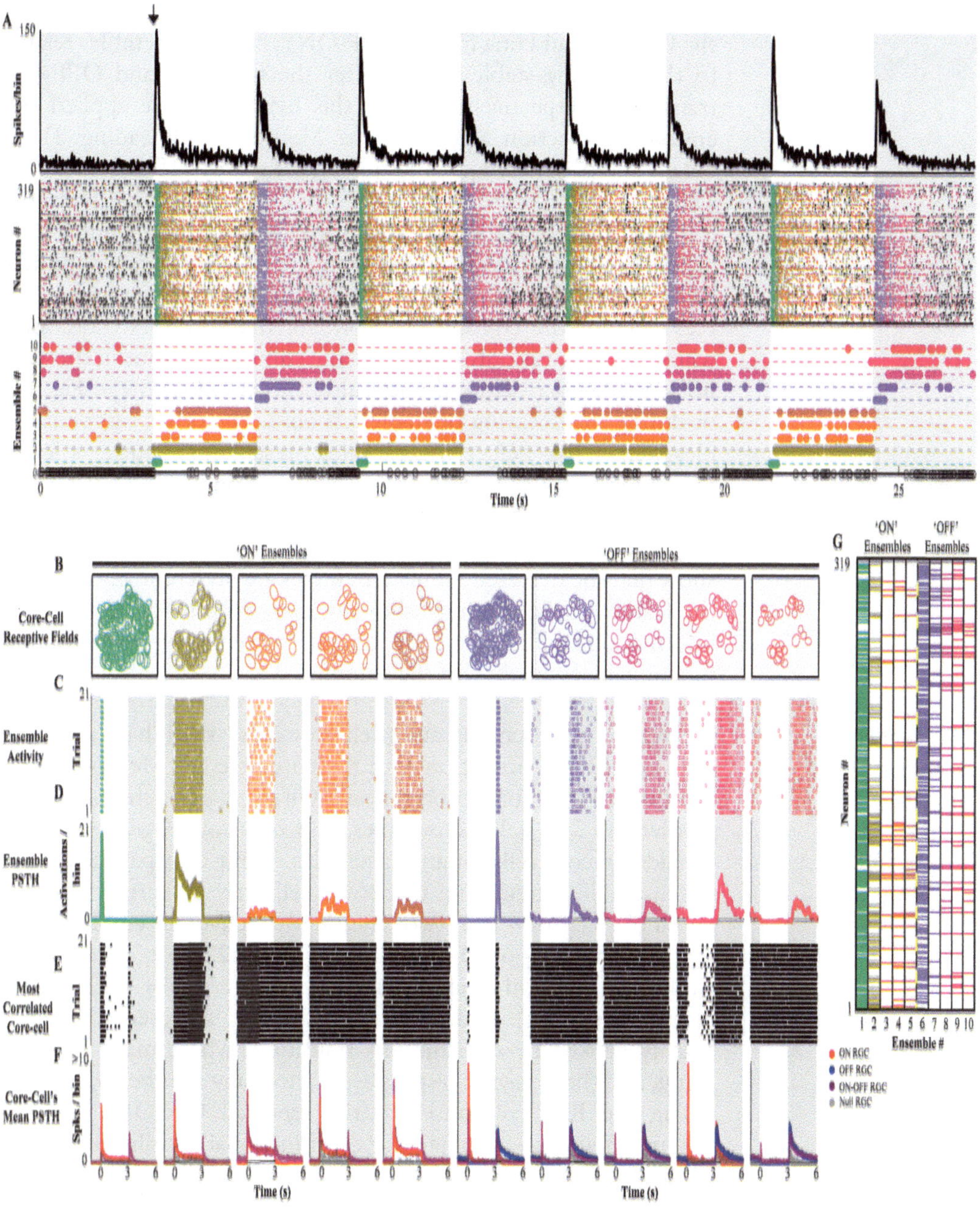

Fig. 5 Stimulus-evoked retinal ensembles. (**a**) *Top.* Population activity (spikes per bin) of 319 RGC's in time during ON light stimulation (white background) and OFF (gray background). The black arrow shows the ON stimulus that started. As expected for the retina, population activity increases transiently when the stimulus changes, then decays and stabilizes. *Middle.* Each population vector is colored according to the ensemble to which it belongs. *Bottom.* The activation of each ensemble in time is represented by a colored dot (corresponding to the color code of the population vector). The method detects consistent stimulus-locked ensemble activity with ON (ensembles of 1 to 5) and OFF (ensembles of 6 to 10) ensembles. Ensemble 0 represents all population vectors that were discarded according to the rejection criteria (*see* Methods (Subheading 2)). (**b**) The spatial distribution of the estimated receptive fields for the corresponding core cells of each ensemble. (**c**) The stacked trial-by-trial responses of each ensemble in time show their reliable and

types of ensembles: two ensembles related to transitions (one for the ON-OFF and one for the OFF-ON) and two ensembles related to the decaying-stable activity after the ON-OFF and OFF-ON transitions, respectively. To test this hypothesis, we applied our ensemble detection algorithm (*see* Methods (Subheading 2) for details and parameters) to the spiking activity of 319 RGCs during 120 s of MEA recording. We found 10 ensembles that comprise ~68% of the population vectors in the analyzed recording. Their activity was strongly associated with the stimulus (Fig. 5a middle and bottom, and d). We discovered two transiently active ensembles (one for the ON-OFF and one for the OFF-ON transition), whose activity was only evoked by the stimulus transition, showing no activity either before the stimulus start (black arrow in Fig. 5a) or during the decaying or stable response. The other eight ensembles were active before the stimulus presentation, but at a lower rate, and during the decaying or stable response. Notably, four of them (Ens. 7–10) are preferentially active in the OFF stimulus, and the other four during the ON stimulus (Ens. 2–5). These results show that the detected retinal ensembles are preferentially tuned to the feature of the stimulus, showing that without stimulus-related information, our method can obtain ensembles whose activity is functionally coupled with the stimulus.

2.6 Functional Properties of RGC Ensembles

We detect the ensembles and their core cells, which have a statistically significant correlation with the ensemble (*see* Methods (Subheading 2) for more information). On average, each core cell is involved in 2.7 ± 1.3 ensembles, and only four RGCs were not considered core cells for any ensembles. Three cells participated in six ensembles, suggesting that some cells may be part of multiple ensemble classes. The two transiently active ensembles, Ens. 1 (ON) and Ens. 6 (OFF), have 257 and 222 core cells, respectively, while the other ensembles have an average of 47.2 ± 17.9 core cells (Fig. 5g). This result aligns with the increased population activity evoked by the stimulus transition, where many cells are firing, and the decaying responses, where many cells become silent. Using the RGC responses to the repeated ON-OFF stimulus (21 trials) and an automated RGC functional classification algorithm (*see* Methods (Subheading 2) for details), we obtained

Fig. 5 (continued) selective responses to specific stimulus features. (**d**) The PSTH for each ensemble shows the stimulus preference of the ON and OFF ensembles. The different responses can be classified as transient and sustained, showing detailed stimulus tuning. (**e**) The stacked trial-by-trial responses of the core cell most correlated for each ensemble show that the temporal activity of the ensemble is poorly determined by the activity of its core cells. (**f**) For each ensemble, the average PSTHs of their corresponding core cells grouped by cell type (colors) do not show a clear tuning preference. (**g**) The colored matrix shows the RGCs in rows and the ensembles in columns. Each column is colored in the entries corresponding to their core cells, according to the ensemble to which they belong (**a**). (This figure is from a previous publication in PlosOne [17])

44 ON, 23 OFF, 205 ON-OFF, and 47 null (no significant preference) RGCs. All ensembles were composed of ON-OFF core cells, but the ON ensembles were mainly composed of ON RGCs, whereas the OFF ensembles were composed of OFF RGCs. Null cells showed significant participation in some ensembles, despite their lack of preference according to the classification algorithm. In fact, many core cells participated in multiple ensembles. For each ensemble, the spatial distribution of the estimated receptive fields (*see* Methods (Subheading 2)) of their corresponding core cells is shown in Fig. 5b, revealing spatial overlap between the ensembles. In this simple setup, we found that the ensemble classification is consistent with the dominant classes of their corresponding core cells.

The detection of core cells allows us to inquire if the tuning preferences of the detected ensembles are inherited from their core cells or if the ensembles have a specific tuning preference as a functional unit. We recall that the core cells are not perfectly correlated with the activation times of their corresponding core cells. In other words, the activation times of the ensemble cannot be completely determined by the spike trains of its core cells, so the tuning preferences of the ensemble could differ from the tuning preferences of individual cells. To qualitatively evaluate this for each ensemble, the peristimulus time histogram (PSTH) of each ensemble (Fig. 5c, d) was calculated and compared with the PSTH of their corresponding core cells (Fig. 5e, f). Furthermore, the core cell most representative for each ensemble (highest correlation with the activity of the ensemble, Fig. 5e) showed responses that differ strongly from the responses of the ensemble. Finally, for each ensemble, we averaged the PSTH of the core cells grouped by RGC class Fig. 5f, obtaining one average PSTH per cell class, which also shows the correspondence between the preference for ensemble tuning and the dominance of cell types. Since ON-OFF cells are present in all ensembles, the core cells of each ensemble, as a group, prefer both light transitions, despite the precise stimulus tuning of the ensembles (Fig. 5d).

As a proof of concept, we tested our method on retinal data. We did not conduct a deep be completely derived from the tuning preference of its core cells, providing preliminary evidence in favor of neuronal ensembles as whole functional units in the early visual system.

2.7 Discussion

In this chapter, we present a previously used method for detecting neuronal ensembles from spiking data that is computationally efficient and scalable [17]. Using simple dimensionality reduction, clustering techniques [47], and suitable statistical tests, we have described a method that is fast, accurate, and easy to use. Our example of mouse RGCs provides evidence of a causal relationship between stimuli and RGC ensembles, while the simulated data examples demonstrate that our method is effective for a wide range of neural activity scenarios, outperforming existing tools for ensemble detection in terms of accuracy and speed. Our method considers four general properties of the ensembles: transient activation (spontaneous and/or stimuli evoked), the presence of a core, the same neurons may participate in different ensembles, and the correlation within the ensemble should be higher than the rest of the population. Other ensemble detection methods have different criteria, such as finding communities between neurons that break apart over time [30, 51], or using activity related to events [48]. Our analysis relies only on grouping the population vectors with no event-related information, allowing us to segment a spiking network's temporal activity under both spontaneous and stimuli-evoked conditions. Compared to the SVD-based method [12, 52], our method is more computationally efficient (for relatively small sample sizes, Fig. 3) and requires less parameter tuning. Additionally, our method does not require linear independence between clusters and has an explicit implementation for the evaluation of the within-ensemble correlation, which is essential for distinguishing between a noisy cluster and an ensemble cluster.

Our method has room for improvement. For instance, non-stationarity and nonlinear spike correlations may lead to inaccurate results in ensemble detection methods that rely on PCA. Our approach can be adapted to include other measures of spike train similarity in addition to linear correlation values [53] or alternative surrogate data methods to provide a null hypothesis for correlations that maintain the temporal characteristics of spike trains, such as their interval between spikes or autocorrelation [54]. However, for generality, we use the least constrained null hypothesis, i.e., to destroy all the spike train temporal structures to test the significance of correlation values. For simplicity, in our example of RGCs, we used the standard bin size for mammalian RGCs, but the bin size used to create the spike trains may affect the detected ensembles, and thus different bin sizes can yield different results. It is important to note that our method is designed for data already binarized, so we consider the binning procedure as a preprocessing step, and, as we did, it should be informed by prior knowledge of the temporal resolution of the recorded neural population [49, 50].

Our approach does not assess the statistical significance of the ensemble sequences detected, which has been suggested to be physiologically meaningful [38].

The extension to clustering temporal sequences of population vectors is straightforward, and we consider its validation and evaluation as a task to be undertaken. Our contribution is to present a simple, fast, accurate, and scalable algorithm to detect synchronous neural ensembles that, due to its simplicity and code availability, can be applied in multiple scenarios and extended in various ways.

Another interesting research direction is based on the idea of using mathematical models of neuronal activity that explicitly include the anatomical structure [55] so that we can disentangle the role the anatomical structure plays in the functional assemblies that we observe. We provide evidence to support the presence of ensembles of retinal that are functionally related to stimuli. Our aim was to illustrate our method in a biological spiking network rather than explain the potential mechanisms involved in the activity of retinal ensembles or their functional implications. We believe that the study of retinal ensembles is an exciting new research area that needs to be explored to understand vision, sensory systems, and, more generally, the nervous system. In line with this perspective, we describe a method that is general enough to be applied to any multivariate binary data set. In addition, it is intuitive and can generate results that are easy to visualize, which should facilitate its comprehension and general use by the scientific community.

3 Notes

3.1 Code

The GUI and the codes used to validate the method using synthetic and real data can be found at https://github.com/brincolab/NeuralEnsembles for Matlab and for Python as well as a GUI that implements the method described in this chapter.

3.2 Animals and RGC Recordings

The mice used in the experiment were housed at the Animal Facility of the Universidad de Valparaíso, in a temperature range of 20–25 ° C, with a 12-h light-dark cycle and free access to food and water. For the present work, only one recording was used, which was previously described in [56]. The procedure involved euthanizing the animals under deep isoflurane or halothane anesthesia, extracting both eyes, and then cutting one of them into quarters. The other eye was stored in oxygenated (95% O_2 and 5% CO_2) AMES at 32 °C in the dark for further experiments. The same AMES medium was used for continuous perfusion during extracellular recordings. MEA recording (MEA USB-256, 20 kHz sample, Multichannel Systems GmbH, Germany) was done by placing a piece of retina onto a dialysis membrane and then into a ring device mounted on a traveling (up/down) cylinder, which was moved to contact the electrode surface of the MEA recording array. Data were processed offline using the SpyKING-CIRCUS spike-sorting algorithm [46] with default parameters.

3.3 Visual Stimuli

Custom software created with PsychoToolbox (Matlab) was used to generate visual stimuli, which were projected onto the retina with a LED projector (PLED-W500, Viewsonic, USA) connected to an inverted microscope (Lens 4×, Eclipse TE2000, NIKON, Japan). The image was 380 × 380 pixels, each covering 5μm^2. A checkerboard stimulus (visual white noise) with a block size of

50μm was presented at a rate of 60 Hz for 20 min, and each block independently took 0 or 255 (max value) on the pixel value scale. To classify the RGC, a green ON-OFF light stimulus was presented, with each part lasting 3 s and repeated 21 times. The first trial was discarded for classification analysis.

3.4 Automated RGC Classification

We classified RGCs as ON, OFF, ON-OFF, or null based on their response to a light stimulus. To do this, we calculated the peristimulus time histogram (PSTH) for each RGC and compared the maximum activity in the ON and OFF parts of the stimulus. We then set a threshold based on the average of the PSTH plus two standard deviations. If only the maximum in the ON (OFF) part was above this threshold, we classified the RGC as ON (OFF). If both maxima were above the threshold, we classified the RGC as ON-OFF. If neither maximum was above the threshold, the RGC was classified as null due to its lack of preference for the stimulus, as described in [57].

3.5 RGC Receptive Field Estimation

The reverse correlation method was used to calculate the spike-triggered average (STA) of each retinal ganglion cell (RGC). This was done by aggregating the 18 frames preceding any spike emitted in a matrix, as described in Methods section 1.2 (Subheading 2), using a checkerboard stimulus. The STA matrix was then broken down using singular value decomposition (SVD), which estimates the temporal and spatial components of the receptive field. The temporal component is the average fluctuation of the stimulus before a spike, while the spatial component is the preferred location of the RGC in the stimulus space. Then, an ellipse was fitted to the spatial component to estimate the receptive field of the RGC, as described in [52].

Acknowledgments

Grant from ANID Fondecyt 1200880 and FONDEQUIP EQM190032 to AGP and Fondecyt 1230170 to MJE.

References

1. Hebb DO (1949) The organization of behavior: a neuropsychological theory, A Wiley book in clinical psychology. Wiley, New York
2. Harris KD, Csicsvari J, Hirase H, Dragoi G, Buzsaki G (2003) Organization of cell assemblies in the hippocampus. Nature 424(6948): 552–556. https://doi.org/10.1038/nature01834
3. Buzsaki G (2010) Neural syntax: cell assemblies, synapsembles, and readers. Neuron 68(3):362–385. https://doi.org/10.1016/j.neuron.2010.09.023
4. Maass W (1997) Networks of spiking neurons: the third generation of neural network models. Neural Netw 10(9):1659–1671
5. Russo E, Durstewitz D (2017) Cell assemblies at multiple time scales with arbitrary lag constellations. elife 6. https://doi.org/10.7554/eLife.19428
6. Varela F, Lachaux JP, Rodriguez E, Martinerie J (2001) The brainweb: phase synchronization and large-scale integration. Nat Rev Neurosci 2(4):229–239. https://doi.org/10.1038/35067550

7. Uhlhaas PJ, Pipa G, Lima B, Melloni L, Neuenschwander S, Nikolic D et al (2009) Neural synchrony in cortical networks: history, concept and current status. Front Integr Neurosci 3:17. https://doi.org/10.3389/neuro.07.017.2009
8. Stringer C, Pachitariu M, Steinmetz N, Reddy CB, Carandini M, Harris KD (2019) Spontaneous behaviors drive multidimensional, brainwide activity. Science 364(6437):255. https://doi.org/10.1126/science.aav7893
9. Yuste R (2015) From the neuron doctrine to neural networks. Nat Rev Neurosci 16(8): 487–497. https://doi.org/10.1038/nrn3962
10. Ringach DL (2009) Spontaneous and driven cortical activity: implications for computation. Curr Opin Neurobiol 19(4):439–444. https://doi.org/10.1016/j.conb.2009.07.005
11. Carrillo-Reid L, Yuste R (2020) Playing the piano with the cortex: role of neuronal ensembles and pattern completion in perception and behavior. Curr Opin Neurobiol 64:89–95. https://doi.org/10.1016/j.conb.2020.03.014
12. Carrillo-Reid L, Lopez-Huerta VG, Garcia-Munoz M, Theiss S, Arbuthnott GW (2015) Cell assembly signatures defined by short-term synaptic plasticity in cortical networks. Int J Neural Syst 25(7):1550026. https://doi.org/10.1142/S0129065715500264
13. Buzsaki G (2004) Large-scale recording of neuronal ensembles. Nat Neurosci 7(5): 446–451. https://doi.org/10.1038/nn1233
14. Einevoll GT, Franke F, Hagen E, Pouzat C, Harris KD (2012) Towards reliable spike-train recordings from thousands of neurons with multielectrodes. Curr Opin Neurobiol 22(1): 11–17. https://doi.org/10.1016/j.conb.2011.10.001
15. Watanabe K, Haga T, Tatsuno M, Euston DR, Fukai T (2019) Unsupervised detection of cell-assembly sequences by similarity-based clustering. Front Neuroinform 13:39. https://doi.org/10.3389/fninf.2019.00039
16. Escobar MJ, Reyes C, Herzog R, Araya J, Otero M, Ibaceta C et al (2018) Characterization of retinal functionality at different eccentricities in a diurnal rodent. Front Cell Neurosci 12:444. https://doi.org/10.3389/fncel.2018.00444
17. Herzog R, Morales A, Mora S, Araya J, Escobar MJ, Palacios AG et al (2021) Scalable and accurate method for neuronal ensemble detection in spiking neural networks. PLoS One 16(7): e0251647. https://doi.org/10.1371/journal.pone.0251647
18. Harris KD (2005) Neural signatures of cell assembly organization. Nat Rev Neurosci 6(5):399–407. https://doi.org/10.1038/nrn1669
19. Carrillo-Reid L, Yang W, Kang Miller JE, Peterka DS, Yuste R (2017) Imaging and optically manipulating neuronal ensembles. Annu Rev Biophys 46:271–293. https://doi.org/10.1146/annurev-biophys-070816-033647
20. See JZ, Atencio CA, Sohal VS, Schreiner CE (2018) Coordinated neuronal ensembles in primary auditory cortical columns. elife 7. https://doi.org/10.7554/eLife.35587
21. Carrillo-Reid L, Miller JE, Hamm JP, Jackson J, Yuste R (2015) Endogenous sequential cortical activity evoked by visual stimuli. J Neurosci 35(23):8813–8828. https://doi.org/10.1523/JNEUROSCI.5214-14.2015
22. Wenzel M, Hamm JP, Peterka DS, Yuste R (2019) Acute focal seizures start as local synchronizations of neuronal ensembles. J Neurosci 39(43):8562–8575. https://doi.org/10.1523/JNEUROSCI.3176-18.2019
23. Hamm JP, Peterka DS, Gogos JA, Yuste R (2017) Altered cortical ensembles in mouse models of schizophrenia. Neuron 94(1): 153–67 e8. https://doi.org/10.1016/j.neuron.2017.03.019
24. Wenzel M, Han S, Smith EH, Hoel E, Greger B, House PA et al (2019) Reduced repertoire of cortical microstates and neuronal ensembles in medically induced loss of consciousness. Cell Syst 8(5):467–74 e4. https://doi.org/10.1016/j.cels.2019.03.007
25. Fang WQ, Yuste R (2017) Overproduction of neurons is correlated with enhanced cortical ensembles and increased perceptual discrimination. Cell Rep 21(2):381–392. https://doi.org/10.1016/j.celrep.2017.09.040
26. Eichenbaum H (2018) Barlow versus Hebb: when is it time to abandon the notion of feature detectors and adopt the cell assembly as the unit of cognition? Neurosci Lett 680:88–93. https://doi.org/10.1016/j.neulet.2017.04.006
27. Nicolelis MA, Baccala LA, Lin RC, Chapin JK (1995) Sensorimotor encoding by synchronous neural ensemble activity at multiple levels of the somatosensory system. Science 268(5215):1353–1358. https://doi.org/10.1126/science.7761855
28. Benchenane K, Peyrache A, Khamassi M, Tierney PL, Gioanni Y, Battaglia FP et al (2010) Coherent theta oscillations and reorganization of spike timing in the hippocampal- prefrontal

network upon learning. Neuron 66(6): 921–936. https://doi.org/10.1016/j.neuron.2010.05.013

29. Peyrache A, Benchenane K, Khamassi M, Wiener SI, Battaglia FP (2010) Principal component analysis of ensemble recordings reveals cell assemblies at high temporal resolution. J Comput Neurosci 29(1–2):309–325. https://doi.org/10.1007/s10827-009-0154-6
30. Lopes-dos-Santos V, Conde-Ocazionez S, Nicolelis MA, Ribeiro ST, Tort AB (2011) Neuronal assembly detection and cell membership specification by principal component analysis. PLoS One 6(6):e20996. https://doi.org/10.1371/journal.pone.0020996
31. Singh A, Humphries MD (2015) Finding communities in sparse networks. Sci Rep 5:8828. https://doi.org/10.1038/srep08828
32. Bruno AM, Frost WN, Humphries MD (2015) Modular deconstruction reveals the dynamical and physical building blocks of a locomotion motor program. Neuron 86(1):304–318. https://doi.org/10.1016/j.neuron.2015.03.005
33. Torre E, Picado-Muino D, Denker M, Borgelt C, Grun S (2013) Statistical evaluation of synchronous spike patterns extracted by frequent item set mining. Front Comput Neurosci 7:132. https://doi.org/10.3389/fncom.2013.00132
34. Torre E, Canova C, Denker M, Gerstein G, Helias M, Grun S (2016) ASSET: analysis of sequences of synchronous events in massively parallel spike trains. PLoS Comput Biol 12(7): e1004939. https://doi.org/10.1371/journal.pcbi.1004939
35. Yegenoglu A, Quaglio P, Torre E, Grün S, Endres D (2016) Exploring the usefulness of formal concept analysis for robust detection of spatio-temporal spike patterns in massively parallel spike trains. In: Haemmerlé O, Stapleton G, Faron Zucker C (eds) Graph-based representation and reasoning ICCS 2016, Lecture notes in computer science, vol 9717, pp 3–1. https://doi.org/10.1007/978-3-319-40985-6_1
36. Quaglio P, Yegenoglu A, Torre E, Endres DM, Grun S (2017) Detection and evaluation of spatio-temporal spike patterns in massively parallel spike train data with SPADE. Front Comput Neurosci 11:41. https://doi.org/10.3389/fncom.2017.00041
37. Onken A, Liu JK, Karunasekara PP, Delis I, Gollisch T, Panzeri S (2016) Using matrix and tensor factorizations for the single-trial analysis of population spike trains. PLoS Comput Biol 12(11):e1005189. https://doi.org/10.1371/journal.pcbi.1005189
38. Carrillo-Reid L, Yuste R (2020) What is a neuronal ensemble? Oxford Research Encyclopedia, Neuroscience. https://doi.org/10.1093/acrefore/9780190264086.013.298
39. Euler T, Haverkamp S, Schubert T, Baden T (2014) Retinal bipolar cells: elementary building blocks of vision. Nat Rev Neurosci 15(8): 507–519. https://doi.org/10.1038/nrn3783
40. Shekhar K, Lapan SW, Whitney IE, Tran NM, Macosko EZ, Kowalczyk M et al (2016) Comprehensive classification of retinal bipolar neurons by single-cell Transcriptomics. Cell 166(5):1308–1323.e30. https://doi.org/10.1016/j.cell.2016.07.054
41. Vlasits AL, Euler T, Franke K (2019) Function first: classifying cell types and circuits of the retina. Curr Opin Neurobiol 56:8–15. https://doi.org/10.1016/j.conb.2018.10.011
42. Masland RH (2012) The neuronal organization of the retina. Neuron 76(2):266–280. https://doi.org/10.1016/j.neuron.2012.10.002
43. Gollisch T, Meister M (2010) Eye smarter than scientists believed: neural computations in circuits of the retina. Neuron 65(2):150–164. https://doi.org/10.1016/j.neuron.2009.12.009
44. Real E, Asari H, Gollisch T, Meister M (2017) Neural circuit inference from function to structure. Curr Biol 27(2):189–198. https://doi.org/10.1016/j.cub.2016.11.040
45. Tikidji-Hamburyan A, Reinhard K, Seitter H, Hovhannisyan A, Procyk CA, Allen AE et al (2015) Retinal output changes qualitatively with every change in ambient illuminance. Nat Neurosci 18(1):66–74. https://doi.org/10.1038/nn.3891
46. Yger P, Spampinato GL, Esposito E, Lefebvre B, Deny S, Gardella C et al (2018) A spike sorting toolbox for up to thousands of electrodes validated with ground truth recordings in vitro and in vivo. elife 7. https://doi.org/10.7554/eLife.34518
47. Rodriguez A, Laio A (2014) Machine learning. Clustering by fast search and find of density peaks. Science 344(6191):1492–1496. https://doi.org/10.1126/science.1242072
48. Montijn JS, Olcese U, Pennartz CM (2016) Visual stimulus detection correlates with the consistency of temporal sequences within stereotyped events of V1 neuronal population activity. J Neurosci 36(33):8624–8640. https://doi.org/10.1523/JNEUROSCI.0853-16.2016
49. Greschner M, Shlens J, Bakolitsa C, Field GD, Gauthier JL, Jepson LH et al (2011)

Correlated firing among major ganglion cell types in primate retina. J Physiol 589(Pt 1): 75–86. https://doi.org/10.1113/jphysiol.2010.193888

50. Schneidman E, Berry MJ 2nd, Segev R, Bialek W (2006) Weak pairwise correlations imply strongly correlated network states in a neural population. Nature 440(7087):1007–1012. https://doi.org/10.1038/nature04701
51. Humphries MD (2017) Dynamical networks: finding, measuring, and tracking neural population activity using network science. Netw Neurosci 1(4):324–338. https://doi.org/10.1162/NETN_a_00020
52. Chichilnisky EJ (2001) A simple white noise analysis of neuronal light responses. Network 12(2):199–213
53. Humphries MD (2011) Spike-train communities: finding groups of similar spike trains. J Neurosci 31(6):2321–2336. https://doi.org/10.1523/JNEUROSCI.2853-10.2011
54. Sebastien Louis CBSG (2010) Generation and selection of surrogate methods for correlation analysis. In: Analysis of parallel spike trains, Springer series in computational neuroscience (NEUROSCI), vol 7. Springer, Cham, pp 359–382
55. Luppi AI, Cabral J, Cofre R, Destexhe A, Deco G, Kringelbach ML (2022) Dynamical models to evaluate structure-function relationships in network neuroscience. Nat Rev Neurosci 23(12):767–768. https://doi.org/10.1038/s41583-022-00646-w
56. Palacios-Munoz A, Escobar MJ, Vielma A, Araya J, Astudillo A, Valdivia G et al (2014) Role of connexin channels in the retinal light response of a diurnal rodent. Front Cell Neurosci 8:249. https://doi.org/10.3389/fncel.2014.00249
57. Carcieri SM, Jacobs AL, Nirenberg S (2003) Classification of retinal ganglion cells: a statistical approach. J Neurophysiol 90(3): 1704–1713. https://doi.org/10.1152/jn.00127.2003

Chapter 4

Fast Neuronal Segmentation of Two-Photon Functional Imaging Recordings Using CITE-On

Marco Brondi, Luca Sità, Sebastiano Curreli, and Tommaso Fellin

Abstract

Two-photon imaging is a powerful method to record the activity of neuronal ensembles in the intact animal brain. Here, we describe a protocol to run a recent analytical approach, CITE-On, specifically designed for online morphological cell detection, segmentation, and signal extraction in two-photon functional imaging recordings. A background introduction and the description of the main challenges CITE-On was conceived to overcome are presented, together with a detailed experimental procedure to install CITE-On and run it on two-photon calcium imaging time-series (t-series). By performing fast image segmentation, CITE-On will facilitate the establishment of efficient close-loop experimental approaches, including all-optical two-photon imaging and perturbation experiments.

Key words Two-photon functional imaging, Genetically encoded calcium indicators (GECIs), Deep neural networks, Online image segmentation, Neuronal ensembles, All-optical experiments, Closed-loop

1 Introduction

1.1 Neuronal Ensemble Monitoring Using Two-Photon Functional Microscopy

In vivo two-photon calcium imaging is a widely used technique to study brain cell activity in animal models [1–6]. In combination with genetically encoded calcium indicators (GECIs) such as GCaMPs [7–9], two-photon microscopy allows measurement of calcium signals in neurons with subcellular spatial resolution [1, 2]. Different approaches to perform in vivo two-photon calcium imaging include acute recordings, longitudinal protocols, large field-of-view (FOV) imaging, deep-brain recording using micro-endoscopes, and volumetric imaging [10–16]. Importantly, under most experimental conditions, neuronal calcium signals are used as a proxy of spiking activity [6–9, 17]. The spatial and temporal resolution of the calcium imaging data varies, depending on several

The original version of this chapter was previously published without open access. A correction to this chapter is available at https://doi.org/10.1007/978-1-0716-4208-5_13

Luis Carrillo-Reid (ed.), *Identification, Characterization, and Manipulation of Neuronal Ensembles*, Neuromethods, vol. 215, https://doi.org/10.1007/978-1-0716-4208-5_4, © The Author(s) 2025, Corrected Publication 2025

factors, among which differences in the acquisition setups, the level of GCaMP expression, and the shape and density of imaged cells [18]. Moreover, the various experimental configurations lead to differences in the signal-to-noise ratio of calcium transients and susceptibility to motion artifacts [17, 19–23].

Two-photon calcium imaging data are typically used to reconstruct the activity of neuronal networks and identify neuronal ensembles. Providing such a description of network activity aids the understanding of the mechanisms through which the brain encodes sensory stimuli [24–29], of how behavioral outputs are mirrored by network dynamics [30–34], and of what are the cellular and circuital phenotypes associated with pathological conditions [35, 36]. Furthermore, thanks to advancements in two-photon optogenetics [3, 37–43], it is now possible to interfere in real time with ongoing neuronal activity, driving changes to the membrane potentials of identified neurons in the FOV with near cellular resolution, or modifying experimental contingencies on the basis of imaging readouts [44]. This closed-loop strategy is crucial to causally test the role of specific neuronal ensembles in a given brain function under study [45]. However, this experimental approach requires precise manipulation of specific sets of neurons with minimal temporal lags. Thus, this approach needs accurate segmentation of and functional trace extraction from two-photon calcium imaging t-series to be performed in real-time, ideally on a frame-by-frame basis. This chapter describes CITE-On, a deep learning algorithm, which has been specifically conceived and developed to this aim (Fig. 1).

1.2 Challenges in Fast Neuronal Segmentation of Two-Photon Imaging Datasets

Two-photon calcium imaging data are typically processed offline, i.e., after completion of the imaging t-series acquisition. The prototypical analysis pipeline includes the following: (i) the compensation of motion artifacts; (ii) the identification of regions of interest (ROIs), a process called image segmentation; (iii) the extraction of functional fluorescence trace from each ROI; and (iv) the subtraction of background and neuropil signals. Many of these steps are significantly time consuming, because of the amount of computation they require [46–49]. However, closed-loop two-photon imaging and perturbation experiments require fast methods for image analysis to identify neurons and extract their functional signature. Such fast analytical methods should be able to track the identity of each imaged cell during the ongoing acquisition and return functional traces to be updated at each frame to instruct the closed-loop protocol. Ideally, fast analytical approaches should therefore be: (i) compliant with the motion artifacts typically encountered in experiments with awake head-restrained behaving animals; (ii) applicable to data with low signal-to-noise ratio, which characterizes individual frames in two-photon imaging t-series; (iii)

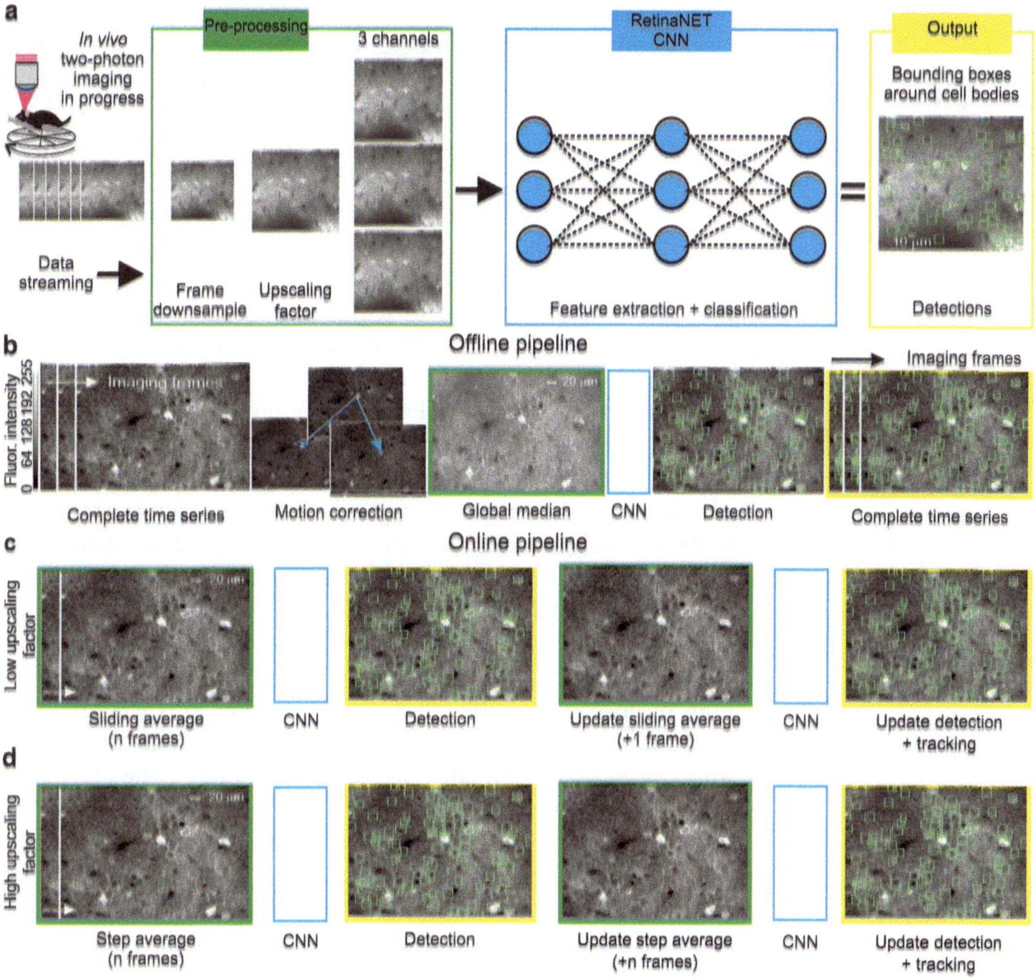

Fig. 1 CITE-On pipeline. (**a**) Schematic of the image detection process in CITE-On: individual frames are transferred to CITE-On as they are acquired. Temporal downsampling, image upscaling, and image triplication (green square) are performed before input to the CNN. The CNN output is the detection of neuronal somata in the form of bounding boxes (green squares in the grayscale image on the right) for each input image. (**b**) CITE-On offline pipeline. From left to right: correction of motion artifacts on the whole t-series, global median projection, bounding box detection, and projection of identified bounding boxes on the complete t-series (yellow). (**c-d**) CITE-On online pipeline using a sliding average approach (top) or a step average approach (bottom). (Reproduced from [61])

applicable to data acquired with different resolution, frame rate, cellular density, and different calcium indicators; and (iv) capable of detecting cells even when they are inactive.

1.3 Current Approaches to Analyze Two-Photon Calcium Imaging T-Series

Motion artifacts can significantly influence the reliability and performance of data analysis of two-photon imaging recordings. Several factors may contribute to generating motion artifacts including animal's respiration, heartbeat, and movement and laser-induced heating of the brain tissue [50–52]. Magnitude and frequency of motion artifacts in two-photon imaging data can be variable and are typically considered compensable when they are <30 μm of planar drift per tens of frames acquired at 30 Hz. Several approaches have been proposed for offline compensation of movement artifacts, including the use of cross-correlation-based alignment [53], block matching algorithms [54], and optical flow-based techniques [55]. All these offline methods are generally time-consuming. Other approaches have been developed for online compensation, including the use of a motion-tracking system [56], a piezoelectric stage for motion correction [50], and image processing algorithms [21, 53, 57, 58]. For example in [59], the authors proposed a method to track the movement of putative cells as the motion of contiguous regions of the FOV surrounding cell somata. The estimated translation is then applied independently to each cell separately. The method corrects for motion artifacts caused by both rigid and non-rigid motion. The authors demonstrate the effectiveness of the method on in vivo two-photon calcium imaging data of the mouse cortex, showing that it improved the quality of extracted functional traces and increased the number of cells, which could be reliably tracked across frames. However, this method requires a significant amount of computational resources, including a high-performance GPU, to run in real time. Furthermore, this method requires an initial delay during which a certain amount of frames are used to estimate the planar drift before the algorithm runs online. More recently, Cai et al. developed FIOLA, an online preprocessing pipeline that implements online rigid motion correction [60]. Alternative methods include the use of a planar motion artifact compensation, which tracks bounding boxes identifying the position of cells across frames using intersection over union (IoU) to follow the identity of the boxes and a Kalman filter to track boxes in time [61–63]. The advantage of using a tracking method based on IoU and a Kallman filter is that it allows for real-time tracking of the position and motion of the cell bodies, which can be used to compensate for motion artifacts in real-time imaging. This method is less computationally intensive than other tracking methods that require the calculation, for example, of dense optical flow fields or the use of deep learning algorithms [55, 64, 65]. Additionally, this method is highly robust to occlusion, which can occur when multiple cells are in close proximity one to the other and their movement may lead to merging of bounding boxes of individual neurons into a single bounding box including multiple cells. However, it is important to note that the IoU and Kalman filter-based tracking method may not be as accurate as other more computationally demanding approaches. Additionally, the IoU and Kalman filter-based tracking method is more suitable for tracking individual cells rather than large populations of cells. Therefore, the choice of tracking method should be based on the specific experimental needs and constraints.

Another major challenge in the analysis of imaging experiments is segmentation, i.e. the process by which pixels in each t-series are classified as belonging to individual cells or to background structures. Currently, different approaches are used to perform segmentation, including manual or model-based approaches and machine learning-based methods (e.g., [46–49, 60, 61, 66–76]). Manual segmentation is still largely considered the gold standard for segmenting neuronal somata from two-photon calcium imaging data. This is typically done using specialized software packages such as ImageJ (https://imagej.net/ij/), Fiji [77], or similar implementations under Matlab (https://www.mathworks.com/) or Python (https://www.python.org/) environments. Although manual segmentation can be accurate, it is time-consuming and can be prone to human error, especially when dealing with large datasets. In recent years, several model-based approaches have been developed for segmenting neuronal somata from two-photon calcium imaging data [46–49, 60, 61, 66–76]. These methods typically use mathematical models to describe the shape, size, and intensity of individual cells in the FOV. One popular example is the active contour model, which uses deformable curves to delineate the contours of individual cells [78, 79]. Other examples include level set methods, which use partial differential equations to model the contours of cells [80], and graph cut methods, which use graph theory to segment cells based on their intensity and spatial relationships [81]. Several recent automatic segmentation approaches instead rely on PCA or CNMF to define the somata footprints, combining morphological and functional (i.e., the coherent temporal fluctuations of fluorescence intensity in the collection of pixels) features extracted from a t-series [46, 48, 49]. These powerful methods return well-defined somata shapes but require long processing time, work best on active cells, and demand, in many cases, some level of manual curation. Importantly, CNMF-based methods are available for online applications downstream of online motion correction, producing neuronal segmentation typically updated every hundreds of frames while the data acquisition continues [48]. The computational cost of such methods scales up with the number of processed cells and with the length (i.e., the number of individual frames) of the t-series and requires a priori setting of initialization parameters [48]. Moreover, machine learning-based approaches have been developed for segmenting neuronal somata in two-photon calcium imaging data. These methods typically use convolutional neural networks (CNNs), which are a type of deep learning algorithm that can automatically learn to recognize complex patterns in large datasets [61]. CNNs are trained to learn the collection of image features associated with cellular somata using large sets of examples built from manual or model-based segmentations, with the aim to perform unsupervised segmentations on new datasets [61]. CNN-based segmentation methods have shown promise in accurately segmenting neuronal somata from two-photon calcium imaging data. One example of a CNN-based segmentation method is U-Net, which was originally developed for biomedical image segmentation [82]. U-Net uses an encoder-decoder architecture, where the input image is progressively down-sampled and then upsampled to produce a segmentation map. Another example is Mask R-CNN, which uses a combination of region proposal and instance segmentation to identify individual cells in the FOV [83]. Despite the big impact and large promise of machine learning-based segmentation approaches for two-photon imaging analysis, there are still significant challenges that need to be addressed. One major challenge is the lack of ground truth datasets for training and evaluating these methods. Creating accurate ground truth datasets is time-consuming and requires significant expertise, especially when dealing with large datasets of complex neuronal structures. Additionally, there are often significant variations in experimental conditions, such as changes in GCaMP expression levels, cell shape, and background noise, which can make it difficult to generalize segmentation algorithms across different datasets.

To extract functional traces from ROIs identified in two-photon calcium imaging data, various methods are used. One common approach calculates the relative change in fluorescence over time with respect to a basal fluorescence value ($\Delta F/F_0$), operating on the set of segmented pixels assigned to each cell. This approach requires the estimation of the basal fluorescence from the whole t-series or from small subsamples of the t-series [84, 85]. Although simple and computationally fast, this method relies on precise segmentation and accurate motion correction of the t-series, and it can be susceptible to noise and contamination from background signals [58, 86].

In the remaining part of the chapter, we present the materials and methods necessary to implement CITE-On [61], a recent analytical approach, which we developed to perform online morphological cell detection, segmentation, and functional trace extraction in two-photon imaging t-series, including those from mesoscopic two-photon calcium imaging experiments. CITE-On is based on the publicly available CNN RetinaNet [87] (*see* **Note 1**), an image detector we used to identify neuronal somata.

2 Materials

2.1 Dataset

Our dataset was composed of two-photon fluorescence t-series acquired in different brain areas, with different indicators, and under various acquisition conditions in head-fixed mice [61]. All recordings adhered to the guidelines of the European Communities Council Directive and were approved by the National

Council on Animal Care of the Italian Ministry of Health (authorization #34/2015-PR, #1134/2015-PR, and #61/2019-PR). More specifically, the dataset included the following:

- 131 t-series from layer IV neurons of the primary somatosensory cortex (S1) expressing green-emitting genetically encoded calcium indicators (either GCaMP6s, GCaMP6f, or GCaMP7f). T-series were recorded in 29 C57BL/6 head-fixed anesthetized mice during stimulation of mystacial vibrissae with an air puff.
- 33 t-series from hippocampal neurons of the cornu Ammonis 1 (CA1) region. Neurons co-expressed GCaMP6f and the red-emitting genetically encoded calcium indicator jRCaMP1a [88]. T-series were recorded in two head-fixed awake C57BL/6 mice navigating in a virtual corridor.
- 9 t-series from neurons of the ventral posteromedial (VPM) nucleus of the thalamus. Neurons expressed GCaMP6s. T-series were recorded in head-restrained anesthetized C57BL/6 mice implanted with an aberration-corrected microendoscope [13].

A ground truth was established for all recordings by combining the annotations from two expert graders. The dataset can be found at https://doi.org/10.48557/TRGQOD.

We used the S1 and the CA1 datasets for both training and testing. When using a dataset for training and testing, the data used for training was not used for testing in order to prevent overfitting. The VPM dataset, publicly available datasets from the Neurofinder challenge (https://github.com/codeneuro/neurofinder), datasets from the Allen Brain Observatory (https://observatory.brain-map.org/visualcoding), and datasets from mesoscopic two-photon imaging experiments [12] were used for testing CITE-On performance. More specifically:

- For the publicly available Allen Brain Observatory (ABO), we used 19 t-series, https://observatory.brain-map.org/visualcoding. T-series identification numbers: 501271265, 501484643, 501574836, 501704220, 501729039, 501836392, 502115959, 502205092, 502608215, 503109347, 504637623, 510214538, 510514474, 510517131, 527048992, 531006860, 539670003, 540684467, 545446482.
- For the publicly available Neurofinder (NF) challenge dataset, we used 28 t-series, https://github.com/codeneuro/neurofinder. T-series identification numbers: neurofinder.00.00, neurofinder.00.01, neurofinder.00.02, neurofinder.00.03, neurofinder.00.04, neurofinder.00.05, neurofinder.00.06, neurofinder.00.07, neurofinder.00.08, neurofinder.00.09, neurofinder.00.10, neurofinder.00.11, neurofinder.01.00,

neurofinder.01.01, neurofinder.02.00, neurofinder.02.01, neurofinder.03.00, neurofinder.04.00, neurofinder.04.01, neurofinder.00.00.test, neurofinder.00.01.test, neurofinder.01.00.test, neurofinder.01.01.test, neurofinder.02.00.test, neurofinder.02.01.test, neurofinder.03.00.test, neurofinder.04.00.test, neurofinder.04.01.test.

- For two-photon mesoscopic imaging, we used a single t-series from the reference [12].

The training and testing were done using the original Keras-RetinaNet repository (https://github.com/fizyr/keras-retinanet/tree/24769ffcdb7a0d9a23bba541ed58e8e70494d0af). The final model was exported as a TensorFlow SavedModel.

2.2 Hardware

- Dell Precision 7920 workstation equipped with: Intel Xeon Silver 4110 CPU, 32GB ECC RAM, Nvidia Quadro RTX 5000 GPU, and 512 GB NVMe SSD.

2.3 Software

- The final CITE-On package (https://gitlab.iit.it/fellin-public/cite-on) is a complete Python package (https://www.python.org/) for inference, tracking, and trace extraction. It relies on TensorFlow [89], Numpy [90], and Scikit [91]) for computations, PIL (https://web.archive.org/web/20201121102218/http://www.pythonware.com/products/pil/) and OpenCV [92] for I/O and image correction, and Matplotlib [93] and Holoviews (https://doi.org/10.5281/zenodo.596560) for visualization and data exploration. All dependencies are handled by a Conda (https://docs.anaconda.com/) environment.

3 Methods

Here we provide instructions, which will guide new users through the following: (i) CITE-On installation, (ii) data handling and preprocessing, (iii) adjustment of cell detector parameters, and (iv) optimization of performance for online data analysis. The software architecture of CITE-On (*see* **Notes 1–4**) requires minimal parameter tuning, prompting users to adjust three parameters: upscaling factor, confidence score threshold (score threshold), and temporal downsampling. Upscaling factor determines the scaling transformation of input images to the first layer of the CITE-On object detector (*see* **Notes 1** and **2**). CITE-On object detector outputs a set of putative cell identities, each one encoded as the coordinates of a bounding rectangle surrounding the cell soma (bounding box). Each cell identity is accompanied by a confidence score, which indicates how confident the model is about whether the object is present inside the box (*see* **Notes 3** and **4**). The confidence score threshold determines which bounding boxes will be retained in the analysis based on a threshold criterion. Lastly, the temporal downsampling modalities regulate the frequency of cell detection updates.

Below, we provide a detailed step-by-step protocol to initiate the analysis with CITE-On.

3.1 Installation

- Clone the repository by running:

```
git clone https://gitlab.iit.it/fellin-public/
   cite-on
```

- Enter the folder and create the Conda environment by running:

```
cd ./cite-on
conda env create -f env.yml
```

- In case there is access to gpu acceleration, use:

```
conda env create -f env_gpu.yml.
```

- Once the environment is created, build the parts of the Cython code and install the CITE-on package by running:

```
conda activate cite-on.
python setup.py build_ext -inplace
pip install .
```

- Register the newly created kernel in your existing Jupyter installation by running:

```
python -m ipykernel install --user
```

3.2 Input Data and Preprocessing

- Select the input t-series to be analyzed. We used uncompressed TIFF files, imported using PIL (https://web.archive.org/web/20201121102218/http://www.pythonware.com/products/pil/). Different data formats should be converted to TIFF format ahead of import.
- Select the path to a directory where you want to save output and intermediate files. If such a folder does not exist, it will be automatically created.
- Compute the median temporal projection of the first N (e.g., $N = 500$–1000) frames of the t-series or of the whole t-series. Optionally, a parallelized implementation of the moco motion correction algorithm [21] can be run on the TIFF stack using the aforementioned median projection as a reference image for the alignment procedure (default moco precision `pix_precision=10`). The moco-aligned t-series is then processed to return a median projection (`np.median(img_moco, axis=2)`). Both the median projection and the moco-corrected projection are displayed as gray-scale images. The linear shift

from moco is reported along with the relative median projection. Displayed images are saved in the path as SVG files.

```
fname='/home/jovyan/scratch/1209.tif'

save='outputs/demo'

os.makedirs(save,exist_ok=True)


im=Image.open(fname)

img=np.ndarray((im.size[1],im.size[0],im.n_frames),dtype=np.uint16)

for i in tqdm(range(im.n_frames)):

    im.seek(i)

    fr=np.array(im)

    img[:,:,i]=fr.copy()

median=np.median(img,axis=2)

show_image(median, os.path.join(save,'raw_median.svg'), format='svg')


moco=utils.MpMotionCorrection(

    pix_precision=10,

    register_stack=True,

    workers=16

)

img_moco=img.copy()

shift_vector=moco.motion_corr(img_moco, ref_image=median)

show_plot(

    (shift_vector[:,0]**2+shift_vector[:,1]**2)**0.5),

    os.path.join(save,'RMS_shift.svg'), format='svg'

)
```

```
median=np.median(img_moco,axis=2)

show_image(

    median,

    os.path.join(save,'median.svg'), format='svg'

)
```

3.3 Draw Boxes

- Estimate the most appropriate upscaling factor for optimal performance (*see* **Note 2**) of the object detector (RetinaNet, *see* **Note 1**). The upscaling factor depends on the ratio between the FOV area and the bounding box average area (Fig. 2). The user can consult Fig. 2a to estimate the optimal upscaling factor for new datasets.
- Select the *BoxEdit* tool in the toolbar:
- Double-click to start drawing a new box at one corner and double-click to complete the box at the opposite corner.
- Click and drag on an existing box to move it.

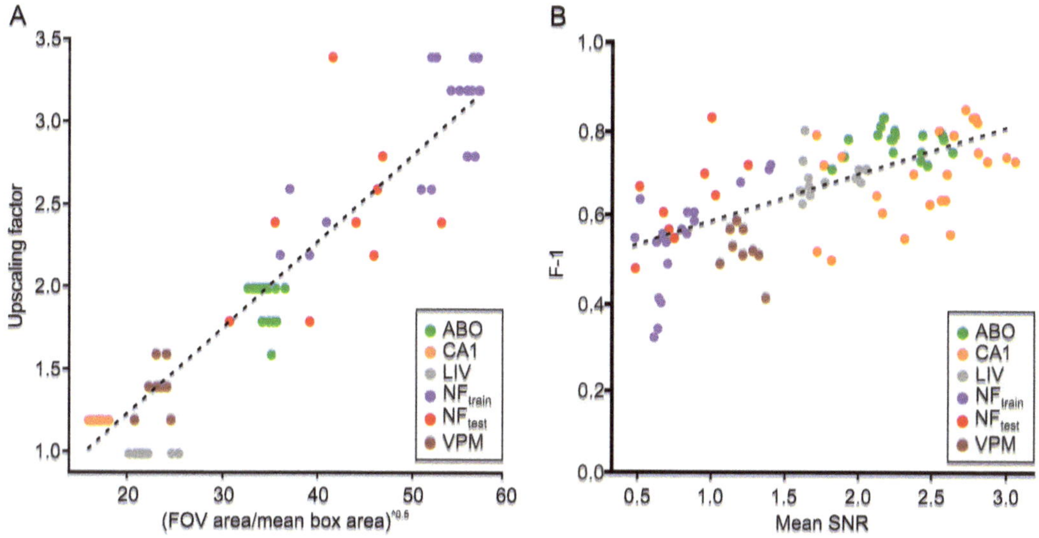

Fig. 2 CITE-On performance vs. SNR. (**a**) Upscaling factor as a function of the ratio between the FOV area and the mean bounding box area for all acquisitions in the validation datasets (color-coded). Dashed line indicates linear fit: upscaling factor = 0.052*((FOV area/mean box area)$^{0.5}$) + 0.182. (**b**) CITE-On detection performance (F-1) as a function of the mean signal-to-noise ratio (SNR) of the t-series. (Reproduced from [61])

- Select a box and press the backspace or delete keys (depending on the operating system) to delete it (Fig. 3 and **Note 3**).
- Edit the box coordinates in the table to resize it:

```
scale=3

boxes = hv.Rectangles([]).opts(

    width=median.shape[1]*scale,

    height=median.shape[0]*scale,

    responsive=False,

    color=None,

    line_color='lime'

)

box_annotator = hv.annotate.instance()

hv_img=hv.Image((np.arange(median.shape[1]),np.arange(median.shape[0]),median

)).opts(invert_yaxis=True, cmap='gray')

box_annotator(hv_img * boxes, annotations=[], name="Cell Annotator")

sample_boxes=box_annotator.annotated.dframe().values[:,:4]

avg_box_area=np.mean((sample_boxes[:,2]-

sample_boxes[:,0])*(sample_boxes[:,1]-sample_boxes[:,3]))

fov_area=img.shape[0]*img.shape[1]

upscaling=((fov_area/avg_box_area)**0.5)*0.052069301959865250+0.18241277070053
896

print('The suggested upscaling factor is: {:,1f}'.format(upscaling))
```

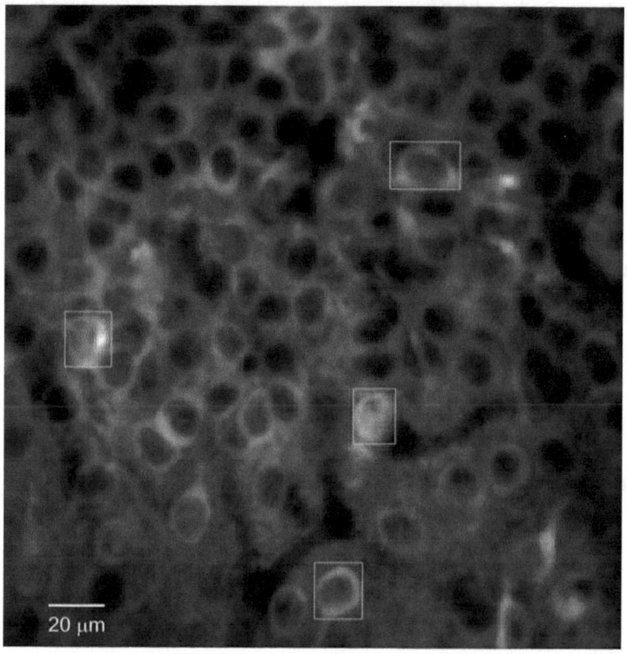

Fig. 3 Drawing bounding box. Two-photon image showing four representative bounding boxes (green rectangles). Initially, bounding boxes around neuronal somata should be carefully drawn for a few cells in order to obtain a good estimate of the average box surface and to accordingly set the upscaling factor

3.4 Run the Object Detector and Fine-Tune CITE-On Parameters for Offline Analysis

- Using the right-hand slider, observe how changing upscaling factor and score threshold influences the performance of the object detector. Changing the score threshold updates the number of detected boxes, which are displayed. Changing the upscaling factor also influences the performance of the object detector and consequently changes the number of detected and displayed boxes for a given score threshold (Fig. 4 and **Notes 3** and **4**). Please note that changing the upscaling factor requires the entire network to be reallocated; thus, at least a couple of seconds are needed to respond to the input change.
- Increase or decrease the upscaling factor to fit bounding boxes to neuronal cell bodies.
- Increase the score threshold to reduce false positives.

```
detector=Detector('./weights/', gpu='0', init_shape=img.shape[:2],
init_upscaling=upscaling)

def explore(Score, Upscaling):
    width  = median.shape[1]
    height = median.shape[0]
    detections=detector.detect(median,Score,Upscaling)
    return hv.Image((np.arange(width),np.arange(height),median)).opts(
        title='',
        invert_yaxis=True,
        cmap='gray',
        width=median.shape[1]*scale,
        height=median.shape[0]*scale) * hv.Path(
            [hv.Bounds(tuple(r)) for r in detections[:,:4]]
        ).opts(color='lime')

dmap=hv.DynamicMap(explore, kdims=['Upscaling','Score'])
dmap.redim.values(
    Score=np.arange(0.1,1,0.05),
    Upscaling=np.arange(1,3,0.1)
)
```

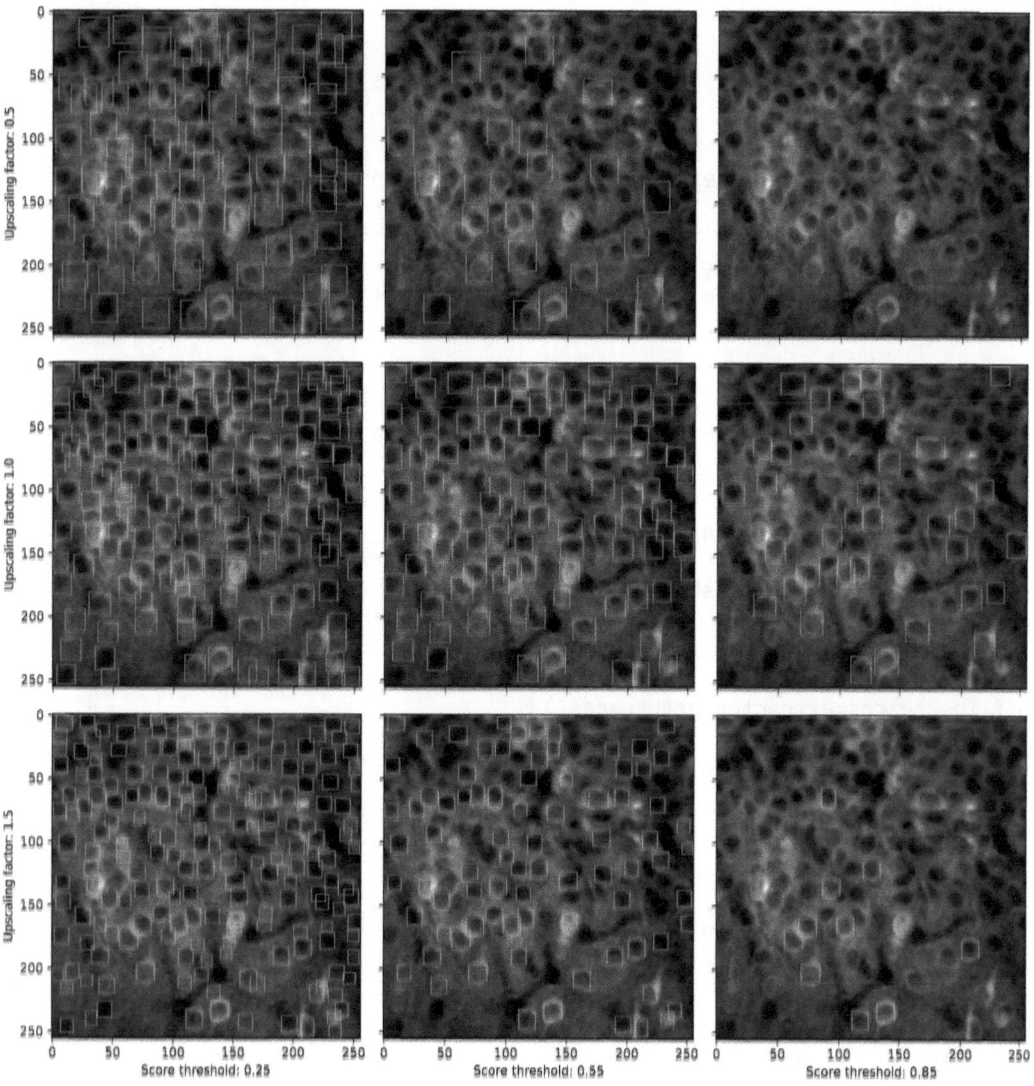

Fig. 4 Changing upscaling factor and score threshold modifies the number of detected cells. The same t-series is analyzed with CITE-On using nine different combinations of upscaling factor and score threshold (nine different panels in the figure divided into three rows). The central panel (mid panel in the central row) displays the detection results using optimized values of upscaling factor and score threshold

3.5 Run Trace Extraction

- Enter the parameters optimized in the previous step (*see* above).
- Run the trace extraction routine.
- Plot results.
- Select the number of workers (i.e., the number of processes per task in a parallel batch) for the trace extraction according to the performance of your computer:

```
upscaling=1.2
offline_score=0.45

offline_boxes=detector.detect(median, upscaling=upscaling,
threshold=offline_score)
np.save(os.path.join(save, 'offline_boxes.npy'),offline_boxes)
# array of shape (num_boxes,5), where each box is represented as
[xmin,ymin,xmax,ymax,score]

extractor=Extractor(workers=16 bg_corr=True)
for i in tqdm(range(img.shape[2])):
    extractor.extract(img[:,:,i], offline_boxes[:,:4])
offline_traces=extractor.get_traces()
np.save(os.path.join(save, 'offline_traces.npy'),offline_traces)
# an array of shape (num_frames,num_boxes) with the functional trace for each
box

# plot results
draw_annotations(median, offline_boxes, save=os.path.join(save,
'offline_detections.svg'))
multiline_plot(offline_traces.T[:20,:], save=os.path.join(save,
'sample_traces.svg'))
clustermap(np.corrcoef(offline_traces.T),save=os.path.join(save,
'offline_clustermap.svg')
```

3.6 Grid Search for Optimal Performance of Online Analysis

In the absence of available ground truth, CITE-On detection performance and quality of the extracted traces should be evaluated by experts in the analysis of calcium imaging recordings. Evaluation of results can be done comparing CITE-On outputs against manual segmentation or against third-party methods. At the end of the offline analysis described above, optimized values of the upscaling factor and the score threshold for a representative t-series are identified. Since the upscaling factor is mostly influenced by morphological constraints (i.e., cell size and FOV size), the output of the offline CITE-On pipeline can be directly used for the online CITE-On analysis of similar data (i.e., same pixel size, same FOV dimension, similar signal-to-noise ratio). The score threshold value obtained from offline analysis should be used as a starting point for its optimization in the online pipeline. We suggest that parameter optimization for the online analysis is performed on the same t-series used for the offline pipeline and identified parameter values are then used with new data (with similar features as the one used for offline analysis) without further tuning.

The next step in the online analysis pipeline is to define the optimal combination of the magnitude of the "local temporal downsampling" and the score threshold. To this aim, the user indicates a range of frames to be averaged in order to evaluate CITE-On performance as a function of different temporal downsampling in the online analysis. The user then indicates a range of score threshold values, which are similar to the ones obtained in the offline analysis. CITE-On performs detection on every combination of score threshold values and temporal downsampling intervals (Fig. 5).

CITE-On detection performance is returned in terms of precision, recall, and F-1 score comparing online outputs with offline outputs. A grid search returns the optimal combination of temporal downsampling and score threshold to be used in the online pipeline for the considered t-series and for similar new data.

```
averages=np.geomspace(3,80,6).astype(np.int)
os.makedirs(os.path.join(save,'temp'), exist_ok=True)
detector=Detector('./weights/', gpu='0', init_shape=img.shape[:2],
init_upscaling=upscaling)
for a in averages:
    print(f'Processing averages: {a}')
    boxes=[]
    for f in tqdm(range(img.shape[2]-a)):
        med=np.mean(img[...,f:f+a],axis=2)
        detections=detector.detect(med,threshold=0.05,upscaling=upscaling)
        boxes.append(detections)
    with open(os.path.join(save,'temp',str(a)+'.pkl'),'wb') as fd:
        pkl.dump(boxes,fd)
        fd.close()

scores=np.arange(0.25,1,0.05)
stats=np.zeros((len(averages),len(scores),3))
for i,a in tqdm(enumerate(averages),total=len(averages)):
    with open(os.path.join(save,'temp',str(a)+'.pkl'),'rb') as fd:
        boxes=pkl.load(fd)
        fd.close()
    for j,s in enumerate(scores):
        tracker=Tracker(threshold=0.2, frame_buffer=500)
        for f in range(len(boxes)):
```

```
            detections=boxes[f][boxes[f][:,-1]>s]

            tracker.update(detections, align_miss=True)

        res=tracker.get_positions()

        if 'shift_vect' in vars():

            shifted_gt=offline_boxes.copy()

            shifted_gt[[0,2],:]+=shift_vect[-1,0]

            shifted_gt[[1,3],:]+=shift_vect[-1,1]

            stat=utils.compute_stats(shifted_gt, res) if len(res) else np.nan

        else:

            stat=utils.compute_stats(offline_boxes, res) if len(res) else
np.nan

        stats[i,j]=[stat['f1'],stat['precision'],stat['recall']] if len(res)
else [0,0,0]

plot_performance_heatmap(stats, save=os.path.join(save,
'online_performace_heatmap.svg')
```

3.7 Evaluate Performance Over Time

At this point, the user retrieved the optimal values of the needed parameters (i.e., upscaling factor, temporal downsampling, and score threshold). CITE-On can now run an online analysis using these parameters. A movie of the t-series (not motion-corrected) is displayed with online detections, which are updated at each temporal downsampling window. Functional calcium traces are displayed for all detected identities. Precision, recall, and F-1 score are calculated frame-by-frame using the offline pipeline output as ground truth. The user may now save the optimized parameter setting and run CITE-On online on new data:

```
with open(os.path.join(save,'temp',str(averages[y[0]])+'.pkl'),'rb') as fd:
    boxes=pkl.load(fd)
    fd.close()
tracker=Tracker(threshold=0.2, frame_buffer=500)
extractor=Extractor(workers=12, bg_corr=True)
time_stats=np.zeros((len(boxes),3))
for f in range(len(boxes)):
    detections=boxes[f][boxes[f][:,-1]>scores[x[0]]]
    tracker.update(detections, align_miss=True)
    res=tracker.get_positions()
    extractor.extract(img[...,f+averages[y[0]]],res)
    if 'shift_vect' in vars():
        shifted_gt=offline_boxes.copy()
        shifted_gt[[0,2],:]+=shift_vect[f+averages[y[0]],0]
        shifted_gt[[1,3],:]+=shift_vect[f+averages[y[0]],1]
        stat=utils.compute_stats(shifted_gt, res) if len(res) else np.nan
    else:
        stat=utils.compute_stats(offline_boxes, res) if len(res) else np.nan
    time_stats[f]=[stat['f1'],stat['precision'],stat['recall']] if len(res)
else [0,0,0]

online_boxes=tracker.data[1:tracker.current_frame, 1:]
np.save(os.path.join(save,'online_boxes.npy'),online_boxes)
#array of shape (num_updates,num_id, 6) ordered by id and for each one
having[xmin,ymin,xmax,ymax,score(0 if lost), hits]
online_traces=extractor.get_traces()
np.save(os.path.join(save,'online_traces.npy'),online_traces)
```

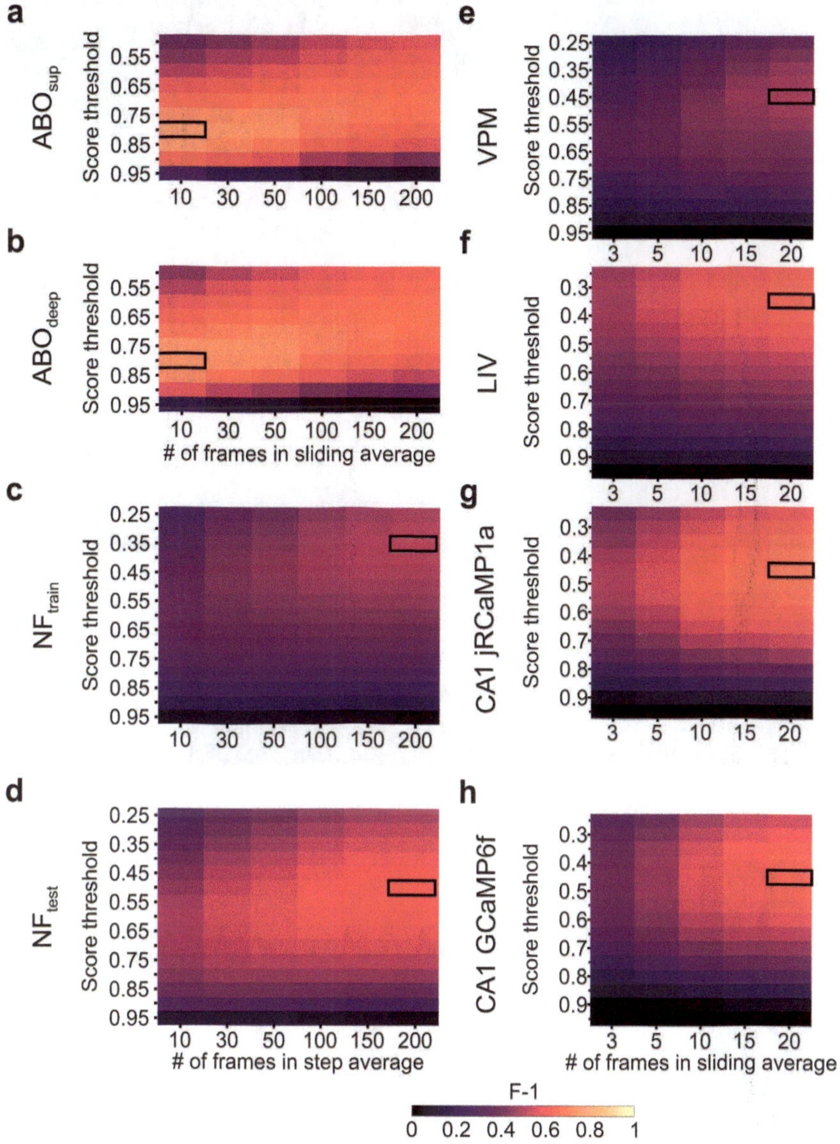

Fig. 5 CITE-On online detection performance on the different datasets. (**a-h**) Relationship between F-1 score (pseudocolor), score threshold (vertical axis), and number of frames in the temporal downsampling (horizontal axis) for ABOsup (**a**), ABOdeep (**b**), NFtrain (**c**), NFtest (**d**), VPM (**e**), LIV(**f**), CA1 jRCaMP1a (**g**), and CA1 GCaMP6f (**h**) datasets. The black rectangle indicates the maximal F-1 score. (Reproduced from [61])

4 Conclusions

In this chapter, we describe a protocol to run CITE-On, a recent analytical approach for morphological cell detection, segmentation, and functional trace extraction in two-photon imaging t-series, including t-series from mesoscopic two-photon calcium imaging experiments (Fig. 6).

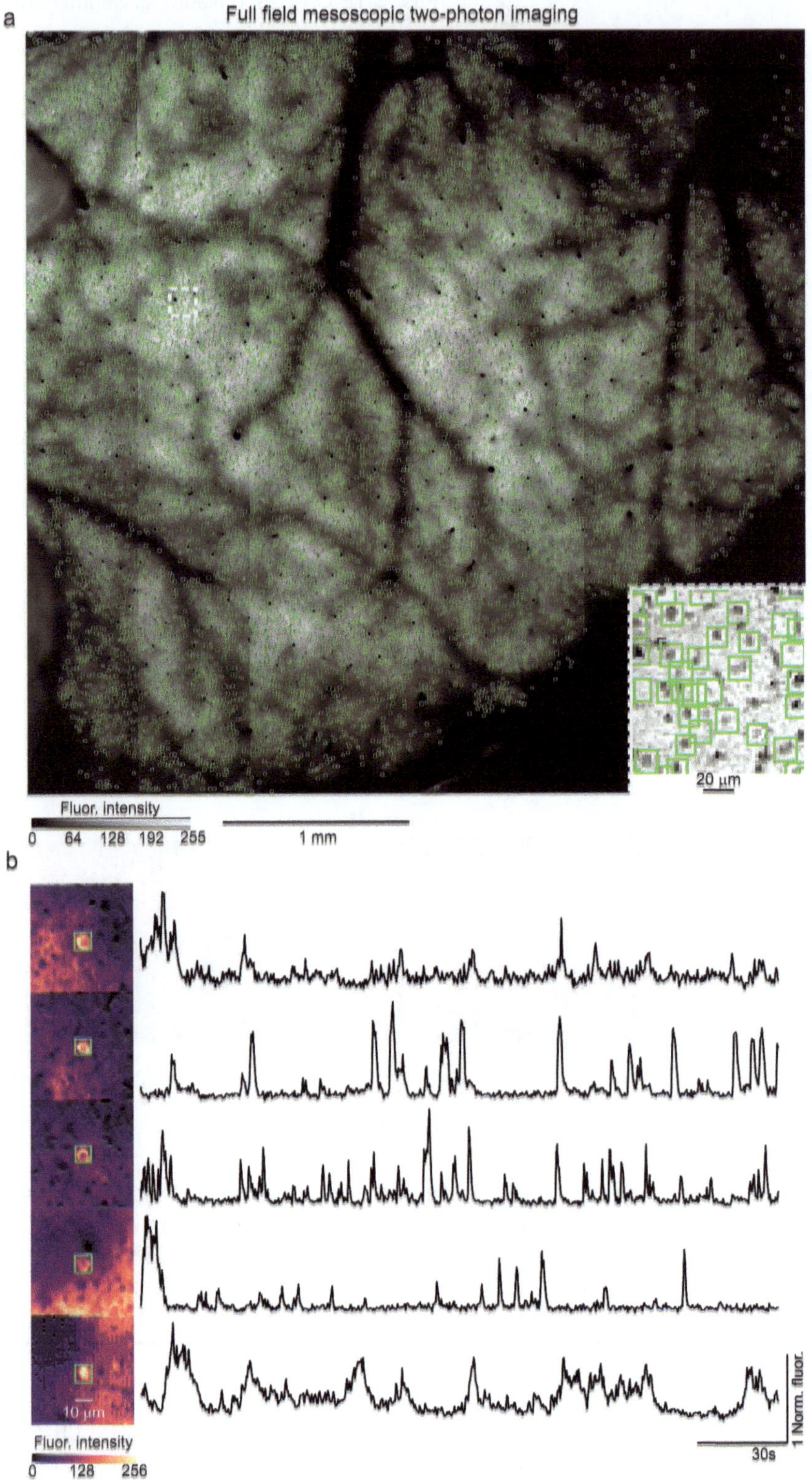

Fig. 6 CITE-On segmentation of two-photon mesoscopic t-series. (**a**) Median projection of a mesoscopic t-series from [12]. CITE-On online detections are shown in green (total number of bounding boxes, 4842). Magnification of a small region of the FOV (white dashed square) is displayed in the insert. (**b**) Fluorescence traces extracted using CITE-On online for five representative cells of the same t-series in (**a**). (Reproduced from [61])

Our approach involves using the convolutional neural network (CNN) RetinaNet (*see* **Note 1**), trained on manually annotated two-photon imaging datasets, to detect cells in the FOV. Our approach offers several advantages compared to other methods. First, it operates in real time, enabling the use of closed-loop optogenetic strategies to manipulate neuronal network activity with single-cell resolution. Second, it is compatible with motion artifacts and works on low signal-to-noise ratio data, making it suitable for in vivo imaging experiments on awake head-restrained animals. Third, it can detect cells even when they are inactive, which is an important aspect to consider when monitoring neuronal populations under different conditions and across days [94]. CITE-On is based on the RetinaNet architecture [87], a one-stage object detection CNN that achieves high accuracy and efficiency using a novel focal loss function. The choice of a CNN for cell detection is based on the observation that CNNs achieve state-of-the-art performance on various image recognition tasks, including object detection and segmentation [95, 96]. Moreover, CNNs can be trained using transfer learning, which involves taking a pretrained model on a large dataset and fine-tuning it on a smaller dataset although with largely dissimilar characteristics [97]. Transfer learning significantly reduces the amount of training data required and improves the performance of the model. Finally, CNNs can be designed to operate in real-time on standard hardware, making them suitable for real-time applications [98].

In conclusion, CITE-On is a fast and accurate analytical tool for cell detection and segmentation in two-photon calcium imaging data. CITE-On will facilitate the establishment of efficient close-loop experimental approaches in all-optical two-photon imaging and perturbation experiments, enabling real-time manipulation of behaviorally relevant neural ensembles during sensory perception and network computations.

5 Notes

1. *CNN and RetinaNet*

   Convolutional neural networks (CNNs) are effective in a variety of image processing tasks including object recognition, image segmentation, and object detection [99, 100]. The main strength of CNNs is their ability to automatically learn features from input data, without the need for manual feature engineering. This makes them well-suited to process large and complex datasets, such as two-photon calcium imaging data. One popular CNN architecture for object detection is RetinaNet [87]. RetinaNet is designed to address the problem of class imbalance in object detection, where the number of background regions greatly exceeds the number of object regions.

RetinaNet uses a feature pyramid network to detect objects at multiple scales and a novel focal loss function to give more weight to hard-to-detect objects. RetinaNet has been shown to achieve state-of-the-art performance on several object detection benchmarks, including COCO and Pascal VOC [101]. It has also been successfully applied in biomedical imaging tasks, such as detecting mitotic cells in microscopy images [102] and segmenting nuclei in histopathology images [103]. In terms of speed, RetinaNet can process images in real time on modern GPUs, making it suitable for real-time applications. The accuracy of RetinaNet can be further improved by using transfer learning, which involves fine-tuning a pretrained model on a new dataset. Transfer learning has been shown to be effective for various image-processing tasks [104, 105] and, given the limited size of available training datasets, transfer learning was used for effective network training in CITE-On [61].

2. *Evaluating Network Performance*

The best practice to test the accuracy, precision, and recall [106] of a given artificial network on the new dataset is to divide the dataset into three subsets: a training set, a validation set, and a test set [107]. The training set is used to train the network, the validation set is used to monitor the training progress and tune the hyperparameters of the network, and the test set is used to evaluate the performance of the network on new, unseen data. To evaluate the performance of the network, accuracy, precision, and recall metrics can be used [106]. Accuracy is the ratio between the number of correctly classified objects and the total number of objects in the dataset. Precision is the ratio between the number of true positives (i.e., correctly classified objects) and the total number of objects classified as positive (i.e., classified as neuronal somata). Recall is the ratio between the number of true positives and the total number of objects that are actually positive (i.e., neuronal somata). Moreover, F1 and mean average precision (mAP) can also be used to evaluate performance. F1 is equal to 2 (Precision*Recall) / (Precision + Recall), while mAP is equal to the average of the IoU. Hyperparameter tuning is an essential step in adapting pretrained networks. Hyperparameters are values that are set prior to the network training, and their values have a significant impact on the network performance. In CITE-On, hyperparameters include, among others, learning rate, batch size, and number of training epochs. Fine-tuning these hyperparameters can significantly improve the network accuracy and speed in detecting neuronal somata. RetinaNet, the CNN network used in CITE-On, has demonstrated exceptional performance on a range of object detection tasks, including the ImageNet dataset. According to [87], the RetinaNet model achieved a mAP of 39.1% on the challenging COCO dataset, outperforming previous state-of-the-art methods such as Faster R-CNN and SSD. The COCO dataset consists of complex images with numerous objects and challenging backgrounds. However, the performance of RetinaNet on two-photon calcium imaging data is likely to be affected by differences in image properties compared to natural images. For example, two-photon calcium imaging data often exhibit low signal-to-noise ratio and are affected by motion artifacts, which can affect the accuracy of the network predictions. Therefore, it is essential to fine-tune the hyperparameters to achieve optimal performance on two-photon calcium imaging data. It is important to note that the performance of RetinaNet, and thus of CITE-On, on two-photon calcium imaging data may vary depending on various factors, such as the quality of the imaging data, the expression level of the fluorescent indicator, and the density of indicator-expressing cells in the tissue. Therefore, it is recommended to evaluate the performance of CITE-On on the specific dataset of interest, using metrics such as accuracy, precision, and recall, based on offline evaluation against manually curated datasets and fine-tune the hyperparameters of the model as necessary to optimize its performance.

3. *Cell Detection*

RetinaNet is an object detection model that outputs a set of bounding boxes around the detected objects in an image. Each bounding box is associated with a confidence score indicating how confident the model is that the object is present in the box. RetinaNet output can be visualized as a set of bounding boxes overlaid on the original image (Figs. 4 and 5). In our approach, an initial fraction of the running imaging t-series is required to generate a first set of bounding boxes, which are then updated at each subsequent incoming frame in a sliding averaging window. More precisely, at relatively high signal-to-noise ratio values (2–3, [61]) the first average of 10 frames is required to generate the first set of bounding boxes. The next average is produced incorporating the following frame with the previous nine and so on. Under our experimental conditions, a delay of 0.3 s was necessary to obtain online cell detection under high signal-to-noise ratio conditions. When dealing with acquisitions with lower signal-to-noise ratio (1–2, [61]) instead, 2–6 s were required. In datasets with further lower signal-to-noise ratio (0.5–1, [61]), the sliding average approach should be replaced with a step average approach, updating cell detection every Δt seconds (where Δt is the duration of the time window used in the step average approach).

4. *Trace Extraction*

When applied to two-photon calcium imaging data, RetinaNet can detect the locations of neuronal somata in real time. However, the output of RetinaNet is not a precise segmentation of individual neurons, as the bounding boxes may include surrounding neuropil regions, as well as fragments of neighboring cells. Therefore, an additional step is required to extract the functional fluorescence traces of individual neurons from the output of RetinaNet. To obtain real-time functional segmentation of neuronal cells, we used the bounding box coordinates to extract the pixel values corresponding to each bounding box from the original two-photon imaging data. Then, these pixel values are used to extract the precise contours of individual neurons within each bounding box. In order to reduce the computational time and be able to properly segment the cells and extract the functional traces, we opted for segmentation refinement based on the instantaneous fluorescence statistics of the pixels inside each bounding box. All the pixels within the interval 80-95th percentile of the fluorescence intensity values were assigned as belonging to the region-of-interest belonging to the cell somata. This procedure returned a "dynamic segmentation," which was updated for each bounding box at each new frame. Finally, the functional fluorescence traces of each neuron were extracted by averaging the pixel values within each region-of-interest in each frame. This process could be performed in real time to provide continuous functional segmentation of neuronal activity. Dynamic segmentation and functional trace extraction operated downstream of the tracking system performed on the bounding boxes. Tracking, dynamic segmentation, and functional trace extraction all operate in real time on a frame-by-frame basis.

Acknowledgments

We are indebted to Pedro Lagomarsino de Leon Roig for his contribution to the development of CITE-On. This work was funded by Horizon 2020 ICT (https://cordis.europa.eu/project/id/101016787, DEEPER), next generation EU (https://fondazione-fair.it/, FAIR), and NIH Brain Initiative (https://braininitiative.nih.gov/, U19 NS107464) to TF.

References

1. Helmchen F, Denk W (2005) Deep tissue two-photon microscopy. Nat Methods 2: 932–940
2. Svoboda K, Yasuda R (2006) Principles of two-photon excitation microscopy and its applications to neuroscience. Neuron 50:823–839
3. Carrillo-Reid L, Yang W, Kang Miller JE et al (2017) Imaging and optically manipulating neuronal ensembles. Annu Rev Biophys 46: 271–293
4. Carrillo-Reid L, Yuste R (2020) Playing the piano with the cortex: role of neuronal

ensembles and pattern completion in perception and behavior. Curr Opin Neurobiol 64: 89–95
5. Lecoq J, Orlova N, Grewe BF (2019) Wide. Fast. Deep: recent advances in multiphoton microscopy of in vivo neuronal activity. J Neurosci 39:9042–9052
6. Brondi M, Moroni M, Vecchia D et al (2020) High-accuracy detection of neuronal ensemble activity in two-photon functional microscopy using smart line scanning. Cell Rep 30: 2567–2580
7. Chen TW, Wardill TJ, Sun Y et al (2013) Ultrasensitive fluorescent proteins for imaging neuronal activity. Nature 499:295–300
8. Zhang Y, Rozsa M, Liang Y et al (2023) Fast and sensitive GCaMP calcium indicators for imaging neural populations. Nature 615: 884–891
9. Dana H, Sun Y, Mohar B et al (2019) High-performance calcium sensors for imaging activity in neuronal populations and microcompartments. Nat Methods 16(7):649–657
10. Kerr JN, de Kock CP, Greenberg DS et al (2007) Spatial organization of neuronal population responses in layer 2/3 of rat barrel cortex. J Neurosci 27:13316–13328
11. Lendvai B, Stern EA, Chen B et al (2000) Experience-dependent plasticity of dendritic spines in the developing rat barrel cortex in vivo. Nature 404:876–881
12. Sofroniew NJ, Flickinger D, King J et al (2016) A large field of view two-photon mesoscope with subcellular resolution for in vivo imaging. elife 5:e14472
13. Antonini A, Sattin A, Moroni M et al (2020) Extended field-of-view ultrathin micro-endoscopes for high-resolution two-photon imaging with minimal invasiveness. elife 9: e58882
14. Jung JC, Schnitzer MJ (2003) Multiphoton endoscopy. Opt Lett 28:902–904
15. Demas J, Manley J, Tejera F et al (2021) High-speed, cortex-wide volumetric recording of neuroactivity at cellular resolution using light beads microscopy. Nat Methods 18:1103–1111
16. Kazemipour A, Novak O, Flickinger D et al (2019) Kilohertz frame-rate two-photon tomography. Nat Methods 16:778–786
17. Huang L, Ledochowitsch P, Knoblich U et al (2021) Relationship between simultaneously recorded spiking activity and fluorescence signal in GCaMP6 transgenic mice. elife 10:e51675
18. Harris KD, Quiroga RQ, Freeman J et al (2016) Improving data quality in neuronal population recordings. Nat Neurosci 19: 1165–1174
19. Theis L, Berens P, Froudarakis E et al (2016) Benchmarking spike rate inference in population calcium imaging. Neuron 90:471–482
20. Pnevmatikakis EA, Giovannucci A (2017) NoRMCorre: an online algorithm for piecewise rigid motion correction of calcium imaging data. J Neurosci Methods 291:83–94
21. Dubbs A, Guevara J, Yuste R (2016) Moco: fast motion correction for calcium imaging. Front Neuroinform 10:6
22. Nadella KM, Ros H, Baragli C et al (2016) Random-access scanning microscopy for 3D imaging in awake behaving animals. Nat Methods 13(12):1001–1004
23. Szalay G, Judak L, Katona G et al (2016) Fast 3D imaging of spine, dendritic, and neuronal assemblies in behaving animals. Neuron 92: 723–738
24. Marshel JH, Kim YS, Machado TA et al (2019) Cortical layer-specific critical dynamics triggering perception. Science 365:eaaw5202
25. Carrillo-Reid L, Han S, Yang W et al (2019) Controlling visually guided behavior by holographic recalling of cortical ensembles. Cell 178:447–457
26. Dalgleish HW, Russell LE, Packer AM et al (2020) How many neurons are sufficient for perception of cortical activity? elife 9:e58889
27. Gill JV, Lerman GM, Zhao H et al (2020) Precise holographic manipulation of olfactory circuits reveals coding features determining perceptual detection. Neuron 108(2): 382–393
28. Carrillo-Reid L, Calderon V (2022) Conceptual framework for neuronal ensemble identification and manipulation related to behavior using calcium imaging. Neurophotonics 9: 041403
29. Perez-Ortega J, Alejandre-Garcia T, Yuste R (2021) Long-term stability of cortical ensembles. elife 10:e64449
30. Liang B, Zhang L, Barbera G et al (2018) Distinct and dynamic ON and OFF neural ensembles in the prefrontal cortex code social exploration. Neuron 100:700–714
31. Hamm JP, Shymkiv Y, Han S et al (2021) Cortical ensembles selective for context. Proc Natl Acad Sci USA 118:e2026179118
32. Remedios R, Kennedy A, Zelikowsky M et al (2017) Social behaviour shapes hypothalamic neural ensemble representations of conspecific sex. Nature 550:388–392
33. Jaidar O, Carrillo-Reid L, Nakano Y et al (2019) Synchronized activation of striatal

direct and indirect pathways underlies the behavior in unilateral dopamine-depleted mice. Eur J Neurosci 49:1512–1528

34. Oliva A (2023) Neuronal ensemble dynamics in social memory. Curr Opin Neurobiol 78: 102654

35. Hamm JP, Shymkiv Y, Mukai J et al (2020) Aberrant cortical ensembles and schizophrenia-like sensory phenotypes in Setd1a(+/−) mice. Biol Psychiatry 88:215–223

36. Hamm JP, Peterka DS, Gogos JA et al (2017) Altered cortical ensembles in mouse models of schizophrenia. Neuron 94:153–167

37. Papagiakoumou E, Anselmi F, Begue A et al (2010) Scanless two-photon excitation of channelrhodopsin-2. Nat Methods 7:848–854

38. Rickgauer JP, Tank DW (2009) Two-photon excitation of channelrhodopsin-2 at saturation. Proc Natl Acad Sci USA 106:15025–15030

39. Rickgauer JP, Deisseroth K, Tank DW (2014) Simultaneous cellular-resolution optical perturbation and imaging of place cell firing fields. Nat Neurosci 17:1816–1824

40. Packer AM, Russell LE, Dalgleish HW et al (2015) Simultaneous all-optical manipulation and recording of neural circuit activity with cellular resolution in vivo. Nat Methods 2: 140–146

41. Forli A, Vecchia D, Binini N et al (2018) Two-photon bidirectional control and imaging of neuronal excitability with high spatial resolution in vivo. Cell Rep 22:3087–3098

42. Forli A, Pisoni M, Printz Y et al (2021) Optogenetic strategies for high-efficiency all-optical interrogation using blue-light-sensitive opsins. elife 10:e63359

43. Russell LE, Dalgleish HWP, Nutbrown R et al (2022) All-optical interrogation of neural circuits in behaving mice. Nat Protoc 17:1579–1620

44. Zhang Z, Russell LE, Packer AM et al (2018) Closed-loop all-optical interrogation of neural circuits in vivo. Nat Methods 15:1037–1040

45. Grosenick L, Marshel JH, Deisseroth K (2015) Closed-loop and activity-guided optogenetic control. Neuron 86:106–139

46. Mukamel EA, Nimmerjahn A, Schnitzer MJ (2009) Automated analysis of cellular signals from large-scale calcium imaging data. Neuron 63:747–760

47. Pnevmatikakis EA, Soudry D, Gao Y et al (2016) Simultaneous denoising, deconvolution, and demixing of calcium imaging data. Neuron 89:285–299

48. Giovannucci A, Friedrich J, Gunn P et al (2019) CaImAn an open source tool for scalable calcium imaging data analysis. elife 8: e38173

49. Pacitariu M, Stringer C, Dipoppa M et al (2017) Suite2p: beyond 10,000 neurons with standard two-photon microscopy. BioRxiv. https://doi.org/10.1101/061507

50. Ryan TM, Hinojosa AJ, Vroman R et al (2020) Correction of z-motion artefacts to allow population imaging of synaptic activity in behaving mice. J Physiol 598:1809–1827

51. Kong L, Little JP, Cui M (2016) Motion quantification during multi-photon functional imaging in behaving animals. Biomed Opt Express 7:3686–3695

52. Podgorski K, Ranganathan G (2016) Brain heating induced by near-infrared lasers during multiphoton microscopy. J Neurophysiol 116:1012–1023

53. Liu W, Pan J, Xu Y et al (2022) Fast and accurate motion correction for two-photon Ca(2+) imaging in behaving mice. Front Neuroinform 16:851188

54. Hattori R, Komiyama T (2022) PatchWarp: corrections of non-uniform image distortions in two-photon calcium imaging data by patchwork affine transformations. Cell Rep Methods 2:100205

55. Flotho P, Nomura S, Kuhn B et al (2022) Software for non-parametric image registration of 2-photon imaging data. J Biophotonics 15:e202100330

56. Griffiths VA, Valera AM, Lau JY et al (2020) Real-time 3D movement correction for two-photon imaging in behaving animals. Nat Methods 17:741–748

57. Chen T, Xue Z, Wang C et al (2010) Motion artifact correction of multi-photon imaging of awake mice models using speed embedded HMM. Med Image Comput Comput Assist Interv 13:473–480

58. Mitani A, Komiyama T (2018) Real-time processing of two-photon calcium imaging data including lateral motion artifact correction. Front Neuroinform 12:98

59. Giovannucci A, Friedrich J, Kaufman M et al (2017) OnACID: online analysis of calcium imaging data in real time. BioRxiv. https://doi.org/10.1101/193383

60. Cai C, Dong C, Friedrich J et al (2023) FIOLA: an accelerated pipeline for fluorescence imaging online analysis. Nat Methods 20:1417–1425

61. Sita L, Brondi M, Lagomarsino de Leon RP et al (2022) A deep-learning approach for online cell identification and trace extraction

in functional two-photon calcium imaging. Nat Commun 13:1529

62. Karagyozov D, Mihovilovic SM, Lesar A et al (2018) Recording neural activity in unrestrained animals with three-dimensional tracking two-photon microscopy. Cell Rep 25:1371–1383
63. Yamaguchi A, Wu R, McNulty P et al (2023) Multi-neuronal recording in unrestrained animals with all acousto-optic random-access line-scanning two-photon microscopy. Front Neurosci 17:1135457
64. Denis J, Dard RF, Quiroli E et al (2020) DeepCINAC: a deep-learning-based python toolbox for inferring calcium imaging neuronal activity based on movie visualization. eNeuro 7:ENEURO.0038-20.2020
65. Wen C, Miura T, Voleti V et al (2021) 3DeeCellTracker, a deep learning-based pipeline for segmenting and tracking cells in 3D time lapse images. elife 10:e59187
66. Soltanian-Zadeh S, Sahingur K, Blau S et al (2019) Fast and robust active neuron segmentation in two-photon calcium imaging using spatiotemporal deep learning. Proc Natl Acad Sci USA 116:8554–8563
67. Bao Y, Soltanian-Zadeh S, Farsiu S et al (2021) Segmentation of neurons from fluorescence calcium recordings beyond real-time. Nat Mach Intell 3:590–600
68. Xu Z, Wu Y, Guan J et al (2023) NeuroSeg-II: a deep learning approach for generalized neuron segmentation in two-photon Ca(2+) imaging. Front Cell Neurosci 17:1127847
69. Dursun G, Bijelic D, Aysit N et al (2023) Combined segmentation and classification-based approach to automated analysis of biomedical signals obtained from calcium imaging. PLoS One 18:e0281236
70. Stringer C, Wang T, Michaelos M et al (2020) Cellpose: a generalist algorithm for cellular segmentation. Nat Methods 18(1):100–106
71. Giovannucci A, Friedrich J, Kaufman M et al (2017) OnACID: Online analysis of calcium imaging data in real time. In: Advances in neural information processing systems. MIT Press, Cambridge
72. de Kraker L, Seignette K, Thamizharasu P et al (2022) SpecSeg is a versatile toolbox that segments neurons and neurites in chronic calcium imaging datasets based on low-frequency cross-spectral power. Cell Rep Methods 2:100299
73. Reynolds S, Abrahamsson T, Schuck R et al (2017) ABLE: an activity-based level set segmentation algorithm for two-photon calcium imaging data. eNeuro 4: ENEURO.0012-17.2017
74. Petersen A, Simon N, Witten D (2018) Scalpel: extracting neurons from calcium imaging data. Ann Appl Stat 12:2430–2456
75. Kang S, Park J, Kim K et al (2022) ICoRD: iterative correlation-based ROI detection method for the extraction of neural signals in calcium imaging. J Neural Eng 19. https://doi.org/10.1088/1741-2552/ac84aa
76. Zhu F, Grier HA, Tandon R et al (2022) A deep learning framework for inference of single-trial neural population dynamics from calcium imaging with subframe temporal resolution. Nat Neurosci 25:1724–1734
77. Schindelin J, Arganda-Carreras I, Frise E et al (2012) Fiji: an open-source platform for biological-image analysis. Nat Methods 9: 676–682
78. Xu J, Gong L, Wang G et al (2019) Convolutional neural network initialized active contour model with adaptive ellipse fitting for nuclear segmentation on breast histopathological images. J Med Imaging (Bellingham) 6: 017501
79. Hatamizadeh A, Sengupta D, Terzopoulos D (2020) End-to-end trainable deep active contour models for automatic image segmentation: delineating buildings in aerial imagery. arXiv. https://doi.org/10.48550/arXiv.2007.11691
80. Pang J, Ozkucur N, Ren M et al (2015) Automatic neuron segmentation and neural network analysis method for phase contrast microscopy images. Biomed Opt Express 6: 4395–4416
81. Li R, Zhu M, Li J et al (2019) Precise segmentation of densely interweaving neuron clusters using G-Cut. Nat Commun 10:1549
82. Ronneberger O, Fischer P, Brox T (2015) U-net: convolution networks for biomedical image segmentation. arXiv. https://doi.org/10.48550/arXiv.1505.04597
83. He K, Gkioxari G, Dollar P et al (2020) Mask R-CNN. IEEE Trans Pattern Anal Mach Intell 42:386–397
84. Stosiek C, Garaschuk O, Holthoff K et al (2003) In vivo two-photon calcium imaging of neuronal networks. Proc Natl Acad Sci USA 100:7319–7324
85. Grienberger C, Giovannucci A, Zeiger W et al (2022) Two-photon calcium imaging of neuronal activity. Nat Rev Methods Primers 2:67
86. Keemink SW, Lowe SC, Pakan JMP et al (2018) FISSA: a neuropil decontamination

toolbox for calcium imaging signals. Sci Rep 8:3493
87. Lin TY, Goyal P, Girshick R et al (2020) Focal loss for dense object detection. IEEE Trans Pattern Anal Mach Intell 42:318–327
88. Dana H, Mohar B, Sun Y et al (2016) Sensitive red protein calcium indicators for imaging neural activity. elife 5:12727
89. Abadi M, Barham P, Chen J et al (2016) TensorFlow: a system for large-scale machine learning. arXiv. https://doi.org/10.48550/arXiv.1605.08695
90. Harris CR, Millman KJ, van der Walt SJ et al (2020) Array programming with NumPy. Nature 585:357–362
91. Pedregosa V, Varoquaux G, Gramfort A et al (2011) Scikit-learn: machine learning in python. J Mach Learn Res 12:2825–2830
92. Bradski G (2000) The OpenCV library. Dr Dobb's J Softw Tools 120:122–125
93. Hunter JD (2007) MatPlotlib: a 2D graphics environment. Comput Sci Eng 9:90–95
94. Sheintuch L, Rubin A, Brande-Eilat N et al (2017) Tracking the same neurons across multiple days in Ca(2+) imaging data. Cell Rep 21:1102–1115
95. He K, Zhang X, Ren S et al (2016) Deep residual learning for image recognition. In: Proceedings of the IEEE computer society conference on computer vision and pattern recognition. IEEE, Piscataway
96. Ren JSJ, Xu L, Yan Q et al (2015) Shepard convolutional neural networks. In: Advances in neural information processing systems. MIT Press, Cambridge, pp 901–909
97. Pan SJ, Yang Q (2009) A survey on transfer learning. IEEE Trans Knowl Data Eng 22: 1345–1359
98. Ren S, He K, Girshick R et al (2016) Faster R-CNN: toward real-time object detection with region proposal networks. arXiv. https://doi.org/10.48550/arXiv.1506.01497
99. Sharma N, Jain V, Mishra A (2018) An analysis of convolutional neural networks for image classification. Procedia Comput Sci 132:377–384
100. Chauhan R, Ghanshala KK, Joshi RC (2019) Convolutional neural network (CNN) for image detection and recognition. First international conference on secure cyber computing and communication
101. Wei Q, Hu X, Wang X et al (2023) Improved retinanet target detection model. 2022 2nd international conference on algorithms, high performance computing and artificial intelligence (AHPCAI)
102. Buggenthin F, Buettner F, Hoppe PS et al (2017) Prospective identification of hematopoietic lineage choice by deep learning. Nat Methods 14:403–406
103. Wang JM, Hong R, Demicco EG et al (2023) Deep learning integrates histopathology and proteogenomics at a pan-cancer level. Cell Rep Med 4:101173
104. Shin HC, Roth HR, Gao M et al (2016) Deep convolutional neural networks for computer-aided detection: CNN architectures, dataset characteristics and transfer learning. IEEE Trans Med Imaging 35:1285–1298
105. Yosinski J, Clune J, Bengio J et al (2014) How transferrable are features in deep neural networks? Adv Neural Inf Proces Syst 27: 3320–3328
106. Juba B, Le HS (2023) Precision-recall versus accuracy and the role of large datasets. Proc AAAI Conf Artif Intell 33:4039–4048
107. Tarca AL, Carey VJ, Chen XW et al (2007) Machine learning and its applications to biology. PLoS Comput Biol 3:e116

Chapter 5

Identification of Neuronal Ensembles from Similarity Maps Using Singular Value Decomposition

Ricardo Velazquez-Contreras and Luis Carrillo-Reid

Abstract

Neuronal ensembles, defined as groups of neurons with coordinated activity, have been proposed as the basic motifs for brain computations such as perception, memory, and movement. This chapter delves into the implementation of Singular Value Decomposition (SVD) applied to similarity maps of neuronal population vectors as an analytical method to identify neuronal ensembles from calcium imaging recordings. At its core, SVD is a mathematical tool that allows the factorization of data matrices. In the context of neuronal ensemble identification, SVD could be applied to similarity maps constructed from multidimensional population vectors, where the dimensionality of the array is defined by the total number of observed neurons; thus similar population vectors are extracted as different SVD factors. Population vectors depict neuronal activity states as screenshots of active and inactive neurons at different time windows. The goal of this chapter is to provide the conceptual intuition for the implementation of SVD from similarity maps to identify neuronal ensembles. The identification of neuronal ensembles could shed light to understand the role of neuronal population dynamics in brain function.

Key words Neuronal ensembles, Singular value decomposition (SVD), Similarity maps, Calcium imaging, Population vectors

1 Introduction

Over the years, numerous theories have emerged as a conceptual framework to study how the brain works: from Golgi's interpretation of the brain as a continuous structure [1] to Cajal's notion of a linear information path [2] and more recently Donald Hebb's proposal of cell assemblies. Neuronal ensembles represent groups of neurons with coordinated and recurrent activity [3]. They serve as the building blocks of brain computations, orchestrating the symphony of neuronal firing patterns that encode perceptions, memories, and movements [4–7]. In this chapter, we adopt the

The original version of this chapter was previously published without open access. A correction to this chapter is available at https://doi.org/10.1007/978-1-0716-4208-5_13

Luis Carrillo-Reid (ed.), *Identification, Characterization, and Manipulation of Neuronal Ensembles*, Neuromethods, vol. 215, https://doi.org/10.1007/978-1-0716-4208-5_5, © The Author(s) 2025, Corrected Publication 2025

concept of neuronal ensembles to refer to sets of neurons exhibiting coordinated activity within a defined time frame, rather than sequential activity patterns that have been observed in neuronal population activity.

To identify neuronal ensembles, it is necessary to measure the activity of hundreds of neurons simultaneously. The recent development of optical techniques has allowed the recording and manipulation of population activity with single cell resolution of molecularly identified neuronal populations [8–10]. Two-photon in vivo recordings using transgenic animal models or viral injections to express genetically encoded calcium indicators represent growing techniques to study neural microcircuits in behaving animals [8, 11]. With this technique, it is possible to identify the activity of hundreds of neurons and its location in space with high spatial resolution during several days [12–15].

There are different methods and conceptual approaches that have been used to identify neural ensembles. SVD has been successfully used in primary visual cortex to identify neuronal ensembles that encode different features of visual stimuli [7, 16]. In this chapter, we describe the implementation of SVD from similarity maps as a method to identify predefined neuronal ensembles from simulated data.

The SVD method is a mathematical tool that allows the decomposition of a matrix into its constituent parts, where each factor represents specific features of the original data. For a square matrix, SVD represents the factorization of a matrix $\boldsymbol{M}$ into the product $\boldsymbol{M} = \boldsymbol{V\Sigma V^T}$, where $\boldsymbol{\Sigma}$ represents the singular values and $\boldsymbol{V^T}$ is the transpose of $\boldsymbol{V}$.

These singular values are usually denoted as $\sigma_1, \sigma_2, \sigma_3 \ldots$ in decreasing order and indicate the importance of each feature of the data. The larger the singular value, the more important is that factor in explaining the features of the data. In practice, SVD could be used as a dimensionality reduction method where only the top-k singular values and their associated components are used to extract the most important information from multidimensional datasets.

An intuitive approach to understand SVD is through its application in image compression. In this way, an image can be viewed as a matrix whose elements are defined by pixels (Fig. 1a). Image size can be reduced by creating a compact version of this matrix using a subset of singular values derived from SVD (Fig. 1b). The level of detail in the compressed image varies based on the number of singular values used (Fig. 1c). A smaller number of singular values results in a loss of finer details (Fig. 1d), while a larger number preserves more details (Fig. 1e). Optimal compression is achieved by selecting the most rapidly decaying singular values, as this minimizes the number of components required for the image's reconstruction while retaining the most significant details (Fig. 1f).

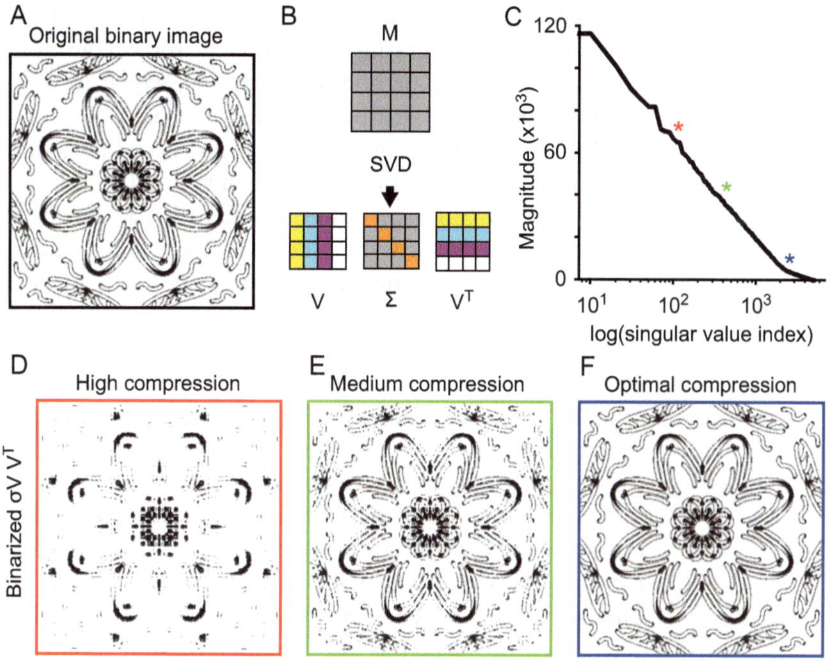

Fig. 1 Conceptualization of SVD algorithm. (**a**) Example of a binary image that shows repeated features. (**b**) Schematic representation of SVD performed on a matrix M, corresponding to the image in A. (**c**) Magnitude and number of singular values selected from SVD. (**d**) Compressed reconstruction of the original image using 9 singular values. (**e**) Compressed reconstruction of the original image using 35 singular values. (**f**) Compressed reconstruction of the original image using 231 singular values. Note that in all compressed images, repeated image features are highlighted and preserved

The identification of neuronal ensembles using SVD requires as a starting point a binary matrix representing the activity (ones) or inactivity (zeros) of all the neurons observed simultaneously. This could be represented as multidimensional population vectors. The most similar population vectors can describe the activation of a given neural ensemble [3, 7, 17, 18]. However, the total and specific moments of activation of each ensemble could be difficult to observe from the similarity matrix. Taking the largest singular values from the SVD, factorization of the similarity map has been used to identify neuronal ensembles [7, 16].

1.1 Experimental Framework

SVD has been used to identify neuronal ensembles encoding different orientations of drifting gratins in primary visual cortex of mice from two-photon calcium imaging recordings [7, 16]. The algorithm has been used in experimental recordings that encompass a Field of View (FOV) of approximately 300 μm by 300 μm containing about 100 active neurons and with a sampling resolution of 4 Hz in head-fixed mice. In a set of experiments, mice were visually stimulated with drifting gratings or natural scenes in passive conditions without any behavioral requirements. In a different set of experiments, mice were trained in a go/no-go task in which the animal produces a specific behavior, such as a licking at a designated port, when presented with a specific orientation, such as the display of vertical drifting-gratings. Conversely, they avoid doing so when presented with the orthogonal stimulus, which in this example is given by horizontal drifting gratings [7, 16].

From calcium imaging population recordings, different software tools can be used to automatically identify each neuron in the FOV, such as CaImAn [14] or suite2p [15]. To accomplish this task, the analysis software uses several key steps, including initial motion correction, the identification of the soma of individual neurons as regions of interest (ROI), the determination of the changes in ROI fluorescence related to the changes in electrical activity of the neuron, and finally the inference of neuronal activity [14, 15]. The creation of a binary matrix representing neuronal activity may involve measuring changes in the fluorescence signal through approaches like the first derivative of the signal, deconvolution, or different fitting techniques. Activity inference is typically binarized into two different states: an inactive state or an active state for each neuron. Then, neuronal activity data is transformed into a binary matrix denoted as $\boldsymbol{A}$ with size $n \times t$, where n is the number of active neurons and t is the number of time points. A practical working method of binarization calcium transients is the thresholding of the first time derivative to fluorescence signals of each cell throughout the recording (see **Note 1**).

The identification of neuronal ensembles from similarity maps using SVD factorization has been implemented in MATLAB (available at https://github.com/luiscareid/Stoixeion) [7]. Such MATLAB implementation uses as input argument the binary activity matrix depicting all the simultaneously observed neurons and the frames to analyze.

To provide an intuition for tuning different initial parameters to perform ensemble identification, we show examples using simulated data with predefined neuronal ensembles. In this chapter, we used simulated data to highlight specific scenarios commonly encountered in real data.

These simulations involve a predefined number of ensembles, along with their temporal profiles. The number of neurons participating in each ensemble was also chosen, so ensemble activity could be validated across the neurons within a specific ensemble. Additionally, a controlled level of noise was introduced to all simulated neurons to test the robustness of the approach.

Simulated data used in this chapter depicts the average level of neuronal activity, the number of neurons within ensembles, the activity level of each ensemble, the total number of neurons, and the duration of the recordings, as previously reported in experimental data [7, 17].

1.2 Important Concepts and Variable Names

- Matrix $\boldsymbol{A}$—Represents the original binary population activity matrix.
- Significant Population Vectors—Represents population vectors where several neurons have coordinated activity above chance levels.
- Matrix $\boldsymbol{A}_{\mathbf{sig}}$—Represents the population matrix containing only significant population vectors.
- Matrix $\boldsymbol{M}$—Represents the similarity map where m_{ij} is the cosine similarity of the vectors $\boldsymbol{a}_{\mathbf{sig}i}$ and $\boldsymbol{a}_{\mathbf{sig}j}$ in the matrix $\boldsymbol{A}_{\mathbf{sig}}$.
- Vector $\boldsymbol{m}_i$—Represents a column vector of the similarity map $\boldsymbol{M}$
- Matrix $\boldsymbol{\Sigma}$—Represents a diagonal matrix in the SVD decomposition of the matrix $\boldsymbol{M}$
- Matrix $\mathbf{V}$—Represents a matrix in the SVD decomposition of the matrix $\boldsymbol{M}$
- Core Neurons—Group of neurons with recurrent activity that define a neuronal ensemble

2 Methods

2.1 Peaks of Synchronous Activity

The activity matrix $\boldsymbol{A}$ of size $n \times t$ is a collection of t population vectors v, each of "n" dimensions (Fig. 2a). These population vectors contain the activity of all simultaneously recorded neurons at a specific time point [18]. Some of these population vectors may represent the moment of activation of different ensembles, but not all of them, in fact some population vectors may have no activity at all [7, 18, 19]. Only population vectors with more neurons firing together above chance levels should be considered for the analysis.

The minimum activity threshold can be determined by using a shuffling process with the activity matrix [18, 20]. In this process the neuronal activity across time is shuffled while preserving the activity interval, and multiple permutations of this shuffling process are performed to find the number of active neurons in a population vector that is higher than the activity expected by chance (see **Note 2**).

Following this step, a new matrix of size $n \times t_s$, for n neurons and t_s significant vectors, is obtained. This matrix contains only population vectors that contain neurons with coordinated activity (peaks of activity) and will be referred to as A_{sig} (Fig. 2b).

2.2 Term-Frequency Inverse Document Frequency Normalization

Not all cells contribute equally to the definition of each ensemble [7, 17]. Some neurons belong to only one ensemble, whereas other neurons could belong to several or all the other identified neuronal ensembles. The neurons that are consistently active have little influence on the definition of an ensemble because they are also active in different ensembles. To normalize the influence of

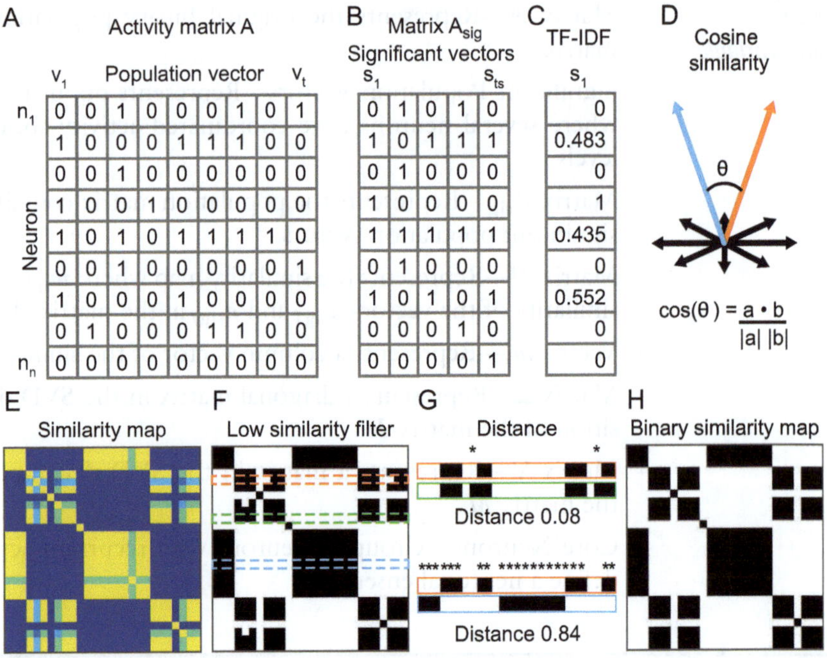

Fig. 2 Representation of neuronal activity as multidimensional population vectors. (**a**) Binarized matrix of population activity where each row represents a neuron and each column represents a time frame from calcium imaging recordings. (**b**) Schematic representation of significant population vectors with at least 3 neurons active at a given time. (**c**) Representation of the TF-IDF normalization. (**d**) Cartoon of population vectors in a multidimensional space. The angle between population vectors is used to construct similarity maps. (**e**) Similarity map depicting the cosine similarity between all possible pairs of significant population vectors. (**f**) Binary version of the similarity map, considering the highest values of similarity. (**g**) Schematic representation of the distance between different rows of the similarity map from F, demonstrating the distance between rows. Asterisks show dissimilar points. (**h**) Binary similarity map after denoising

neurons that are active in all the ensembles, the binary matrix of population activity could be subject to Term Frequency-Inverse Document Frequency (TF-IDF) normalization. Note that this normalization should only be applied when the goal is to exclude the cells that are consistently active, which may be interneurons in certain recordings [7, 18].

TF-IDF normalization is a common technique initially used by web search engines to disregard irrelevant words in a web page or document [21, 22]. This normalization considers the logarithm of the ratio between the total number of documents considered and the number of documents containing the term of interest. In this function, if a term is present in most documents, then that word would have little impact on the definition of each document, but if a term is present in only a few documents, then that term would be more relevant to the meaning of those documents. In the context of neuronal activity, the total number of available documents corresponds to the number of significant population vectors. Thus, the more active a neuron is at different moments, the less relevant it might be for defining an ensemble (Fig. 2c).

After this step (see **Note 3**), a new matrix with dimensions $n \times t_s$ is obtained, where n is the number of neurons in the recording and t_s is the number of significant population vectors.

2.3 Implementation of Similarity Maps from Population Vectors

Using the $A_{\mathbf{sig}}$ matrix with the activity of each neuron in the significant population vectors, similarity maps can be computed. According to the ensemble definition, similar population vectors typically belong to the same ensemble since a similar group of cells would be active at different times [3, 7, 19].

The simultaneous activity of many neurons at different times could be understood as multidimensional population vectors pointing to a specific location in a multidimensional space. To calculate the distance between all the possible pairs of population vectors, the cosine of the angle between each pair of vectors was used. When the vectors point to the same location in space, their angle is 0, rendering $\cos(0) = 1$, yielding the highest similarity. The further apart these vectors are, the lower the similarity value will be, reaching the lowest similarity value when the two vectors are orthogonal [7, 18, 23–25]. This calculation is expressed as $\cos(\theta) = \frac{\boldsymbol{v} \cdot \boldsymbol{u}}{|\boldsymbol{v}| \times |\boldsymbol{u}|}$ where v and $\boldsymbol{u}$ are vectors of equal size. The result of this operation is a single continuous number that gives the proportion of how similar the vectors are to each other, by measuring how far apart these population vectors are within the n-dimensional space (Fig. 2d). The most similar vectors would form clusters in the high-dimensional space, being closer together than the rest of the vectors, which are more different [7, 18].

By applying this procedure between each pair of population vectors, the result will be a new symmetric similarity matrix $\boldsymbol{M}$ with dimensions $[t_s, t_s]$, where t_s is the number of significant population vectors (Fig. 2e). In this matrix, each value m_{ij} represents the cosine similarity between vectors $\boldsymbol{v}_i$ and $\boldsymbol{v}_j$, calculated as $m_{ij} = \frac{\boldsymbol{v}_i \cdot \boldsymbol{v}_j}{|\boldsymbol{v}_i| \times |\boldsymbol{v}_j|}$. This matrix depicts the different patterns of similarity among all the different population vectors, revealing recurrent activity of neuronal ensembles at different times.

Not all similarities shown in the similarity matrix indicate the moments of ensemble activation. Only the most similar and recurrent vectors effectively encode the moment of activation of an ensemble [7, 17]. To facilitate the identification of robust neuronal ensembles, the similarity matrix is binarized [7, 18, 26, 27]. The threshold represents the percentage of coactive neurons at different times above what is expected by chance (Fig. 2f).

This threshold can be manually adjusted according to the data under analysis (see **Note 4**). Alternatively, it can be calculated by randomizing the activity in the population vectors and evaluating the probability of each number of coactive neurons occurring in each population vector [7]. After this step, the result is a new binary symmetric matrix denoted as M indicating a degree of high similarity between each pair of population vectors.

To account for small dissimilarities that may be encoding the same information, the algorithm computes how much a similarity vector must change to be identical to another similarity vector (Fig. 2g).

After this procedure, a new continuous symmetric matrix is generated, which subsequently requires binarization to select only the vectors that require small changes as part of the activity of the same ensemble (Fig. 2h). The identification of neuronal ensembles using SVD requires that the binarization of the similarity matrix reflects known experimental conditions. To find the most appropriate threshold values for the binarization of the similarity matrix (*see* **Note 5**), it is recommended to first design an experiment subjected to controlled experimental conditions, for example, spontaneous activity vs. sensory evoked activity (*see* **Note 6**).

2.4 Singular Value Decomposition of Similarity Maps

The binary representation of neuronal population activity in the recordings may not readily reveal discernible activity patterns (Fig. 3a). However, the similarity matrix M captures the main patterns of population activity derived from the original data [7, 26, 27] (Fig. 3b). It is important to emphasize that the matrix M contains the similarity between population vectors and not between individual neurons. SVD could be applied to similarity maps as an effective technique for pattern extraction through the singular values. These singular values reflect in descending order the relevance of the factors considered in the matrix. By extracting the most significant singular values, the original matrix could be represented with high accuracy [7].

To select the number of singular values, the most rapidly decaying factors in the Σ matrix are considered (Fig. 3c). Such singular values represent most of the variance of the data and define the number of identified neuronal ensembles.

The matrix M can be represented as $M = V\Sigma V^T$, since M is a symmetric matrix. To determine the number of significant singular values to be used to describe the number of neuronal ensembles that better reconstruct the similarity matrix and ultimately the activity matrix, a randomization process can be employed. This process helps identify the number of values that decay more rapidly than what would be expected by chance. The maximum number of ensembles that can be identified using SVD depends on the number of recorded neurons but is generally a fraction of that number [7, 17, 18, 23].

It is also possible to define the proportion of time that an ensemble should be active to be considered an ensemble; this parameter hovers around 5% of the total recorded time (*see* **Note 7**). After this process, the optimal number of ensembles could be obtained, which aligns with the number of singular values selected.

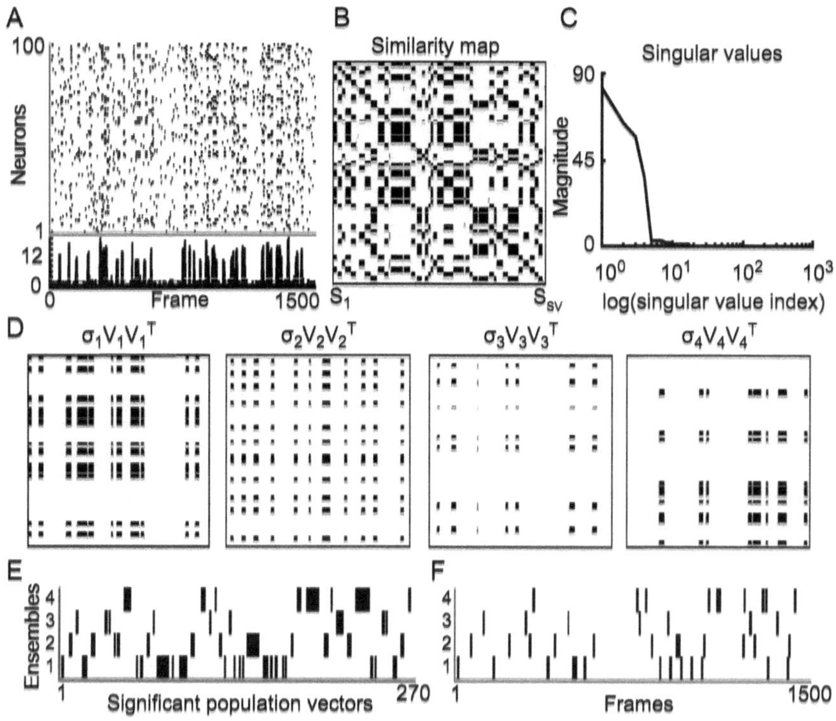

Fig. 3 Extraction of the temporal course of neuronal ensembles from the factorized similarity map. (**a**) Raster plot depicting the binarized activity of simultaneously recorded neurons, where each row represents the activity of an individual neuron. The temporal histogram at the bottom illustrates the number of coactive neurons over time. The dashed line marks the threshold value used to determine significant population vectors. (**b**) Binarized similarity map of the significant population vectors. Black squares indicate high similarity. (**c**) Magnitude of the singular values obtained from SVD, highlighting the cutoff singular value used for extracting the number of neuronal ensembles corresponding to a significant change in slope in the magnitude of contiguous singular values. (**d**) The first four factors from SVD generated by the product of the selected singular value and its corresponding components in V. (**e**) Temporal course of neuronal ensembles extracted from the factorized significant population vectors. (**f**) Temporal course of identified neuronal ensembles rearranged in time

Then a reduced version of the original similarity matrix can be reconstructed for each singular value selected. This reconstruction involves multiplying the first selected singular values in **Σ** by the corresponding components of V and V^T (Fig. 3d).

With the reduced version of the similarity map, the time course of each neuronal ensemble identified can be reconstructed (Fig. 3e). This activity can be further extended to the original recording timescale, by locating each significant vector in the original activity matrix A (Fig. 3f). The final result of this step is the identification of each significant population vector involved in each ensemble from the original raster plot (*see* **Note 8**).

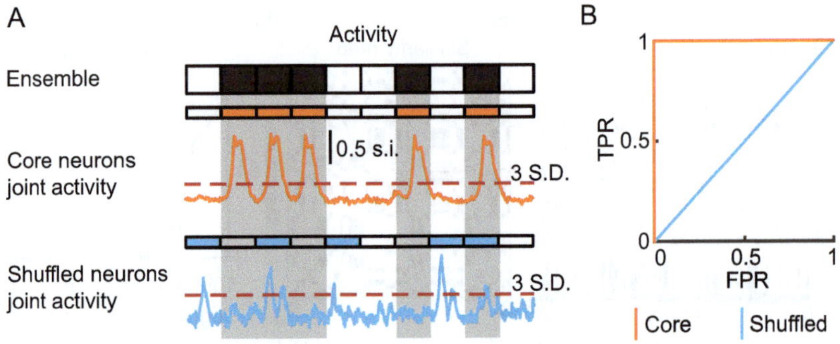

Fig. 4 Identification of the most representative neurons from each neuronal ensemble. (**a**) Similarity between ensemble activity (top) and the combined activity of different sets of neurons. The colored squares indicate activation time course. Core neurons (orange) exhibit high similarity with ensemble activity, whereas a shuffled set of neurons (blue) displays multiple points of mismatch with the ensemble activity. The dashed red lines represent the threshold for the activity of the set of neurons. (**b**) ROC curve demonstrating the different performance of core neurons and the shuffled set of neurons in predicting ensemble activity

2.5 Identification of the Most Representative Cells in each Ensemble

The most representative neurons that define a neuronal ensemble could be conceptualized as the center of mass from the cluster of population vectors representing a given ensemble in a multidimensional space. Intuitively, the neurons that are active in most of the population vectors that define an ensemble represent the core ensemble [7]. To pinpoint such neurons, an iterative approach could be used where neurons that predict ensemble activity are highlighted.

Neurons that are both the most representative and maximize the predictiveness of ensemble activity are denoted as "core neurons" [7, 17, 18]. The ensemble's activity is primarily encoded by the collective behavior of these core neurons. Notably, the joint activity of the core neurons demonstrates greater similarity to the ensemble's activity in comparison to a selection of randomly shuffled neurons from the entire population (Fig. 4a). The precise predictive performance of each neuron set in relation to ensemble activity was evaluated through the Receiver Operating Characteristic (ROC) curve (Fig. 4b).

This last step results in the discovery of the most relevant neurons associated with the activation of an ensemble (*see* **Note 9**).

Once neuronal ensembles and their members are identified, the algorithm implementation displays the spatial maps of each ensemble as well as sequential activity patterns between neuronal ensembles.

3 Conclusions

Ensemble identification using SVD provides a straightforward way to automatically identify the number of neuronal ensembles from

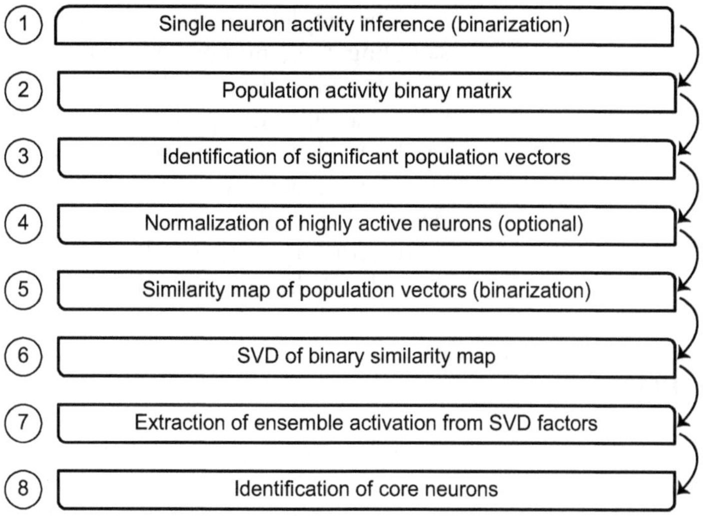

Fig. 5 Schematic representation of key steps in neuronal ensemble identification from similarity maps using SVD

the activation matrix of a population of neurons. Importantly, this can be accomplished even without knowing the temporal presence of a stimulus or specific behavior [7, 17]. This approach empowers researchers to engage in a discovery-driven analysis, enabling the identification and exploration of spontaneous or evoked neuronal ensembles.

The identification process involves several key steps (Fig. 5): firstly, the binarization of neuronal activity by applying a procedure such as the first derivative of the calcium signal and setting an activation threshold. With the binary activation matrix, only the significant population vectors are extracted, with the option to apply the TF-IDF normalization.

Following this, a similarity matrix of the cosine of the angle of each pair of population vectors is computed. At this stage, it is possible to apply a noise reduction technique based on the similarity of the different pairs of vectors. Finally, the SVD is applied on this similarity matrix to extract the most relevant features. The time course of each ensemble is extracted from the SVD of the similarity matrix using the higher singular values and finally extracting the activation of each neuron associated with this ensemble activity.

It has been previously demonstrated that the time course of ensembles identified with this approach accurately mirrors the time course of various presented visual stimuli [7, 16, 17]. Additionally, in spontaneous activity, it is possible to identify various ensembles and, through further analysis, pinpoint sequences of activation. This highlights the method's significant potential in identifying sets of neurons participating in the encoding of different stimuli and characterizing their joint activity over time.

The implementation of the method is independent of the sampling rate, number of neurons, and FOV. The method has been tested in datasets that typically include around 100 neurons within a FOV of approximately 300 μm x 300 μm. Since the method relies on population vectors that contain neurons with coordinated activity at several points in time, the identification of neuronal ensembles from recordings with highly sparse activity or with recordings with less than 30 simultaneously recorded neurons could be challenging (*see* **Note 10**).

4 Notes

1. When utilizing the first derivative of the fluorescence signal, the activity threshold is typically set at 3 standard deviations above the mean of the basal noise level. This threshold is equally applicable to a fast non-negative deconvolution approach [7, 28]. Ideally, ground truth electrophysiological recordings should be performed for different experimental setups.
2. For 100–200 neurons recorded simultaneously in the visual cortex, a threshold to determine representative population vectors is typically 3 or 4 coactive neurons [7]. It is important to note that the higher this threshold is, the fewer population vectors are considered for the analysis. In the code used for this manuscript, this value is set using the variable "pks." If this variable is assigned an empty value using "[ ]," then the automatic threshold selection is performed.
3. To decide whether to perform the TF-IDF normalization, it is necessary to consider the experimental question of interest and experimental conditions. Neurons that participate in all the ensembles could be important for some experimental designs; then the most convenient option is not to perform this normalization. In that case, the function "Ras_tf_idf" that performs this normalization should be excluded from the analysis.
4. The threshold for the low similarity filter is set using the variable "scut." A working value for this threshold typically falls within the range of 0.20 to 0.25. Highly synchronous datasets require higher values ranging from 0.25 to 0.35. If this value is left empty, the automatic selection process is initiated.
5. If the distance threshold is set to the highest value of 1, all similarity vectors will be considered as identical. Conversely, if the value of the threshold is as low as 0, then there will be no changes from the originally binarized similarity matrix. A suitable threshold value for this procedure and experimental conditions falls within the range of 0.20 and 0.25 [7]. Highly

synchronous datasets require higher values ranging from 0.25 to 0.35. This value could be manually set using the variable "hcut."

6. For the implementation of the algorithm, it is important to consider controlled experimental conditions as a ground truth. This could be achieved using optogenetics, electrical stimulation, or pharmacological manipulations. In such conditions, similarity maps must reflects controlled experimental conditions.
7. To avoid the detection of ensembles that were active in a short time window, the algorithm needs a minimum amount of activity for each ensemble.
8. The variable "C_edos" is a binary matrix with dimensions $[t_f \times e]$ where t_f is the number of significant population vectors and e is the number of ensembles. A number 1 indicates the activation time of the corresponding ensemble.
9. The variable "Pools_coords" is a matrix with dimensions $[n \times 3 \times e]$, for n neurons and e ensembles.
10. In performance tests using simulated data with controlled levels of noise and overall activity, better results are observed when recording the activity of more than 30 neurons. Additionally, the neurons should not be active more than the 15% of the entire recording time. The identification of ensembles requires that binary activity represent bursting activity instead of single spikes. Thus, to apply the method to electrophysiological recordings, raster plots should be binarized with a minimum time window of 20 ms.

Acknowledgments

Supported by grants from CONAHCYT (CF6653) and UNAM-DGAPA-PAPIIT (IN213923) to L.C-R., and The Kavli Foundation (NeuroData Discovery Award) to L.C-R. and R.V-C.

References

1. Mazzarello P (1999) Camillo Golgi's scientific biography. J Hist Neurosci 8:121–131. https://doi.org/10.1076/jhin.8.2.121.1836
2. Carrillo-Reid L, Larriva-Sahd J (2020) Contributions of Lorente de Nó to the nonlinear circuitry of the brain. Anat Rec 303:1215–1220. https://doi.org/10.1002/ar.24192
3. Hebb DO (1949) The organization of behavior. Wiley, New York
4. Buzsáki G (2010) Neural syntax: cell assemblies, synapsembles, and readers. Neuron 68:362–385. https://doi.org/10.1016/j.neuron.2010.09.023
5. Dragoi G, Tonegawa S (2011) Preplay of future place cell sequences by hippocampal cellular assemblies. Nature 469:397–401. https://doi.org/10.1038/nature09633
6. Crowe DA, Averbeck BB, Chafee MV (2010) Rapid sequences of population activity patterns dynamically encode task-critical spatial information in parietal cortex. J Neurosci 30:

11640–11653. https://doi.org/10.1523/JNEUROSCI.0954-10.2010

7. Carrillo-Reid L, Miller J-e K, Hamm JP, Jackson J, Yuste R (2015) Endogenous sequential cortical activity evoked by visual stimuli. J Neurosci 35:8813–8828. https://doi.org/10.1523/JNEUROSCI.5214-14.2015
8. Dana H, Sun Y, Mohar B, Hulse BK, Kerlin AM, Hasseman JP, Tsegaye G, Tsang A, Wong A, Patel R, Macklin JJ, Chen Y, Konnerth A, Jayaraman V, Looger LL, Schreiter ER, Svoboda K, Kim DS (2019) High-performance calcium sensors for imaging activity in neuronal populations and microcompartments. Nat Methods 16:649–657. https://doi.org/10.1038/s41592-019-0435-6
9. Grienberger C, Giovannucci A, Zeiger W, Portera-Cailliau C (2022) Two-photon calcium imaging of neuronal activity. Nat Rev Methods Primers 2:67. https://doi.org/10.1038/s43586-022-00147-1
10. Arias A, Manubens-Gil L, Dierssen M (2022) Fluorescent transgenic mouse models for whole-brain imaging in health and disease. Front Mol Neurosci 15:1–21. https://doi.org/10.3389/fnmol.2022.958222
11. Zhang Y, Rózsa M, Liang Y, Bushey D, Wei Z, Zheng J, Reep D, Broussard GJ, Tsang A, Tsegaye G, Narayan S, Obara CJ, Lim J-X, Patel R, Zhang R, Ahrens MB, Turner GC, Wang SS-H, Korff WL, Schreiter ER, Svoboda K, Hasseman JP, Kolb I, Looger LL (2023) Fast and sensitive GCaMP calcium indicators for imaging neural populations. Nature 615:884–891. https://doi.org/10.1038/s41586-023-05828-9
12. Wallace DJ, Kerr JND (2010) Chasing the cell assembly. Curr Opin Neurobiol 20:296–305
13. Yuste R, Katz LC (1991) Control of postsynaptic Ca2+ influx in developing neocortex by excitatory and inhibitory neurotransmitters. Neuron 6:333–344. https://doi.org/10.1016/0896-6273(91)90243-S
14. Kleinfeld D, Giovannucci A, Friedrich J, Gunn P, Kalfon J, Brown BL, Ann Koay S, Taxidis J, Najafi F, Gauthier JL, Zhou P, Khakh BS, Tank DW, Chklovskii DB, Pnevmatikakis EA (2019) Caiman open source tool for scalable calcium imaging data analysis. elife 8: e38173. https://doi.org/10.7554/eLife.38173.001
15. Pachitariu M, Stringer C, Schröder S, Dipoppa M, Rossi LF, Carandini M, Harris KD (2017) Suite2p: beyond 10,000 neurons with standard two-photon microscopy. https://doi.org/10.1101/061507
16. Carrillo-Reid L, Han S, Yang W, Akrouh A, Yuste R (2019) Controlling visually guided behavior by holographic recalling of cortical ensembles. Cell 178:447–457.e5. https://doi.org/10.1016/j.cell.2019.05.045
17. Carrillo-Reid L, Han S, O'Neil D, Taralova E, Jebara T, Yuste R (2021) Identification of pattern completion neurons in neuronal ensembles using probabilistic graphical models. J Neurosci 41:8577–8588. https://doi.org/10.1523/JNEUROSCI.0051-21.2021
18. Carrillo-Reid L, Tecuapetla F, Tapia D, Hernández-Cruz A, Galarraga E, Drucker-Colin R, Bargas J (2008) Encoding network states by striatal cell assemblies. J Neurophysiol 99: 1435–1450. https://doi.org/10.1152/jn.01131.2007
19. Shmiel T, Drori R, Shmiel O, Ben-Shaul Y, Nadasdy Z, Shemesh M, Teicher M, Abeles M (2005) Neurons of the cerebral cortex exhibit precise interspike timing in correspondence to behavior. Proc Natl Acad Sci 102:18655–18657. https://doi.org/10.1073/pnas.0509346102
20. Mao B-Q, Hamzei-Sichani F, Aronov D, Froemke RC, Yuste R (2001) Dynamics of spontaneous activity in neocortical slices. Neuron 32:883–898. https://doi.org/10.1016/S0896-6273(01)00518-9
21. Lan M, Tan CL, Jian S, Yue L (2009) Supervised and traditional term weighting methods for automatic text categorization. IEEE Trans Pattern Anal Mach Intell 31:721–735. https://doi.org/10.1109/TPAMI.2008.110
22. Garbarine E, DePasquale J, Gadia V, Polikar R, Rosen G (2011) Information-theoretic approaches to SVM feature selection for metagenome read classification. Comput Biol Chem 35:199–209. https://doi.org/10.1016/j.compbiolchem.2011.04.007
23. Carrillo-Reid L, Tecuapetla F, Ibáñez-Sandoval O, Hernández-Cruz A, Galarraga E, Bargas J (2009) Activation of the cholinergic system endows compositional properties to striatal cell assemblies. J Neurophysiol 101: 737–749. https://doi.org/10.1152/jn.90975.2008
24. Sasaki T, Matsuki N, Ikegaya Y (2007) Metastability of active CA3 networks. J Neurosci 27: 517–528. https://doi.org/10.1523/JNEUROSCI.4514-06.2007
25. Schreiber S, Fellous JM, Whitmer D, Tiesinga P, Sejnowski TJ (2003) A new correlation-based measure of spike timing reliability. Neurocomputing 52–54:925–931. https://doi.org/10.1016/S0925-2312(02)00838-X

26. Victor JD, Purpura KP (1996) Nature and precision of temporal coding in visual cortex: a metric-space analysis. J Neurophysiol 76: 1310–1326. https://doi.org/10.1152/jn.1996.76.2.1310
27. Levy N, Horn D, Meilijson I, Ruppin E (2001) Distributed synchrony in a cell assembly of spiking neurons. Neural Netw 14:815–824. https://doi.org/10.1016/S0893-6080(01)00044-2
28. Vogelstein JT, Packer AM, Machado TA, Sippy T, Babadi B, Yuste R, Paninski L (2010) Fast nonnegative deconvolution for spike train inference from population calcium imaging. J Neurophysiol 104:3691–3704. https://doi.org/10.1152/jn.01073.2009

Chapter 6

Detection of Overlapping Neuronal Assemblies from Activity Recordings of Large Neuronal Populations by Means of Non-orthogonal Low-Dimensional State Spaces

Sebastián A. Romano and Germán Sumbre

Abstract

Neuronal populations in vivo are characterized by the recurring coactivation of specific groups of neurons, referred to as neuronal assemblies. These assemblies are thought to constitute fundamental units for brain computations. Identifying neuronal assemblies through statistical analysis of simultaneous recordings of large neuronal populations is crucial. Here, we describe a computationally fast algorithm based on dimensionality reduction techniques, developed to detect neuronal assemblies and analyze their dynamics. Importantly, it allows for overlap between the detected assemblies, where neurons are able to transiently participate in multiple assemblies. We show that this method has been successfully applied for the analysis of calcium imaging experiments involving thousands of simultaneously recorded neurons, and its accuracy and scalability is supported by benchmarking studies on simulated data.

Key words Neuronal populations, Neuronal assemblies, Population coding, Dimensionality reduction

1 Introduction

In 1949, Donald Hebb proposed that anatomically distributed groups of neurons act together in transient synchrony, forming neuronal assemblies that underlie the neuronal processing of sensory stimuli, the generation of motor behaviors, and support internal cognitive mechanisms [1]. This extended network of interconnected neurons would be constituted and reinforced through synaptic plasticity mechanisms associated with learning processes. Neuronal assemblies would in turn be interconnected between them, such that external stimulation or internal engagement of one of these assemblies would trigger a reverberating circuit involving the sequential activation of a set of assemblies. The appeal of the "cell

The original version of this chapter was previously published without open access. A correction to this chapter is available at https://doi.org/10.1007/978-1-0716-4208-5_13

Luis Carrillo-Reid (ed.), *Identification, Characterization, and Manipulation of Neuronal Ensembles*, Neuromethods, vol. 215, https://doi.org/10.1007/978-1-0716-4208-5_6, © The Author(s) 2025, Corrected Publication 2025

assembly hypothesis" is that this extended period of brain activation provides a plausible neuronal substrate for bridging the gap between stimulus and behavioral response. Moreover, they are a cornerstone for distributed population coding in neuronal circuits [2–4]. However, as this book testifies, more than 70 years after Hebb's proposal, and despite the technological advances that have allowed an unprecedented explosion in the scale of the neuronal populations simultaneously recorded, identifying neuronal assemblies from these large neuronal datasets is still an open challenge.

A neuronal assembly can thus be depicted as a group of neurons that present tightly coordinated activity [5]. This temporal coordination takes the form of episodic and transient coincident coactivations that are consistently repeated throughout time. To define the timescale for this temporal coincidence, we will take a reader-defined stance [4]. In this view, assemblies are composed of neurons that fire within the integrating time window of a downstream reader neuron, which can be estimated on the order of membrane time constants (e.g., ~20–50 ms) [6]. Importantly, this means that neurons can belong to an assembly irrespectively of whether they are synaptically interconnected or not.

Assemblies can thus be conceived as neuronal processing modules associated with discrete and elemental brain operations performed with distributed population codes. It is essential to recognize that these distributed codes are dynamic, multifunctional, and instantiated by partially overlapping neuronal populations [7–9]. Thus, a neuronal assembly could be involved in a specific computation at a given episode and have other functional role at a different moment. Hence, assemblies are not disjoint sets, and neurons could belong to multiple assemblies (i.e., nonexclusive assemblies), playing diverse computational roles. Indeed, it is in neuronal assemblies' flexible and distributed nature wherein lies one of the major difficulties in identifying them [10, 11]. Finally, there is no ground truth data on which to train algorithms designed to detect assemblies from in vivo neuronal recordings, so unsupervised learning algorithms capable of revealing hidden structure in unlabeled data are needed.

Here, we describe a computationally fast, unsupervised algorithm capable of detecting nonexclusive assemblies (allowing for potential neuronal overlap across assemblies) that are engaged and disengaged on a moment-to-moment basis in episodes of synchronous activity. The method has been applied with success to study neuronal assemblies in two-photon fluorescence calcium imaging data [11–14], but it can be also used to analyze multielectrode recordings. We have reported a computational toolbox implemented in Matlab (Mathworks) that includes this algorithm in a computational module (https://github.com/zebrain-lab/Toolbox-Romano-et-al) [11] and an accompanying protocol [15]. We describe the algorithm step by step, discuss its potential and shortcomings, and provide a pseudo-code for implementation in any programming language.

2 Methods

The present algorithm is designed to detect neuronal assemblies through the analysis of activity recordings of neuronal populations. For N single neurons recorded along T time samples, a matrix F with T rows and N columns must be provided. When working with spike-sorted multielectrode recordings, each column of matrix F should represent the activity of a neuron (e.g., binned spike counts, smooth spike rates, etc.). If calcium imaging data is used, columns in F should represent either the relative fluorescence change ($\Delta F/F$) traces of the imaged neurons, or the neuronal spike-rate traces inferred through a suitable computational algorithm. In both cases, and as discussed above, to calculate the neuronal activity traces, it is desirable to use time bins or smoothing kernels of a temporal scale compatible with membrane time constants.

2.1 Preprocessing Two-Photon Calcium Imaging Data

In the case of using calcium imaging data to infer neuronal activity, here we discuss necessary preprocessing procedures. The starting point for the analysis of calcium imaging data involves essentially two preprocessing steps: video registration to correct for lateral displacements of the imaged plane and image segmentation into regions of interest (ROIs) corresponding to single neurons (either manually or automatically). There are a variety of techniques for the former [16–18] and the latter [11, 19, 20] steps that we have discussed before [11] and will not be covered here.

The following procedure is the extraction of fluorescence traces to calculate the $\Delta F/F$ of each ROI. Importantly, the neuropil fluorescence signal can substantially contaminate the somatic signal. To correct for this contamination, the local neuropil fluorescence signal estimated from a ~ 40 μm diameter spot centered on each ROI (that excludes the remaining ROIs) can be subtracted from the ROI signal. This correction becomes much more important when trying to identify neuronal assemblies, since neuropil fluorescence contamination can drastically impact on the measurement of temporal correlations between the ROI fluorescence traces (Fig. 1). We have previously introduced algorithms for automated correction of this kind of contamination [11].

It is important to keep in mind that the fluorescence $\Delta F/F$ transients measured via calcium imaging are a low-passed and nonlinearly distorted transformation of the underlying spiking activity, as it reflects the comparatively slower cellular dynamics of cellular calcium and the fluorescent calcium indicators. Therefore, there is the optional step of inferring neuronal spiking rates from fluorescence traces in calcium imaging data. Several computational methods have been proposed for this [20–23]. A recent benchmark on a variety of algorithms revealed that, when compared to ground-truth spiking activity obtained with electrophysiology, the spike-

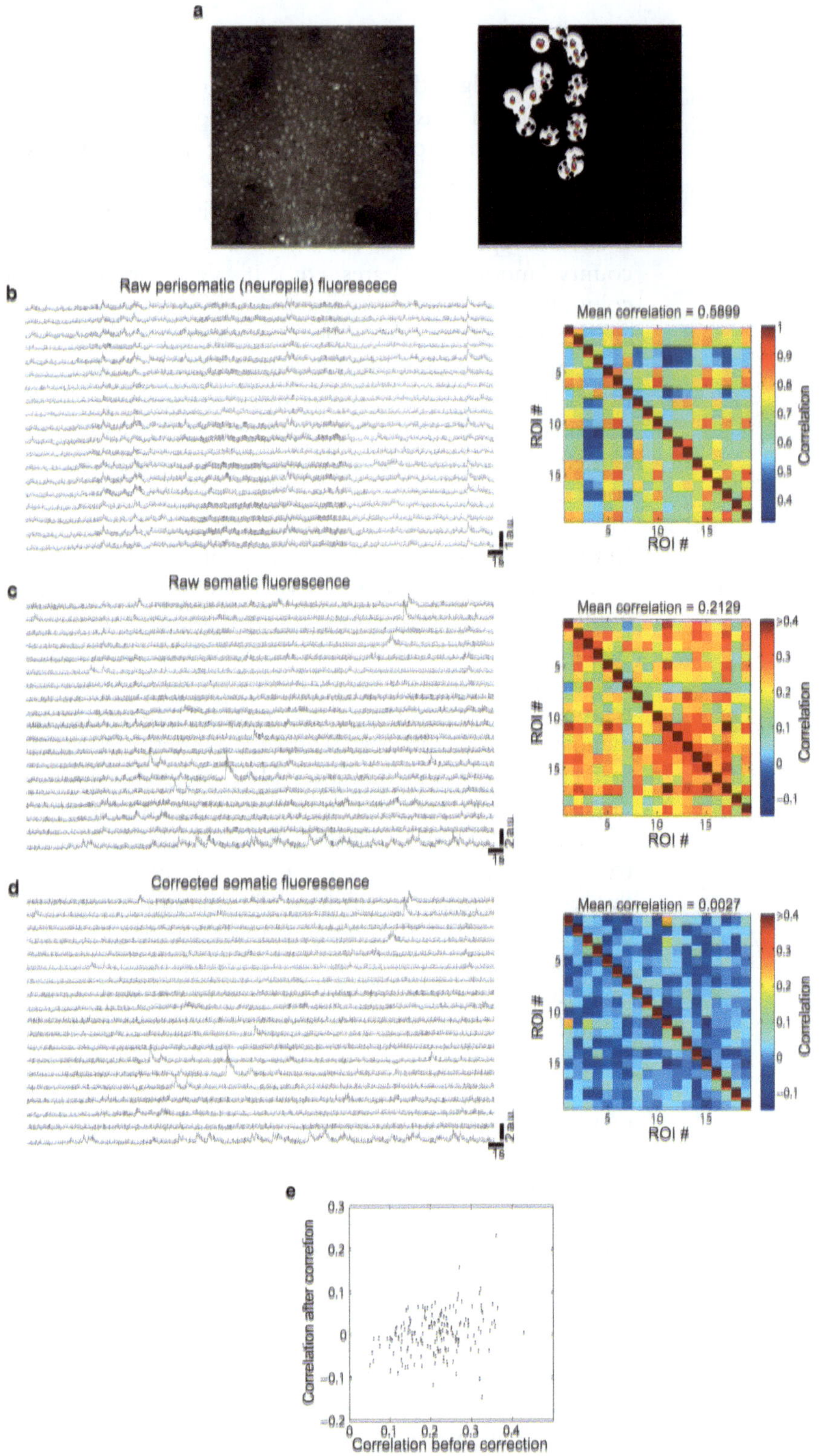

Fig. 1 In calcium imaging data contamination from neuropil, fluorescence can corrupt activity correlations between neurons, potentially precluding the correct identification of neuronal assemblies (**a**) Left: optical plane imaged with two-photon microscopy of the mouse somatosensory cortex. Right: examples of the

rate traces inferred from two-photon imaging data correlate well with the true spike rates for slower imaging sampling rates ($r \sim 0.7$ for 3.125 Hz sampling, where r is the Pearson correlation coefficient), but perform poorer for faster imaging rates ($r \sim 0.4$ for 25 Hz sampling) [24].

For the *PCA-promax* procedure that we describe here, the user can choose to work either with $\Delta F/F$ fluorescence traces or with inferred spike rate traces. In case of working directly with fluorescence traces, better results are obtained by removing fluorescence fluctuations due to noise in the measurement. We included a computational module designed for this in the calcium imaging processing toolbox, where a noise model is fitted to the $\Delta F/F$ data and fluorescence signals that do not significantly differ from the noise model are set to 0 [11].

2.2 Procedure for Detecting Assemblies

The complete procedure for identification of assemblies is depicted in pseudo-code in *Appendix 1*. It involves a main *PCA-promax* algorithm, along with five additional subroutines. In this section, we will conceptually describe and discuss the steps of the complete procedure.

First, the matrix F of the recorded neuronal activity is represented in a space of reduced dimensionality through principal component analysis (*PCA*). In order to standardize the data variance across neurons and avoid biasing the principal components (*PCs*) towards the most active neurons (i.e., the neurons with largest variance), the matrix F is z-scored before performing *PCA*. This standardization of variances will induce *PCA* towards representing neuronal activity covariance rather than variance.

In principal component space, each neuron will be represented by a point, whose coordinates are called *PC* "loadings". Neurons with correlated activity would typically appear occupying similar regions of this PC space, separated from a cloud of uncorrelated neurons that sit at the origin (Fig. 2a). The *PCA-promax* algorithm aims at finding these regions in a data-driven manner, to delineate clusters of correlated neurons. But first, we need to determine the right dimensionality of *PC* space that best describes the concerted neuronal activity in the recorded data.

Fig. 1 (continued) detected ROIs (red) and their circular perisomatic masks used to calculate the local neuropil signals (white). Perisomatic masks are overlapping, but each ROI is associated with a single circular mask. Black holes inside the perisomatic masks are other detected ROIs not included in the local neuropil signal calculation. (**b**) Left: raw fluorescence traces obtained with the perisomatic masks shown in (**a**) (neuropil signal). Right: pairwise correlation matrix for the signals shown in the left. Note the high temporal correlation across the traces. (**c**) Same as (**b**), for the raw fluorescence traces of the ROIs shown in (**a**) (somata). (**d**) Same as (**c**), for the ROI fluorescence traces automatically corrected with the computational toolbox [11], obtained by subtracting the traces shown in (**b**) from the corresponding traces shown in (**c**). Note the reduction in the temporal correlations, compared to those found in (**c**), despite the small changes of the individual fluorescence traces. (**e**) Relationship between the pairwise correlations shown in (**c**) and (**d**)

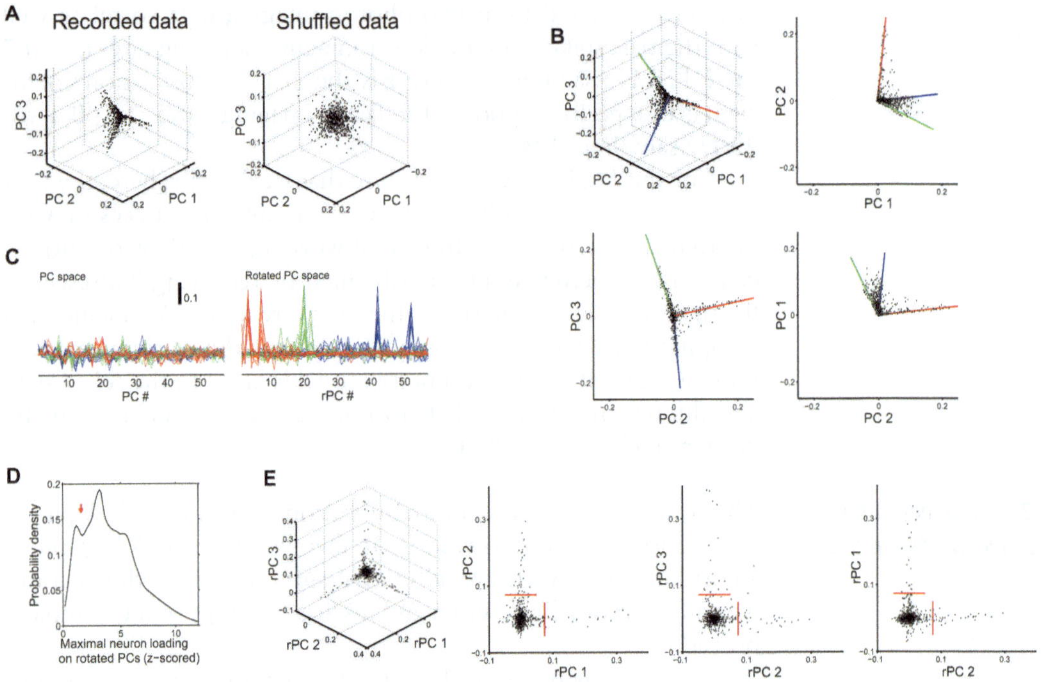

Fig. 2 Representing neurons in *PCA-promax* allows detection of assemblies. (**a**) Examples of neuronal population activity observed in principal component (*PC*) space. Only the first three most significant *PCs* are shown (see (**c**) for a representation of all the significant *PCs*). Left: Each black dot depicts a neuron; its coordinates are given by the entire neurons' spontaneous activity during a 1-hr recording during a typical experiment. Notice the dots extending along straight lines projecting away from the origin. Right: same as right panel, but for null model data where activity correlations between neurons were eliminated by shuffling the neuronal activation times. Note how dots cluster around the origin without a clear structure. (**b**) The axes that describe the projections observed in the recorded data shown in A are obtained by *promax* rotation of the first three *PCs* (*rPCs* are shown as red, green, and blue lines). (**c**) The component loadings of neurons from three assemblies (red, green, and blue curves) over all the significant *PCs* (left) and *rPCs* (right). Each curve depicts a neuron, representing all its loadings over the corresponding component axes. Note the overlap between similar color curves, indicating that neurons with similar loadings in the *rPC* space belong to the same assembly. Structure in neuronal loadings is enhanced in *rPC* space. (**d**) Probability density of the maximal neuronal z-scored loadings on *rPCs* of a representative experiment. These distributions typically show a first peak that encompasses the cloud of points scattered around the origin of the *rPC* space. The threshold loadThresh for *rPC* loadings that discriminate assemblies (red lines in **e**) is manually selected from this distribution (red arrow). Its value is usually around 2, and small variations in its definition did not affect the validity of the assemblies, as measured through clustering validation techniques. (**e**) The data shown in A is represented over *rPC* space, and the threshold loadThresh is shown in red (neurons with higher loading over an *rPC* are grouped in an assembly)

2.3 Implementing a Null Model to Detect PCs Associated with Correlated Neuronal Activity

The first step is to reduce the dimensionality of the dataset, by keeping data dimensions that describe the correlated activity of neuronal assemblies and throwing away the rest. This can be obtained by only keeping the top *PCs* which are associated with eigenvalues large enough so that they cannot be explained by the

PCs of a null model of the data where assemblies are absent. From these null models, a threshold to the eigenvalues of informative *PCs* must be obtained, λ_{TH}, and two possible null model types are considered, from which the experimenter must choose.

The first approach, initially implemented by Peyrache and colleagues [25] and used to detect assemblies in neuronal data [12–14, 26, 27], relies on an analytical bound from the theory of random matrices, given by the Marcenko-Pastur distribution. This distribution describes the eigenvalues of a correlation matrix computed for statistically independent variables (in our case, neurons with independent activity) and uses this eigenvalue distribution as a null model of the data. The theoretical upper limit provided by the Marcenko-Pastur distribution can be used as a lower bound to identify significant eigenvalues associated with correlated firing in the neuronal data. This threshold value was derived for the asymptotic case of $N \rightarrow \infty$ and $T \rightarrow \infty$, with $T/N >= 1$. Thus, to account for the case of matrices of finite size, the Tracy-Widom correction [28] is included in the estimation of λ_{TH}:

$$\lambda_{TH}^{M-P} = \left(1 + \sqrt{\frac{N}{T}}\right)^2 + N^{\frac{-2}{3}}$$

where $N^{-2/3}$ is the bias correcting term introduced for finite matrices (this result applies for matrices of unit variance, which in our case holds true due to the z-score standardization of F). This threshold value can be used to reduce data dimensionality by discarding *PCs* associated with eigenvalues lower than $\lambda_{TH}{}^{\text{M-P}}$.

The second approach relies on constructing a null model of F that destroys activity correlations across neurons, while preserving two important statistics of the recorded neuronal activity: the mean neuronal activity and the activity autocorrelation of each neuron. Namely, the matrix F can be written as a column-wise concatenation of N row vectors of the form $f_i = [f_{i,1}, f_{i,2}, \ldots f_{i,T}]$ (written in row form) containing the activity time series of each neuron i, *with* $i = 1, 2, \ldots, N$. Thus, for each neuron i, the proposed null model preserves the sample mean activity:

$$\bar{f_i} = \frac{1}{T} \sum_{t=1}^{T} f_{i,t}$$

and the sample autocorrelation of the neuronal activity:

$$xcorr_i(\tau) = \sum_{t=1}^{T-\tau} \left(f_{i,t+\tau} - \bar{f_i}\right)\left(f_{i,t} - \bar{f_i}\right)$$

The procedure for constructing this null model F_{CS} is to perform independent random circular time shifts of each column f_i that compose the recorded matrix F [29]. A fast implementation of this kind of cyclic permutation is to choose a random time sample independently for each neuron i and use that random sample to divide the corresponding f_i column in two nonoverlapping segments which are then switched in position. This procedure does not interfere with neither the mean activity levels nor the neuronal autocorrelations. However, since segments are switched at different random time samples for each neuron, it disrupts temporal coordination across neurons. The resulting F_{CS} obtained this way is an appropriate null model for a neuronal population devoid of assemblies (which, in *PC* space, would look similar to the shuffled data shown Fig. 2a). This procedure is iterated to obtain 500 different F_{CS} matrices, the *PCA* eigenvalues are calculated for each iteration, and the eigenvalues are pooled together in a distribution $eigenvalues_{CS}$. The threshold value ${\lambda_{TH}}^{CS}$ for a significant eigenvalue is determined by the 95th percentile of this distribution:

$$\lambda_{TH}^{CS} = percentile95\left(eigenvalues_{CS}\right)$$

This threshold value is then used to reduce data dimensionality by discarding *PCs* associated with eigenvalues lower than ${\lambda_{TH}}^{CS}$.

Obtaining a threshold value for significant eigenvalues through the circular shifts approach is evidently much slower than the analytic result by Marcenko-Pastur since it involves the generation of several surrogate null models. Previous studies have found that reducing the dimensionality of F through ${\lambda_{TH}}^{CS}$ yields similar results to the ones obtained through ${\lambda_{TH}}^{M\text{-}P}$ [29], while others working on calcium imaging data have found significant improvements when using ${\lambda_{TH}}^{CS}$ (depending on how performance was quantified) [30] (see "Performance on simulated data" below). Ultimately, the best method will depend on the firing statistics in F and how performance will be evaluated, and we encourage experimenting with both methods.

2.4 Finding Overlapping Assemblies Through Non-orthogonal Factor Rotation

After obtaining the significant *PCs* through the dimensionality reduction approach explained above, one could use these components to identify clusters of correlated neurons [25]. However, *PCA* imposes a condition of orthogonality between *PCs*. Thus, if we were to use the *PCs* to define clusters, it would result in clusters that do not share neurons (exclusive or disjoint clusters). We want to relax this condition, since we are interested in the more general case of overlapping assemblies that could share neurons (i.e., neurons could be assigned to multiple assemblies).

Therefore, to delineate the assemblies, we turn to a space of obliquely rotated components, through the *promax* algorithm [31]. An example of rotated *PCs* (*rPCs*) found by *promax* is shown in Fig. 2b. *Promax* is a technique from Factor Analysis, which rotates components to obtain a "simple structure" of rotated factors (*rPCs*) sparsely populated by few high loadings, with the rest of the loadings being close to zero (i.e., each *rPC* will represent a sparse group of neurons) (Fig. 2c). Since *promax* performs an oblique (non-orthogonal) rotation, it allows for correlated *rPCs*, unlike the orthogonal *PCs*.

To delineate clusters using the *rPCs*, we impose a data-driven threshold *loadThresh* obtained from the distribution of neuron loadings along the *rPCs*. For this, first each *rPC* is normalized according to their euclidean 2-norm and then sign-flipped so that its most extreme loading value is positive. We obtain the maximum loading of each neuron in the z-scored *rPCs*, estimate the (smoothed) distribution of these values, and determine *loadThresh* as the loading value that corresponds to the first robust minima in this distribution (Fig. 2d). In this way, *loadThresh* separates the cloud of neurons sitting at the origin in *rPC* space from neurons that significantly project along the *rPC* axes that define neuronal assemblies (Fig. 2d). Typically, users manually select *loadThresh* from visual inspection of the distribution of loadings mentioned above. Neurons are assigned to an assembly if their z-scored loading along the respective *rPC* exceeds *loadThresh* (Fig. 2e).

Following this procedure, one will obtain as many assemblies as *rPCs* found in the data. Since the relaxation of the orthogonality condition could change the relevant dimensionality of the data (the number of latent dimensions obtained through *PCA* and λ_{TH}), some rPCs could potentially extensively overlap. This would result in redundant clusters. Therefore, there is the optional step of merging pairs of assemblies that are determined by two *rPCs* if the dot product between both *rPCs* exceeds a threshold value (typically 0.6).

2.5 Calculation of Time Series of Assembly Activations

Once assemblies are defined, the final step is to infer the time series of assembly activations. For this, we use the *engagement* index (that we called before matching index [12]) that quantifies the degree of similarity, at the neuronal level, between any given pair of population activation patterns. In this sense, the *engagement* index can be used to quantify the match of neuronal activations between the time series of observed activation patterns and given target patterns consisting of the neuronal assemblies found by the *PCA-promax* algorithm. This index is similar to the Dice coefficient, which was previously used to evaluate the degree of similarity in the connection patterns of brain regions [32, 33]. The *engagement* index is defined as

$$engagment_{i,t=j} = 2\frac{|Pat_i \cap Pat_j|}{|Pat_i| + |Pat_j|}$$

where both Pat_i and Pat_j are two binary neuronal activation patterns (a 1 x *N* binary vector representing the recorded neuronal population, with ones indicating active neurons and zeros indicating silent neurons). Norms are equal to the number of ones in each vector. In the present case of studying the temporal dynamics of activation of a given neuronal assembly, Pat_i is the binary pattern of assembly i (with ones labeling assembly neurons and zeros labeling the rest of the neuronal population), and Pat_j is the binary population activity pattern of time sample j (the row j of matrix F converted to logical values). Thus, $engagement_{i,t=j}$ is the engagement of assembly i at time sample j, and it will quantify the proportion of neuronal activations that are common to the assembly and to the population activity observed at time sample j. It is valued between 0 (no overlap in activations) and 1 (perfect overlap in activations). The *engagement* index aims to estimate the number of "hits" between neuronal activations (common ones in the vectors) without specifically considering as a "hit" a match between silent neurons (common zeros), opposed to the popularly used Hamming distance, which can become problematic when considering the match of sparse binary vectors.

To estimate the significance across the time series of values obtained with the *engagement* index, the hypergeometric distribution is used to obtain the *engSignificance* index. Under the null hypothesis of independent neuronal activations, this is a discrete distribution that describes the probability of having k "hits" (common neuronal activations) with n target neuronal activations in a population of N neurons showing K activations at a given moment. Therefore, it allows us to estimate the probability of observing a given activation match by chance, with neurons independently activated. Thus, under this null hypothesis, the time series $engSignificance_{i,t=j}$ gives the p-value of the engagement of assembly i at time sample j, under the null hypothesis that the k hits were the result of chance.

3 Results

3.1 Performance on Simulated Data

There is no ground truth data on which to benchmark the performance of algorithms designed to detect assemblies from in vivo neuronal recordings. Nevertheless, one can test the performance on simulated data. Mölter and collaborators performed such a test for simulated calcium imaging recordings [30] and tested a variety of algorithms. They embedded 10 assemblies into a population of size varying from ~200 to 900 neurons and evaluated algorithms while changing different parameters of the simulation.

The performance of the different algorithms was measured with a "Best Match" score, which evaluates the difference between the set of embedded assemblies and the set of reconstructed assemblies, in a way which takes into account overlaps between the assemblies that compose each set. This score ranges from 0 to 1 for complete mismatch and perfect match, correspondingly, and 0.07 indicates chance level performance for these simulations (i.e., the expected value when comparing two random sets of assemblies).

A significantly better performance was observed for *PCA-promax* using the eigenvalue threshold $\lambda_{TH}{}^{CS}$ obtained with circular shift null models. Under this condition, the Best Match performance score remained stable in the 0.8 to 0.9 range when varying the total number of simulated neurons (from 200 to 900 neurons), the size of assemblies (from 10 to 25 neurons per assembly), the simulation duration (from 1800s to 7200 s), the imaging frame rate (from 2 Hz to 10 Hz), and the calcium indicator half-life $\tau_{1/2}$ (from 0.1 to 2 s). Performance also remained within the 0.8 to 0.9 range when varying signal-to-noise conditions, namely, the average number of activation events that increased the firing rate of neurons (from 9 to 36 per hour), the multiplicative gain for these increases in firing rates (from 4 to 25), and the standard deviation of the gaussian noise added to the simulated fluorescence traces (from 0 to 1.5). Performance significantly degraded for simulation durations shorter than 1800s and more challenging signal-to-noise conditions, but this was true for all the algorithms tested. *PCA-promax* was among the best-performing algorithms, and one of the fastest ones, between one and two orders of magnitude faster than similarly performing algorithms.

Despite these encouraging results, caution should be taken when analyzing real neuronal datasets, where complexities and particularities of each experiment recording could bias performance either way. There is no single algorithm that is guaranteed to work across recording techniques (electrophysiology, calcium, and voltage imaging), signal-to-noise ratios, regimes of network dynamics, and neural connectivity architectures. Moreover, without ground truth data on which to test the algorithms, conclusions on the real performance of algorithms are still speculative. Therefore, testing and comparing results across different algorithms is always recommended.

3.2 Studying Neuronal Assemblies in Real Recordings

To illustrate how the detection of assemblies through *PCA-promax* can reveal biologically relevant information about large neuronal networks, we will present a series of results obtained with different imaging datasets. These results have been previously published [11, 12, 14].

3.2.1 Unsupervised Analysis of Population Dynamics During a Task

We applied the *PCA-promax* procedure on two-photon fluorescence calcium imaging from the mouse somatosensory cortex where neurons expressed GCaMP6s [34, 35]. In this experiment, the superficial barrel cortex was imaged while the mouse performed a single-whisker object localization task. This dataset is freely available at crcns.org (dataset ssc-1).

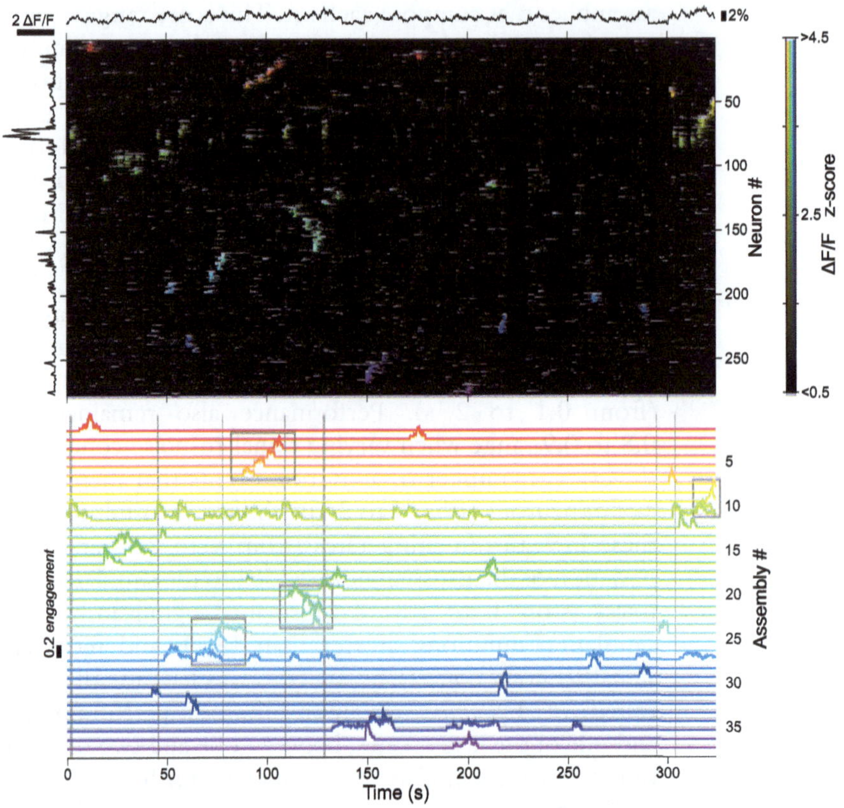

Fig. 3 Inspecting population activity dynamics through the detection of neuronal assemblies in the barrel cortex of mice performing a whisker-guided object localization task. Top panel: raster plot of the z-scored $\Delta F/F$ of 277 neurons distributed in 37 assemblies, from the total imaged population of 1025 neurons. Neurons are sorted and color-coded according to the assembly to which they belong to (color bar on the left). Black trace on top, fluctuations of the percentage of active neurons in the total imaged population; black trace on the left, average neuronal responses to a whisker-object contact (each contact is labeled by a vertical dotted line in the bottom panel). Bottom panel: activation dynamics of the detected assemblies (as measured by the *engagement* index), color-coded as in the raster plot. Vertical dotted lines indicate moments of whisker-object contact. Note that activations of assembly #11 are associated with contacts (neurons from assembly #11 are indicated by light green bar on the right of top panel). Episodes of activity sequentially progressing through assemblies are marked by gray boxes

Activity fluctuations along the experiment were sparse, being limited to <5% of total neuronal population imaged (1025 neurons, see top trace in Fig. 3). However, clustering neurons according to their activities revealed the coordinated activation of several sets of neurons (Fig. 3). Interestingly, one of these neuronal assemblies (assembly #11) consisted of a group of correlated neurons that was responsive to the mouse contacting the whisker with the object. Thus, insightful functional information can be obtained through the completely unsupervised procedure of *PCA-promax*, without requiring any additional information besides the recorded neuronal activity.

Apart from detecting this group of functionally related neurons, the clustering algorithm also exposed the ongoing concerted activity of other assemblies along the task, which may convey additional contextual or task-related signals. For example, the sequential activation of sets of assemblies (as predicted by Hebb's hypothesis) can be observed.

3.2.2 Detecting Assemblies in Large-Scale Neuronal Activity Data

Recent advances in optical methods and optogenetics provide exceptional possibilities for functional imaging of large neuronal populations, and even whole brains [36, 37], with high spatial resolution. In contrast to the important experimental advances in imaging techniques, computational analysis tools are still incipient.

In Fig. 4, we illustrate the result of using the *PCA-promax* method to detect assemblies in brain-wide light-sheet GCaMP5 imaging throughout a volumetrically scanned zebrafish larva, where the activity of more than 80.000 ROIs was analyzed. The *PCA-promax* algorithm revealed several bilaterally symmetric assemblies and unilateral assemblies with their corresponding contralateral counterparts. In particular, we observed a symmetric pair of assemblies distributed over the pre-tectum and the anterior and caudal hindbrain (Fig. 4a panel iv), whose activities were tightly related to visual stimulation that induced the larva to trigger optomotor responses (OMR, an innate visuomotor reflex characterized by the fish swimming in the same direction as a high-contrast visual stimulus) (Fig. 4b), suggesting a visuomotor role for these assemblies. This particular assembly was also apparent when averaging ROI responses during OMR stimulation (Fig. 4c), but the average-response map was much noisier and less anatomically defined than the assemblies found through activity clustering. Moreover, we also found topographically precise clusters that were missed by averaging ROI responses to visual stimulation, like a symmetric pair of assemblies comprising the midbrain torus semicircularis, along with its octavolateral (lateral line and acoustic) afferences from the hindbrain (Fig. 4b panel ii). Finally, clusters (assemblies) found with the standard *k-means* algorithm (Fig. 4b panel viii) also show some spatial organization that is roughly consistent with those found through *PCA-promax* (Fig. 4b panel vii), but assemblies found with PCA-promax expose a much richer and informative fine structure, including overlapping assemblies (that cannot be detected by *k-means*).

3.2.3 Studying Functional Characteristics of Neuronal Assemblies

The *PCA-promax* algorithm has also been used to study spontaneously engaged neuronal assemblies in the visual system of the zebrafish larva [12]. This case study shows how this algorithm, combined with functional studies and careful usage of null models to assess statistical and biological significance, can reveal underlying

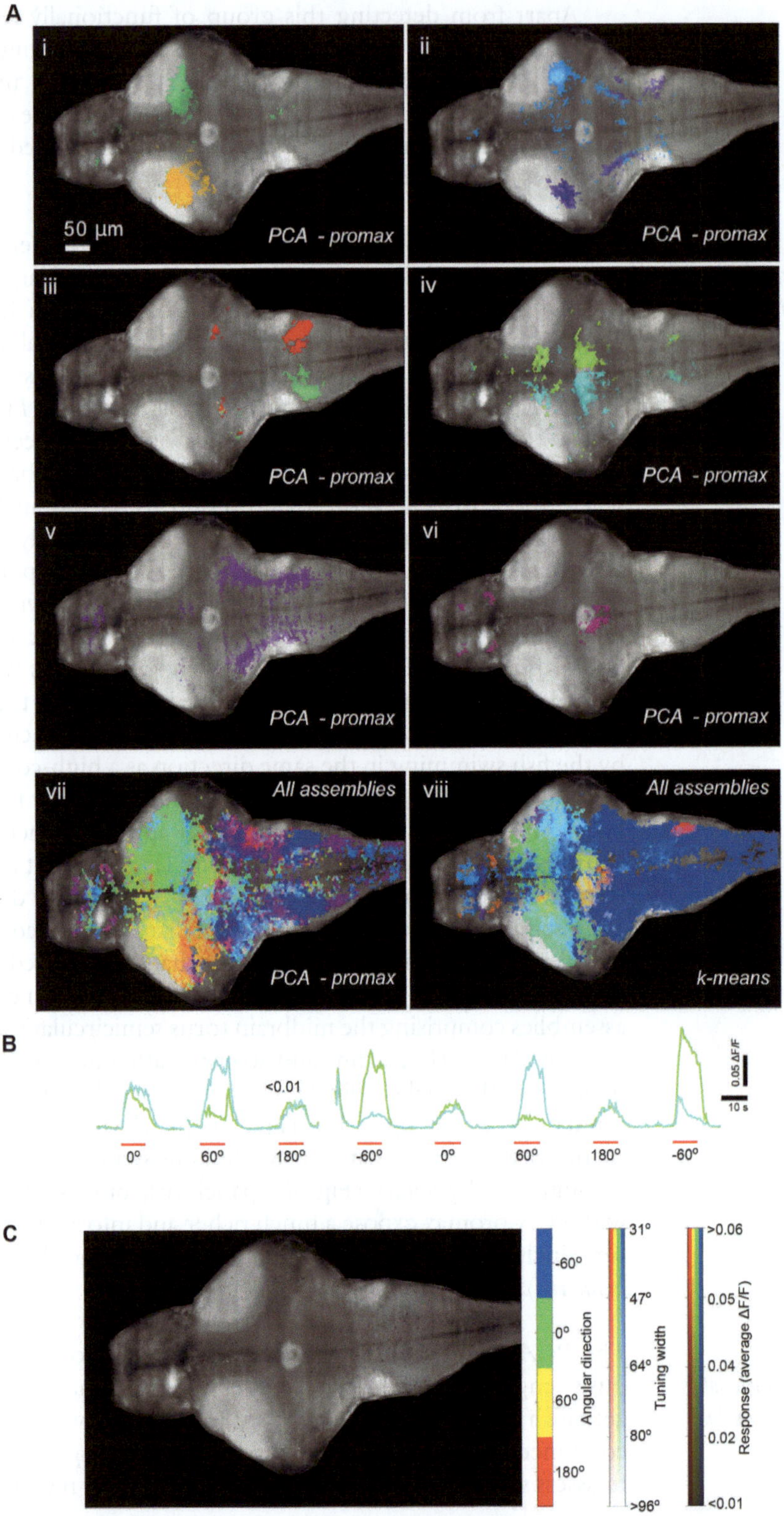

Fig. 4 Neuronal assemblies obtained using whole-brain light-sheet imaging in zebrafish reveal the spatial structure of brain activity. A GCaMP5-expressing zebrafish larva was volumetrically scanned with fast

computational mechanisms and behavioral correlates of neuronal assemblies.

The optic tectum is the largest visual region in the zebrafish and is essential for visually guided prey detection and capture [38]. With two-photon calcium imaging, we monitored the spontaneous activity of ~1000 neurons in the optic tectum in intact, non-anaesthetized, non-paralyzed zebrafish larvae pan-neuronally expressing the genetically encoded calcium indicator GCaMP3. We analyzed the structure of tectal spontaneous neuronal activity within the context of optic tectum's functional role visual object detection and the generation of orienting motor commands [38, 39]. For this, we first recorded the neuronal dynamics of tectal populations for periods of 1 hr., in the absence of sensory stimulation. Then, we visually stimulated the larva with light spots at different positions in the visual field to characterize the visual tuning of tectal neurons with respect to the azimuthal position of the light stimulation.

We applied the PCA-promax procedure to detect assemblies in these recordings and identified spontaneous assemblies consisting of significantly temporally correlated neurons (Fig. 5a), representing 80% ± 19% of the significant neuronal pairwise correlations observed in the tectal population (Fig. 5b). Therefore, these assemblies accounted for a large fraction of the correlational structure of the entire imaged neuronal population.

Spontaneous assemblies tended to be formed by neighboring neurons of the same tectal hemisphere, often resulting in topographically compact neuronal clusters (Fig. 5c). Indeed,

Fig. 4 (continued) multiplane single-photon light-sheet microscopy. Due to the lack of single-neuron resolution in this imaging dataset, a grid of 9-μm-diameter hexagons was imposed over each imaged optical plane. (**a**) (**i–vi**) Some of the ROI assemblies found with the *PCA-promax* method, displayed over the maximal intensity projection of all imaged optical sections. (**i–iv**) Pairs of symmetric unilateral assemblies. (**v–vi**) Single bilaterally symmetric assemblies. (**vii**) All ROI assemblies found with the *PCA-promax* approach, displayed over the maximal intensity projection of all imaged optical sections. Assemblies are colored according to the similarity of their activity dynamics (the more temporally correlated, the more similar the color of the assemblies). This color code was also used in the previous panels (except for the assembly pair shown in **iii**, which was colored differently to facilitate visualization). (**viii**) Same as **vii**, but for assemblies found using k-means, color-coded according to the similarity of their activity dynamics. To allow for a comparison between **vii** and **viii**, the dimensionality of the dataset was reduced through *PCA* before clustering (otherwise, clustering did not converge). (**b**) Single-trial activation dynamics of the assemblies shown in **iv** during one of the blocks of visual stimulation. Periods of moving-grating stimulation are labeled in red, angles indicate the angular direction of gratings. Note that both assemblies are activated by this OMR-inducing stimulation, with the bluish and greenish assemblies preferring 60° and − 60°, respectively. (**c**) Average ROI responses to angular direction of visual moving gratings. Responses are displayed in HSV color code over the maximal intensity projection of all imaged optical sections. Note that, while the tectal neuropil indiscriminately responds to all directions (whitish ROIs), a pair of bilaterally symmetric group of ROIs in the hindbrain responds selectively to either 60° or − 60° (this group is similar to the assembly in panel **iv** shown in **a**)

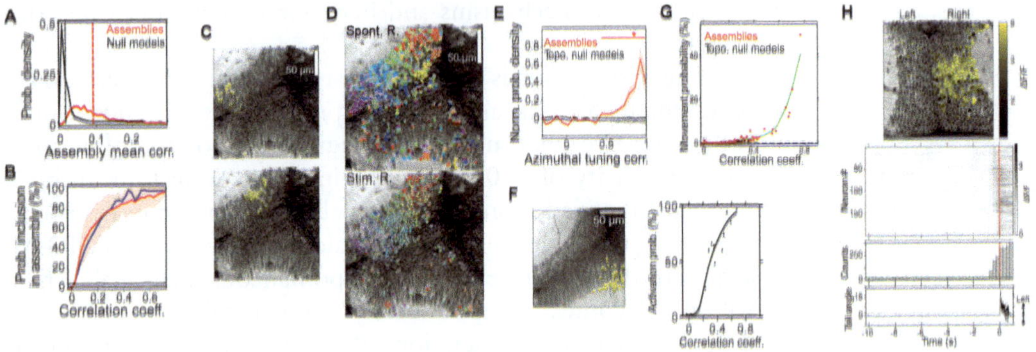

Fig. 5 Spontaneously emerging assemblies are formed by similarly tuned neurons, show nonlinear cooperative dynamics, and present a behavioral correlate (**a**) Probability density distribution of the mean pairwise spontaneous activity correlation coefficient of neurons within each of the spontaneous assemblies (red). To assess significance, the same distribution for null models is presented in black. Dashed lines: the distributions' means (0.09 ± 0.06 for spontaneous assemblies and 0.01 ± 0.01 for null models; means and SD $p < 10^{-5}$). (**b**) Probability of inclusion of a neuronal pair in a spontaneous assembly, as a function of their pairwise spontaneous activity correlation coefficient, averaged across experiments (red), and for a representative single experiment (blue). The same calculation for shuffled sets of surrogate assemblies (null models) is presented in black. Error bars, SD across experiments. (**c**) Topographies of two examples of spontaneous neuronal assemblies. Assembly neurons are labeled in yellow. (**d**) Spontaneously emerging retinotopic-like map in one typical experiment. Top: all the spontaneous assemblies whose centroids lay on the right tectum. Neurons belonging to assemblies with similar topographic centroids were similarly colored. The colors represent the average assemblies' centroid position along the rostro-caudal tectal axis. The transparency represents the SD across centroids. Bottom: same as top panel, but for assemblies found during responses to visual stimulation with light spots. Note the resemblance between both panels, showing a graded change in colors along the rostro-caudal axis. (**e**) Top: probability density distribution of the pooled pairwise correlation coefficients of azimuthal tuning curves of neurons that are grouped by a spontaneous assembly (red). This distribution was normalized by dividing by the distribution resulting from topographical null models. For comparison, we show in black the average normalized distribution for single null model runs. Note that spontaneous assemblies show a bias toward grouping neurons with highly similar tuning curves (significant bias range and median are indicated by top red line and arrowhead). (**f**) Left: topography of a typical spontaneous assembly (assembly neurons in yellow). Right: raw data (dots) and regression fits (black curves) of the activation probabilities of the assembly's neurons. Note the all-or-none-like nature of these curves, suggesting highly recurrent facilitatory cooperative dynamics. (**g**) Probability of a tail movement onset as a function of the correlation between all the assembly patterns and the spontaneous tectal network activity for the imaging frame preceding movement onset (red dots, raw data; green curve, regression fit). Black dots and curve: same calculation for randomized assemblies (null models). Dashed blue line: average probability of a movement onset at any given imaging frame. Analysis for pooled experiments ($n = 5$). (**h**) Example of spontaneous activation of a tectal assembly before a spontaneous tail movement. Bottom to top: tail angle trace (zero, a straight tail; positive, left bends; negative, right bends), histogram of the spontaneously active tectal neurons (green line, threshold for significant population events; red dotted line, tail-flip onset), raster plot of spontaneous activations of tectal neurons active before the movement's onset (red dotted line, tail-flip onset), and topography of the spontaneously activated assembly. Time is relative to tail-flip onset. Note that neurons activated immediately before movement onset are located in the tectal hemisphere contralateral to the initial tail movement direction

spontaneous assemblies were mostly constrained to either tectal hemisphere (71% of them were significantly unilateral), and 51% showed significant topographical compactness. Remarkably, the topographic arrangement of these spontaneous assemblies was reminiscent of the neuronal activation patterns that we recorded in response to visual stimulation (Fig. 5d).

To infer the significance of the measured properties of the assemblies, we controlled against sets of shuffled surrogate assemblies (null models). For this purpose, we shuffled neuronal identities by permuting the indexing of the neuronal population. Then, we built the null models by grouping neurons according to the original assemblies but using the permuted neuronal indexing. This was repeated for 1000 rounds, creating a null model set of mock random clusters with randomized topographies, conserving the rest of the statistics in the data (firing rates, pairwise correlations, etc.). Moreover, we built an additional set of null models with the additional constraint of conserving the relative pairwise physical distances between neurons found in the original assemblies (topographical null models). This method enabled us to destroy the functional specificity of the assemblies while keeping intact their topographic compactness and the tectum's retinotopic map. Using these sets of null models, we were able to test for the significance of the detected assemblies' features, with respect to the shuffled clusters.

We found that spontaneous assemblies had a significant bias to regroup neurons with similar spatial tuning curves, even when compared to the topographic null models that preserve the distribution of pairwise distances (Fig. 5e). Therefore, spontaneous assemblies do not simply derive from the well-known tendency of higher spontaneous correlations between nearby neurons [40, 41]. On the contrary, there is a clear specificity in the composition of spontaneous neuronal assemblies such that they grouped functionally related neurons rather than exclusively neighboring ones.

To study the dynamic principles underlying the emergence of spontaneous assemblies, we analyzed how assembly activation depended on the spontaneously coordinated activations of the neurons within the assembly. For this purpose, we calculated the activation probability of neurons within a given assembly, as a function of the correlation between the instantaneous activity pattern of the tectal population and the topography of the assembly being tested (the latter correlation indicated how well the spontaneous activity at any given moment matched the spontaneous assembly being studied). The obtained activation/correlation profiles showed strong activation probabilities ("facilitations") that sharply increased as the spontaneous activity pattern gradually resembled the topography of the tested assembly (Fig. 5f). This implies that the spontaneous neuronal assembly patterns represent network states that are "preferred" by the assembly neurons, since these neurons are markedly more prone to be activated when the spontaneous network activation better resembles the pattern of the assembly they belong to (up to an ~40-fold increase compared to the null model). This result suggests that the dynamics of assembly neurons are capable of abruptly switching from sparse weak to fully correlated activations.

Finally, we asked whether the spontaneous activation of a tectal assembly can be related to the generation of a behavioral output. For this purpose, we monitored spontaneous tectal activities in head-restrained larvae so we could also monitor the spontaneous tail motor behaviors. Under this condition, larvae produced brief isolated episodes of tail movements. Remarkably, we observed that 11% of these spontaneous movement episodes were immediately preceded by the spontaneous activation of a topographically compact tectal assembly (Fig. 5f). As a control, we used tectal spontaneous activations to predict the onset of tail movements of other larvae. Using this method, we found that only 0.2% of the movement episodes were preceded by compact tectal assemblies. Given the frequency of spontaneous assembly activations and tail movements, the probability of predicting 11% of the episodes by chance is in the order of 10–20. Notably, when the activation of a spontaneous assembly preceded a tail flip with kinematics that indicated a turning movement, in 93% of the cases the assembly corresponded to the contralateral tectal hemisphere (Fig. 5f; 13 out of 14, 7 to each side; $p = 9\times 10^{-4}$). We then calculated the probability of a tail movement onset during the imaging frame *i* as a function of the correlation between the tectal activity pattern at frame *i-1* and the patterns of all the spontaneous neuronal assemblies found. The latter correlation indicates how well the pattern of tectal activity preceding a tail flip matches the topographies of a spontaneous neuronal assembly. We observed a sharp step-like increase of this probability as the spontaneous tectal activity gradually matched the topography of a spontaneous assembly (Fig. 5g).

In summary, these results indicate a behavioral correlate of spontaneous tectal assemblies and reveal that these assemblies are particularly tuned to visual field positions, a vital information computed by the tectum during its role in prey-tracking behavior.

3.2.4 Development of Neuronal Assemblies

From embryonic development to adulthood, newborn neurons are continuously added to the vertebrate brain. This process is an important component of brain plasticity, permitting the formation of new memories and adaptation of neural circuits to the ever-changing sensory environment [42]. Recently born neurons eventually receive synaptic inputs and connect to post-synaptic partners, indicating their integration into neuronal networks [43].

Building on the previous case study, we asked how newborn neurons integrate into neuronal assemblies of the zebrafish optic tectum through [14]. To describe the morphological development of newborn tectal neurons, we used an electroporation approach performed at 4 days postfertilization (dpf), which sparsely labeled newborn neurons with a red fluorescent label (dTomato), and imaged with GCaMP5 tectal activity up to 10 days post electroporation (dpe).

We were able to image and analyze chronically the same newborn neurons for several consecutive days. These chronic experiments showed that the labeled newborn neurons already exhibited spontaneous calcium activity at 1 dpe, and its activity became progressively correlated with the activity of its neighboring mature neurons as development advanced and its dendritic tree became more complex (Fig. 6a).

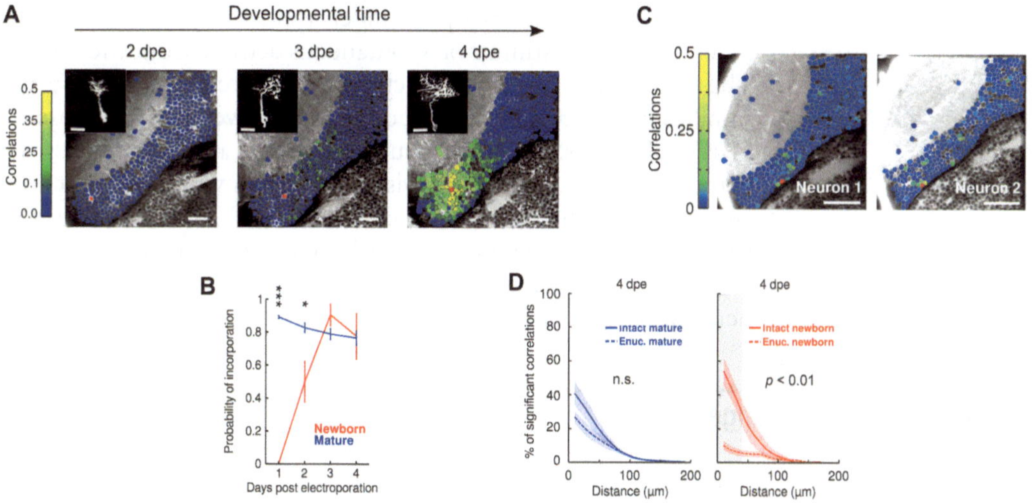

Fig. 6 Neurons need visual experience to integrate into neuronal assemblies. (**a**) Example of the same newborn-labeled neuron chronically imaged at 2, 3, and 4 days post electroporation (dpe). Tectal neurons are color-coded according to their pairwise correlation values with the newborn-labeled neuron (red). The scale bars represent 50 μm. Insets represent the morphology of the labeled neuron. The scale bar represents 20 μm. (**b**) Probability of incorporation of newborn-labeled neurons into the neuronal assemblies (red) and the probability of the functionally mature neurons to belong to one of the neuronal assemblies (blue) for 1–4 dpe (newborn neurons: n = 6, 15, 21, and 15; functionally mature neurons: n = 1962, 3564, 5296, and 3969 from 6, 10, 17, and 11 larvae, respectively).$*$ $p < 0.05$; $***$ $p < 0.001$. The error bars represent SEM. (**c**) Two examples of the pairwise correlations at 4 dpe between newborn-labeled and the rest of the neurons in tecta that were deprived of retinal inputs through eye enucleations. Optical sections of the optic tectum with neurons color-coded (color scale bar) according to their pairwise correlations with the newborn-labeled neuron (red). (**d**) Left: The relationship between the percentage of significant correlations among functionally mature neurons in enucleated (dashed line) and intact (solid line) larvae and the physical distance between the neurons. Note that the two curves are not significantly different. Right: The same as the left panel, but for the correlations between newborn-labeled neurons and the rest of the tectal neurons. Note the significant difference between the two curves for short distances ($p < 0.01$; Mann-Whitney U test). Confidence intervals show the SEM

We then evaluated the probability for a newborn neuron to belong to one of the functionally mature functional spontaneous assemblies of the optic tectum. We observed that, at 1 dpe, newborn neurons did not belong to any of these neuronal assemblies. The probability of belonging increased at 2 dpe and became similar to that of functionally mature neurons at 3 or 4 dpe (Fig. 6b).

To study the role of visual experience on the incorporation of newborn neurons into functionally mature sub-circuits, we performed, just before the electroporation (at 3.5 dpf), unilateral enucleations of the contralateral eye to the monitored tectal hemisphere (in the zebrafish larva, the retina projects exclusively to the contralateral hemisphere). This procedure chronically deprived the optic tectum of all retinal afferents before the labeled newborn neurons became responsive to visually responsive. Under these conditions, the functionally mature population developed under normal retinal conditions (before the enucleations), whereas the newborn neurons developed under retinal deprivation (in the absence of visual stimuli or spontaneous activity from the retina). The correlations between functionally mature neurons were not significantly altered by the enucleations. However, in the absence of retinal inputs, newborn neurons showed a significantly lower percentage of significant pairwise correlations with their functionally mature partners (Fig. 6d, e). These results demonstrate how visual experience shapes functional interactions in the tectal neurons, driving the recruitment of newborn neurons into functional neuronal assemblies.

4 Outlook and Challenges

The recent rapid progress in large-scale neuronal recording techniques currently allows to monitor the activity of hundreds to thousands of individual neurons simultaneously. Together with optogenetic activity manipulations and large-scale reconstruction of neuronal circuits, new opportunities are emerging for finding the neuronal substrates of behavior and cognition. The neuronal assembly is thought to be an essential computational unit that underlies numerous brain operations [4, 6, 10, 44]. In these assemblies, neurons could be associated anatomically (through recurrent synaptic connections between them) and/or functionally (bound together by means of common synaptic inputs that collectively co-activate them during sensory, motor, mnemonic, or cognitive processes). Thus, because of the vast number of the potential member neurons in a functional cell assembly, statistical and computational techniques are crucial to identify assemblies from neuronal recordings. Functional clustering of neuronal data is a complex mathematical procedure still under development. Here we present the *PCA-promax* algorithm that has been extensively tested with success, particularly for calcium imaging experiments. It has also been shown to be robust in benchmark tests performed on simulated data.

Nevertheless, the *PCA-promax* method for detecting assemblies has limitations. The algorithm relies on *PCA*, which implements linear data transformations. Thus, nonlinear neuronal activity correlations (i.e., when plotting the activity of a neuron against another neuron produces a curved cloud of points) could result in spurious assemblies. Most multivariate statistical analysis methods currently applied are also linear, but those that rely on manifold learning algorithms [45–48] could accommodate other nonlinear measures of activity similarity. Furthermore, when using *PCA*, there is no guarantee of a straightforward interpretation of the *PCs* obtained in terms of biological or experimental variables. In principle, this could be improved with the recently proposed *demixed PCA* technique [49], which reduces the dimensionality of the data, taking into account task parameters (e.g., sensory or motor variables). Furthermore, increasingly popular non-negative matrix factorization techniques [50] which, unlike *PCA*, impose a positivity constraint to its components may be more biologically interpretable in terms of positive neuronal activations. Finally, the *PCA-promax* method is particularly suited to delineate discrete clusters of ROIs that are engaged in brief events of collective activation. Thus, for population activations that present a continuous and long-lasting evolution in time (e.g., propagating spatial waves, oscillations, slow and temporally extended patterning), this algorithm would tend to discretize these dynamic phenomena in distinct patterns. For example, a wave would be separated in a progression of distinct assemblies that represent different instances of the moving wave front. Therefore, particular care should be taken in the interpretation of the results in these kinds of scenarios. Nevertheless, this is not a specific problem of the *PCA-promax* algorithm, since time-evolving patterns are usually problematic for all clustering methods. New mathematical techniques to address this issue are eagerly awaited, since unsupervised faithful learning of nonlinear interdependencies across large neuronal populations is an extremely challenging task that remains unresolved.

Acknowledgments

We thank members of the Sumbre lab, especially Verónica Pérez-Schuster and Enrique Hansen. This research was funded by PICT 2017-4023, CONICET PIP 2787, FOCEM-Mercosur (COF 03/11), ERC StG (243106) and ERC CoG (726280).

Appendix 1: Pseudo-code for *PCA-promax* Procedure for the Detection of Neuronal Assemblies

Algorithm 1
PCA-promax ensembles
Main procedure

```
procedure PCA-promax(F,nullModelType)
PCs,eigenvals ← PCA(zscore(F))
λ_TH ← maxNullEigenval(zscore(F),method)
topPC ← lowest argument index i that satisfies eigenvals(i) ≤ λ_TH
PCs ← Keep only top 1,2,...,topPC components in PCs
PCs_rot ← Rotate PCs components with promax oblique rotation
PCs_rot ← normAndFlip(PCs_rot)
neuronLoading ← max(zscore(PCs_rot)^Transpose)
distLoading ← Estimate smooth distribution of neuronLoading values
load_TH ← argument of the first minimum in distLoading
assemblies,PCs_rot ← defineAssemblies(PCs_rot,load_TH)
OPTIONAL: assemblies ← mergeSimilar(assemblies,PCs_rot,load_TH)
engagment,engSignificance ← assemblyEngagement(assemblies,F)
return assemblies,engagment,engSignificance
```

Algorithm 2
maxNullEigenval(F,method)
Procedure to estimate maximum eigenvalue of a null model.

```
T ← number of rows in F
N ← number of columns in F
if method ==" Marcenko - pastur " then
      λ_TH ← (1+√(N/T))^2 + N^(2/3)
else if method == "CircularShift" then
          N_iter ← Number of iterations
          eigenvalsPool ← Empty array
            for k = 1,2,...,N_iter do
              F_CS ← Empty matrix
                for iNeur = 1,2,...,N do
                  sampleSplit ← Random integer ∈ [1,T]
                  segment1 ← F(1 : sampleSplit,iNeur)
                  segment2 ← F(sampleSplit + 1 : end,iNeur)
```

(continued)

```
            shifted ← Row concatenate segment2 and segment1
            F_CS ← Append column shifted
        end for
        eigenvals ← PCA(F_CS)
        eigenvalsPool ← append(eigenvalsPool, eigenvals)
    end for
    λ_TH ← 95^th percentile of eigenvalsPool
end if
return λ_TH
```

Algorithm 3
normAndFlip(PCs_{rot})
Procedure to normalize and flip rotated components.

```
for i = 1,2,..., number of columns in PCs_rot do
        PCs_rot(1 : end,i) ← PCs_rot(1 : end,i)/norm(PCs_rot(1 : end,i))
        if max(PCs_rot(1 : end,i)) <= abs(min(PCs_rot(1 : end,i))) then
            PCs_rot(1 : end,i) ←-PCs_rot(1 : end,i);
    end if
end for
return PCs_rot
```

Algorithm 4
defineAssemblies(PCs_{rot},$load_{TH}$)
Procedure to define assemblies

```
zPCs_rot ← zscore(PCs_rot)
assemblies ← zeros matrix array with the same size as PCs_rot
counter1 ← 0
toDel ← empty array
for i = 1,2,..., number of columns in PCs_rot do
        values,indexes ← sort column i of zPCs_rot in descending order
        if indexes(values >= load_TH) is an empty array then
            todel ← append(todel,i)
            counter1 ← counter1 + 1
```

(continued)

```
        else
            assemblyNeurs ← indexes(values >= load_TH)
            assemblies(assemblyNeurs,i) ← PCs_rot(assemblyNeurs,i)
            assemblies(:,i) ← assemblies(:,i)/norm(assemblies(:,i))
        end if
end for
assemblies ← delete columns toDel in assemblies
PCs_rot ← delete columns toDel in PCs_rot
return assemblies,PCs_rot
```

Algorithm 5
mergeSimilar(assemblies,PCs_{rot},$load_{TH}$)
Procedure to merge similar assemblies

```
sThresh ← set similarity threshold
repeat
    similarity ← assemblies^transpose * assemblies
    similarity ← set upper triangular part of similarity to 0
    simRow,simCol ← indexes that satisfy similarity > sThresh
    if simRow is an empty array then
        break
    end if
    vals ← zero array of same size as simRow
    for i = 1,2,..., number of elements in simRow do
        vals(i) ← similarity(simRow(i),simCol(i))
    end for
    indMax ← Find argmax of vals
    i_merge,j_merge ← simRow(indMax),simCol(indMax)
    summedV ector ← sum columns i_merge and j_merge of PCs_rot
    summedV ector ← summedV ector/norm(summedV ector)
    PCsrot ← delete columns imerge and jmerge of PCsrot
    assemblies ← delete columns i_merge and j_merge of assemblies
    PCs_rot ← append summedV ector as a new column to PCs_rot
    i0 ← indexes that satisfy zscore(summedV ector) < load_TH
    summedV ector ← set summedV ector(i0) to 0
    summedV ector ← summedV ector/norm(summedV ector)
    assemblies ← append column summedV ector to assemblies
until break
return assemblies
```

Algorithm 6
assemblyEngagment(assemblies,F).
Procedure to calculate the activation time series of each assembly.

```
T ← number of rows in F
N ← number of columns in F
A ← number of columns in assemblies
engagment ← zeros matrix array with A rows and T columns
engSignificance ← ones matrix array with A rows and T columns
raster ← convert F into matrix of logical values
popActivations ← sum elements of raster along column dimension
for iAssembly = 1,2,...,A do
        assembly_i ← column iAssembly in assemblies
        assemblyNeurs_i ← indexes in assembly_i with
        non-zero values
        pattern ← zeros array with N columns
        pattern ← set pattern(assemblyNeurs_i) to 1
        p ← sum elements of pattern
        for iSample = 1,2,...,T do
            pSum ← p + popActivations(iSample)
            pIntersection ← pattern AND
            popActivations(iSample)
            pIntersection ← sum elements of
            pIntersection
              engagment(iAssembly,iSample) ← 2 × pIntersection/pSum
            if pIntersection > 0 then
                  x ← array 0,1,2,...,min(p,popActivations(iSample))
                  prob ← hypergeometric prob. dens. function of x for p hits
    with popActivations(iSample) samples from a population of size N
          engSignificance(iAssembly,iSample) ← sum of prob from pIntersection
    to min(p,popActivations(iSample))
        end if
      end for
end for
return engagment,engSignificance
```

References

1. Hebb DO (1949) The organization of behavior. Organ Behav 911:335–335. https://doi.org/10.2307/1418888
2. Pouget A, Dayan P, Zemel R (2000) Information processing with population codes. Nat Rev Neurosci 1:125–132. https://doi.org/10.1038/35039062
3. Averbeck BB, Latham PE, Pouget A (2006) Neural correlations, population coding and computation. Nat Rev Neurosci 7:358–366. https://doi.org/10.1038/nrn1888
4. Buzsáki G (2010) Neural syntax: cell assemblies, synapsembles, and readers. Neuron 68: 362–385. https://doi.org/10.1016/j.neuron.2010.09.023

5. Abeles M, Gerstein GL (1988) Detecting spatiotemporal firing patterns among simultaneously recorded single neurons. J Neurophysiol 60:909–924
6. Harris KD (2005) Neural signatures of cell assembly organization. Nat Rev Neurosci 6: 399–407. https://doi.org/10.1038/nrn1669
7. Briggman KL, Kristan WB (2008) Multifunctional pattern-generating circuits. Annu Rev Neurosci 31:271–294. https://doi.org/10.1146/annurev.neuro.31.060407.125552
8. Anderson ML (2010) Neural reuse: a fundamental organizational principle of the brain. Behav Brain Sci 33:245–266; discussion 266–313. https://doi.org/10.1017/S0140525X10000853
9. Tognoli E, Kelso JAS (2014) The metastable brain. Neuron 81:35–48. https://doi.org/10.1016/j.neuron.2013.12.022
10. Wallace DJ, Kerr JN (2010) Chasing the cell assembly. Curr Opin Neurobiol 20:296–305. https://doi.org/10.1016/j.conb.2010.05.003
11. Romano SA, Pérez-Schuster V, Jouary A et al (2017) An integrated calcium imaging processing toolbox for the analysis of neuronal population dynamics. PLoS Comput Biol 13:1–23. https://doi.org/10.1371/journal.pcbi.1005526
12. Romano SA, Pietri T, Pérez-Schuster V et al (2015) Spontaneous neuronal network dynamics reveal circuit's functional adaptations for behavior. Neuron 85:1070–1085. https://doi.org/10.1016/j.neuron.2015.01.027
13. Pietri T, Romano SA, Pérez-Schuster V et al (2017) The emergence of the spatial structure of Tectal spontaneous activity is independent of visual inputs. Cell Rep 19:939–948. https://doi.org/10.1016/j.celrep.2017.04.015
14. Boulanger-Weill J, Candat V, Jouary A et al (2017) Functional interactions between newborn and mature neurons leading to integration into established neuronal circuits. Curr Biol 27:1707–1720.e5. https://doi.org/10.1016/j.cub.2017.05.029
15. Romano SA, Pérez-schuster V, Jouary A, et al (2017) A computational toolbox and step-by-step tutorial for the analysis of neuronal population dynamics in calcium imaging data. bioRxiv
16. Thévenaz P, Ruttimann UE, Unser M (1998) A pyramid approach to subpixel registration based on intensity. IEEE Trans Image Process 7:27–41. https://doi.org/10.1109/83.650848
17. Guizar-Sicairos M, Thurman ST, Fienup JR (2008) Efficient subpixel image registration algorithms. Opt Lett 33:156–158. https://doi.org/10.1364/ol.33.000156
18. Pnevmatikakis EA, Giovannucci A (2017) NoRMCorre: an online algorithm for piecewise rigid motion correction of calcium imaging data. J Neurosci Methods 291:83–94. https://doi.org/10.1016/j.jneumeth.2017.07.031
19. Mukamel EA, Nimmerjahn A, Schnitzer MJ (2009) Automated analysis of cellular signals from large-scale calcium imaging data. Neuron 63:747–760. https://doi.org/10.1016/j.neuron.2009.08.009
20. Pnevmatikakis EA, Soudry D, Gao Y et al (2016) Simultaneous denoising, deconvolution, and demixing of calcium imaging data. Neuron 89:299–299. https://doi.org/10.1016/j.neuron.2015.11.037
21. Deneux T, Kaszas A, Szalay G et al (2016) Accurate spike estimation from noisy calcium signals for ultrafast three-dimensional imaging of large neuronal populations in vivo. Nat Commun 7:12190. https://doi.org/10.1038/ncomms12190
22. Friedrich J, Zhou P, Paninski L (2017) Fast online deconvolution of calcium imaging data. PLoS Comput Biol 13:e1005423. https://doi.org/10.1371/journal.pcbi.1005423
23. Giovannucci A, Friedrich J, Gunn P et al (2019) CaImAn an open source tool for scalable calcium imaging data analysis. elife 8: e38173. https://doi.org/10.7554/eLife.38173
24. Berens P, Freeman J, Deneux T et al (2018) Community-based benchmarking improves spike rate inference from two-photon calcium imaging data. PLoS Comput Biol 14: e1006157. https://doi.org/10.1371/journal.pcbi.1006157
25. Peyrache A, Benchenane K, Khamassi M et al (2009) Principal component analysis of ensemble recordings reveals cell assemblies at high temporal resolution. J Comput Neurosci 29: 309–325. https://doi.org/10.1007/s10827-009-0154-6
26. Peyrache A, Khamassi M, Benchenane K et al (2009) Replay of rule-learning related neural patterns in the prefrontal cortex during sleep. Nat Neurosci 12:919–926. https://doi.org/10.1038/nn.2337
27. Lopes-dos-Santos V, Conde-Ocazionez S, Nicolelis ML et al (2011) Neuronal assembly detection and cell membership specification by principal component analysis. PLoS One 6: e20996–e20996. https://doi.org/10.1371/journal.pone.0020996

28. Tracy CA, Widom H (1992) Level-spacing distributions and the airy kernel. Commun Math Phys 159:35–35
29. Lopes-dos-Santos V, Ribeiro S, Tort ABL (2013) Detecting cell assemblies in large neuronal populations. J Neurosci Methods 220: 149–166. https://doi.org/10.1016/j.jneumeth.2013.04.010
30. Mölter J, Avitan L, Goodhill GJ (2018) Detecting neural assemblies in calcium imaging data. BMC Biol 16:143. https://doi.org/10.1186/s12915-018-0606-4
31. Hendrickson AE, White PO (1964) PROMAX: a quick method for rotation to oblique simple structure. Br J Stat Psychol 17:65–70. https://doi.org/10.1111/j.2044-8317.1964.tb00244.x
32. Hilgetag C, Kötter R, Stephan K, Sporns O (2002) Computational methods for the analysis of brain connectivity. In: Ascoli G (ed) Computational neuroanatomy – principles and methods. Humana Press, Totowa, pp 295–335
33. Sporns O, Honey C, Kötter R (2007) Identification and classification of hubs in brain networks. PLoS One 2:e1049. https://doi.org/10.1371/Citation
34. Peron SP, Freeman J, Iyer V et al (2015) A cellular resolution map of barrel cortex activity during tactile behavior. Neuron 86:783–799. https://doi.org/10.1016/j.neuron.2015.03.027
35. Guo ZV, Li N, Huber D et al (2014) Flow of cortical activity underlying a tactile decision in mice. Neuron 81:179–194. https://doi.org/10.1016/j.neuron.2013.10.020
36. Ahrens MB, Orger MB, Robson DN et al (2013) Whole-brain functional imaging at cellular resolution using light-sheet microscopy. Nat Methods 10:413–420. https://doi.org/10.1038/nmeth.2434
37. Panier T, Romano SA, Olive R et al (2013) Fast functional imaging of multiple brain regions in intact zebrafish larvae using selective plane illumination microscopy. Front Neural Circuits 7: 65–65. https://doi.org/10.3389/fncir.2013.00065
38. Gahtan E, Tanger P, Baier H (2005) Visual prey capture in larval zebrafish is controlled by identified reticulospinal neurons downstream of the tectum. J Neurosci 25:9294–9303. https://doi.org/10.1523/JNEUROSCI.2678-05.2005
39. Niell CM, Smith SJ (2005) Functional imaging reveals rapid development of visual response properties in the zebrafish tectum. Neuron 45:941–951. https://doi.org/10.1016/j.neuron.2005.01.047
40. Kerr JND, de Kock CPJ, Greenberg DS et al (2007) Spatial organization of neuronal population responses in layer 2/3 of rat barrel cortex. J Neurosci 27:13316–13328. https://doi.org/10.1523/JNEUROSCI.2210-07.2007
41. Smith MA, Kohn A (2008) Spatial and temporal scales of neuronal correlation in primary visual cortex. J Neurosci 28:12591–12603. https://doi.org/10.1523/JNEUROSCI.2929-08.2008
42. Marin-Burgin A, Mongiat LA, Pardi MB, Schinder AF (2012) Unique processing during a period of high excitation/inhibition balance in adult-born neurons. Science 335:1238–1242. https://doi.org/10.1126/science.1214956
43. Toni N, Laplagne DA, Zhao C et al (2008) Neurons born in the adult dentate gyrus form functional synapses with target cells. Nat Neurosci 11:901–907. https://doi.org/10.1038/nn.2156
44. Yuste R (2015) From the neuron doctrine to neural networks. Nat Rev Neurosci 16:487–497. https://doi.org/10.1038/nrn3962
45. Roweis ST, Saul LK (2000) Nonlinear dimensionality reduction by locally linear embedding. Science 290:2323–2326. https://doi.org/10.1126/science.290.5500.2323
46. Tenenbaum JB, de Silva V, Langford JC (2000) A global geometric framework for nonlinear dimensionality reduction. Science 290:2319–2323. https://doi.org/10.1126/science.290.5500.2319
47. van der Maaten L, Hinton G (2008) Visualizing data using t-SNE. J Mach Learn Res 9: 2579–2605
48. McInnes L, Healy J, Melville J (2020) UMAP: uniform manifold approximation and projection for dimension reduction
49. Kobak D, Brendel W, Constantinidis C et al (2016) Demixed principal component analysis of neural population data. eLife 5:1–37. https://doi.org/10.7554/eLife.10989
50. Lee DD, Seung HS (1999) Learning the parts of objects by non-negative matrix factorization. Nature 401:788–791. https://doi.org/10.1038/44565

Chapter 7

Detecting Neural Assemblies Through Similarity Graph Clustering

Jan Mölter and Geoffrey J. Goodhill

Abstract

A hallmark of neural population activity is neural assemblies, groups of neurons that consistently coactivate. Quantitative analysis of these assemblies requires reliable and objective methods for their detection and extraction from recordings of neural population activity, increasingly in the form of calcium imaging data. Here we discuss an algorithm which achieves this goal. The basic idea of this approach is to form a similarity graph of population activity patterns with a high level of coactivity. Methods developed for community detection in graphs can then be applied to obtain a statistical estimate for the number of assemblies, followed by extraction via standard clustering methods. Expanding on the original MATLAB implementation, here we explain the application of this algorithm to example data using a more recent and efficient Python implementation.

Key words Calcium imaging, Spontaneous activity, Neural assemblies, Clustering

1 Introduction

Across many brain regions in many species, a frequently observed feature of neural population activity is the consistent coactivation of certain groups of neurons [1, 2]. These groups of neurons are often referred to as assemblies or ensembles and have been suggested to be fundamental units of neural information processing [3–5]. A variety of qualitative definitions of assemblies have been proposed, and each of these allows many different quantitative interpretations [2, 5]. Here we will use "assembly" to refer to a group of neurons whose activity tends to be coincidentally elevated on some relevant timescale [6]. Note that this is a purely functional definition that makes no assumptions about underlying structural connectivity.

Currently, one of the most important ways to study neural population activity at a large scale is by calcium imaging [7]. Through the use of fluorescent indicators, for example, from the GCaMP family [8, 9], the activity of thousands of neurons can

Luis Carrillo-Reid (ed.), *Identification, Characterization, and Manipulation of Neuronal Ensembles*, Neuromethods, vol. 215, https://doi.org/10.1007/978-1-0716-4208-5_7,

be recorded simultaneously [10]. Detecting neural assemblies from such recordings can be more challenging than from electrophysiological recordings of spiking activity, due to the often low temporal resolution and high levels of background noise of the fluorescence signals.

In the past few years, considerable work has been devoted to the development of algorithms to detect assemblies. The techniques used in these algorithms range from dimensionality reduction and clustering to statistical inference. When applied to the same data, the results can vary substantially, reflecting the different assumptions and methods used in each algorithm [6]. For real data, there is generally no ground truth with which to evaluate these results, but algorithms can be compared using surrogate data of known composition [6].

Here we take a closer look at an algorithm based on a statistical inference approach that has been shown to perform well for calcium imaging data, which we refer to as Similarity Graph Clustering (SGC) [6, 11]. A brief summary of this algorithm is as follows:

1. Starting with the fluorescence time series for a population of neurons (or regions of interest), we first derive a binary indicator signal for each neuron's activity. We refer to the collection of activity states for each neuron in the population as the population's activity pattern at that time.
2. In the absence of assembly activation, these activity patterns are usually sparse. We therefore select only those activity patterns where the level of coactivity exceeds a threshold derived from a statistical null model.
3. We assume that the resulting set of high-activity patterns is due to events of assembly activity and that patterns that correspond to the same assembly are similar. To obtain an estimate for the number of assemblies and perform clustering, we assemble the high-activity patterns into a graph where each node represents a pattern and each edge represents pattern similarity. This is different from many other approaches where one investigates the relation between pairs of individual neurons in terms of correlations, rather than directly the similarity between population-scale activity patterns. In case of the former, one would generally obtain groups of neurons that are pairwise correlated; however, that does not imply coactivity of the neurons across the groups, which we require for an assembly.
4. Since patterns corresponding to the same assembly are similar, this graph exhibits community structure. Thus, by means of a Bayesian inference approach, where the stochastic block model as a particular model of random graph with community structure is fitted to the observed graph of activity patterns, we obtain an estimate for the number of communities.

5. Using spectral clustering methods, we then decompose the graph into its groups of similar activity patterns.
6. From those, assemblies are then reconstructed as follows. We start by constructing preliminary assembly patterns for those groups of similar activity patterns that are not too small. These are then refined by recursively merging groups upon sufficient similarity. This eventually yields a set of refined preliminary assembly patterns. We then take again the entire set of original high-activity patterns and assign each of them to one of those if they are sufficiently similar, or otherwise reject them. This yields a final set of groups of activity patterns. After rejecting particularly small groups, for every remaining group, we obtain the final assembly as the neurons that were active in more than a threshold fraction of its patterns.

2 Recording of Neural Assembly Activity

In order to illustrate the practical application of the SGC algorithm, we consider a recording of neural activity from the optic tectum of the larval zebrafish using calcium imaging [11]. In this experiment, zebrafish-expressing *elavl3:H2B-GCaMP6s* were embedded in agarose, and two-photon calcium imaging was performed from a single plane of the optic tectum at a frequency of 2.2 Hz for up to 1 h (Fig. 1). Imaging was performed in the dark so that purely spontaneous activity was recorded, uncorrupted by visual stimuli.

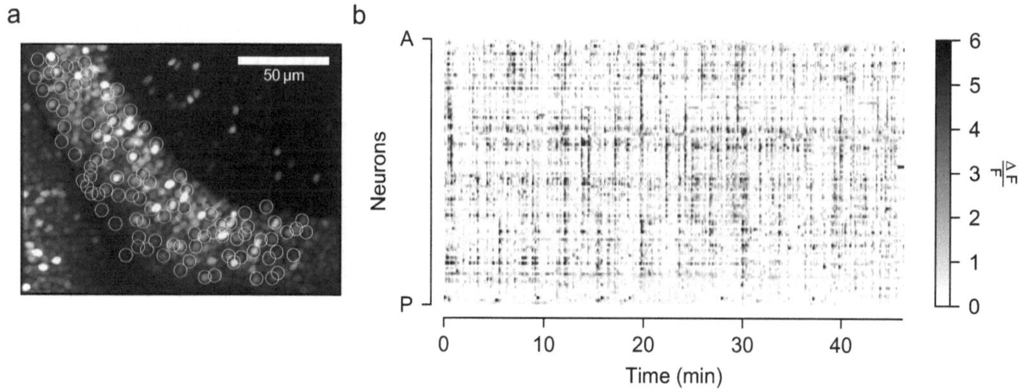

Fig. 1 Calcium fluorescence imaging of a population of neurons. (**a**) Mean two-photon fluorescence image from a single plane through the optic tectum of a larval zebrafish expressing *elavl3:H2B-GCaMP6s*. In the periventricular layer, there is a population of 100 neurons highlighted whose fluorescence activity was considered in this experiment. (**b**) Fluorescence activity of this population ordered by neuron position in the optic tectum from anterior (A) to posterior (P) over 45 min

We here primarily describe the application of a recent Python implementation of the SGC algorithm [12], rather than the original MATLAB implementation [13]. However, both implementations are essentially equivalent and either can be used as desired.

3 Application of the Similarity Graph Clustering Algorithm

Since the original implementation was in MATLAB, the starting point for the algorithm is a MAT-file containing the time series of deflection of calcium fluorescence from baseline level, $\Delta F/F$, for each neuron. Such a file can be created from a NumPy array containing the $\Delta F/F$ time series as shown in the following listing:

```
# import SGC
#
# dF_F = ...
# ↪ T × N array, where T is the length of the time series
and N the size of the neural population

SGC.generate_CALCIUM_FLUORESCENCE_mat('[...]_CALCIUM-
FLUORESCENCE.mat', dF_F)
```

3.1 Data Preprocessing

Starting with the *_CALCIUM-FLUORESCENCE.mat file, the first step is to perform some preprocessing on the calcium fluorescence time series. This preprocessing encompasses the thresholding of each neuron's fluorescence time series to obtain a binary signal for the activity of the neurons, the determination of a threshold for significant coactivity across the population's neurons, and the selection of activity patterns where the coactivity exceeds this threshold (Fig. 2a).

This can be done either interactively within a Python session via

```
# import SGC

SGC.calcium_fluorescence_preprocessing('[...]_CALCIUM-
FLUORESCENCE.mat')
```

or from a shell environment via

```
$ python3 SGC.py preprocessing [...]_CALCIUM-FLUORES-
CENCE.mat
```

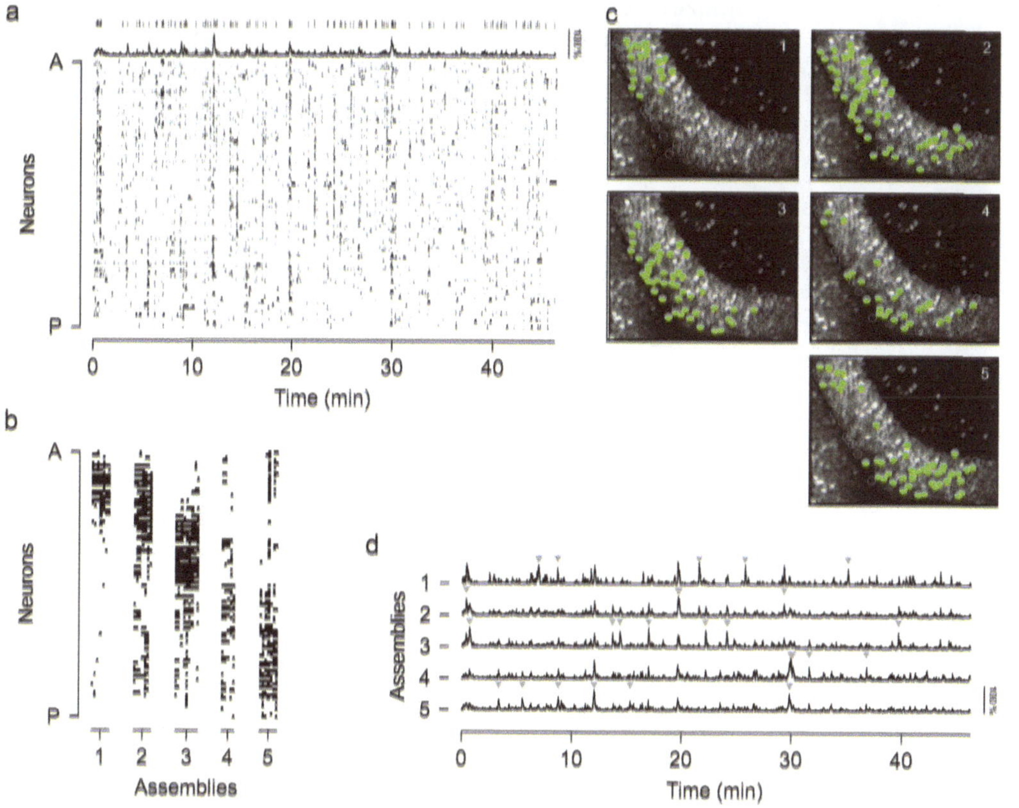

Fig. 2 Detecting assembly activity. (**a**) The raster plot shows the activity indicator signals. Above is the coactivity time series of the population measuring the proportion of neurons in the population which are active, together with the time points of significant coactivity. (**b**) Activity patterns identified by SGC as belonging to five different assemblies. (**c**) Spatial layout of the detected assemblies. It is notable that assemblies are mostly spatially compact. (**d**) Coactivity time series of the different assemblies, with arrows indicating when they are significantly active

This will generate a *_ACTIVITY-RASTER.mat file that contains the results from the preprocessing, i.e., the high-activity patterns with a significant amount of coactivity.

3.2 Assembly Detection

The second step starts with the arrangement of high-activity patterns that have been extracted in the first step, in a graph. The edges in this graph encode the similarity between incident patterns, and it is the key ingredient to this detection algorithm. Following that, the community structure of this graph and, in particular, the most likely number of communities is estimated using a Markov Chain Monte Carlo procedure.

This is the computationally most expensive part of the whole algorithm as it involves fitting a statistical model of a random graph to the "observed" graph of activity patterns in five independent trials of several thousand rounds each. The Python implementation utilizes the original implementation of the inference algorithm [14] as a C extension module, which significantly improves the performance over the MATLAB implementations. It is therefore preferred to use it with this extension module, since otherwise the algorithm falls back to a native Python implementation which is much slower.

Given the estimate for the number of communities, the graph of high-activity patterns is decomposed into groups of similar activity patterns using spectral clustering, from which the assemblies are constructed (Fig. 2b, c).

Again, this can be done either interactively within a Python session via

```
# import SGC

SGC.assembly_detection('[...]_ACTIVITY-RASTER.mat')
```

or from a shell environment via

```
$ python3 SGC.py detection [...]_ACTIVITY-RASTER.mat
```

This will generate a *_SGC-ASSEMBLIES.mat file that contains the results from the assembly detection step.

3.3 Parameters

There are several parameters of the algorithm that have been exposed in the current implementation that can be varied to affect the results. In the following, we will briefly introduce those and explain how changing them from their default values may affect the results.

For the preprocessing step, the activity threshold sets the threshold for the fluorescence time series that determines whether a neuron is considered active. The default value is 2 standard deviations, but the user may want to adjust this depending on the noise levels in the fluorescence signal. The number of rounds of shuffling to generate a null model of coactivity has a default value of 1000. A larger number of shuffling rounds improve the quality of the null model which estimates the level coactivity in the population that can be expected by chance, but at the cost of increased runtime. The significance level α for coactivity determines the threshold for what is considered "high" coactivity and has a default value of 0.05. This threshold is the $(1 - \alpha) \times 100$'th percentile of the null model distribution, so that lower values impose a stricter threshold.

For the assembly detection step, the default for the number of Monte Carlo rounds is 5 and the number steps in the statistical inference procedure is 50,000. Larger values (especially for number of steps) generally improve the estimate of the number of communities in the similarity graph. This might be appropriate when there are particularly many events of high coactivity and the graph is large. The assembly affinity threshold used to construct the assemblies has a default value of 0.2. The higher this value, the stricter the threshold for assembly membership of neurons and, consequently, the sparser the assemblies.

The above parameters can be set either interactively within a Python session via

```
# import SGC

SGC.calcium_fluorescence_preprocessing('[...]_CALCIUM-
FLUORESCENCE.mat', standard_deviations_threshold=[...],
shuffling_rounds=[...], coactivity_significance_level=
[...])
SGC.assembly_detection('[...]_ACTIVITY-RASTER.mat', mon-
tecarlo_rounds=[...], montecarlo_steps=[...], affinity_-
threshold=[...])
```

or from a shell environment via

```
$ python3 SGC.py preprocessing --standard-deviations-
threshold=[...] --shuffling-rounds=[...] --coactivity-
significance-level=[...] [...]_CALCIUM-FLUORESCENCE.mat
$ python3 SGC.py detection --montecarlo-rounds=[...]
--montecarlo-steps=[...] --assembly-affinity-threshold=
[...] [...]_ACTIVITY-RASTER.mat
```

4 Notes

4.1 Algorithm Outputs

The results of assembly detection step are contained in a *_SGC-ASSEMBLIES.mat file, which most importantly includes the detected assemblies. These are saved in two different ways: as lists of indices of the associated neurons and as assembly patterns with components corresponding to the affinity of the respective neuron to the assembly, i.e., the probability the neuron was active in the activity patterns associated with the assembly. The former arises from the latter by selecting only neurons if their affinity exceeded affinity threshold (20% per default), meaning the neuron participated in more than the fraction of the events of this assembly given by the affinity threshold. Thus, if one desires to be stricter in the selection of the assembly neurons, yielding sparser assemblies, one needs to go back to the assembly patterns, as shown in the following listing.

```
# import SGC
# import numpy as np
#
# SGC_ASSEMBLIES = SGC.load_SGC_ASSEMBLIES_mat('[...]
_SGC-ASSEMBLIES.mat')

assemblies = SGC_ASSEMBLIES['assemblies']
assembly_patterns = SGC_ASSEMBLIES['assembly_pattern_de-
tection']['assemblyActivityPatterns']

p_assemblies = lambda p: [np.where(A > p)[0] for A in
assembly_patterns]
# ↪ p_assemblies(p) yields the assemblies when thre-
sholded at an affinity p. Consistently, the following the
expression, comparing the assemblies with assemblies
obtained from p_assemblies(0.2), evaluates to True.

all([np.all(aa[0] == aa[1]) for aa in zip(assemblies,
p_assemblies(0.2))])
```

In addition to the identity of the detected assemblies, another important piece of information is the timing of the assembly activity. In the *_SGC-ASSEMBLIES.mat file, only the indices to activity patterns among the high-activity events are stored for every assembly. Therefore, one needs to combine these with the indices to activity patterns among all the activity patterns that are in turn stored in the *_ACTIVITY-RASTER.mat file. In the following listing, it is demonstrated how this can be done.

```
# import SGC
#
# ACTIVITY_RASTER = SGC.load_ACTIVITY_RASTER_mat('[...]
_ACTIVITY-RASTER.mat')
# SGC_ASSEMBLIES = SGC.load_SGC_ASSEMBLIES_mat('[...]
_SGC-ASSEMBLIES.mat')

activity_raster_peaks = ACTIVITY_RASTER['activity_ras-
ter_peaks']
i_assembly_patterns = SGC_ASSEMBLIES['assembly_pattern_-
detection']['assemblyIActivityPatterns']

[activity_raster_peaks[I] for I in i_assembly_patterns]
# ↪ This yields lists of times of assembly activity.
```

At the end, this leaves one with the identity of the individual assemblies as well as their timing. Those can then be used in further analyses as it is, e.g., done by Avitan et al. [11].

4.2 Data Considerations and Limitations

The notion of assemblies used here is that they consist of neurons that are synchronously active. An alternative concept of assemblies that is also present in the literature is that of neurons that consistently activate in a given sequence [15, 16]. Since temporal relations between activity patterns are not taken into account in the SGC algorithm, it does not address this alternative concept.

An important consideration for applying the algorithm is the scale of the data. On surrogate data, it could, for instance, be shown that recording lengths between 30 and 60 min at 2 Hz are sufficient to obtain reasonably good results, provided that there was sufficient assembly activity. As for the size of the neural population, there was no decline in performance when considering up to almost 1000 neurons [6]. Regardless, there is no a priori limit on the amount of the data that can be handled by the algorithm, given sufficient computing resources and time.

Having applied the algorithm, it is always recommended to compare the results with the original data to confirm that the results are sensible. For instance, an easy sanity check is to compare the fluorescence and activity indicator time series for individual assemblies, where events of coactivity should be obviously visible.

We have found that this algorithm produces good results in many test cases including artificial and real data [6]. However, it is certainly possible that there exists data with characteristics making it ill-suited for the application of SGC, for instance, due to different experimental conditions. In such cases, we recommend investigating other approaches [6].

Acknowledgments

The authors would like to thank Shuyu Zhu and Manning Zhang for their helpful comments on the manuscript. J.M. acknowledges funding from the Deutsche Forschungsgemeinschaft (DFG, German Research Foundation).

References

1. Carrillo-Reid L, Yuste R (2020) What is a neuronal ensemble? In: Oxford research encyclopedia of neuroscience. Oxford University Press. https://doi.org/10.1093/acrefore/9780190264086.013.298
2. Gerstein GL, Bedenbaugh P, Aertsen AMHJ (1989) Neuronal assemblies. IEEE Trans Biomed Eng 36:4–14. https://doi.org/10.1109/10.16444
3. Buzsáki G (2010) Neural syntax: cell assemblies, synapsembles, and readers. Neuron 68:362–385. https://doi.org/10.1016/j.neuron.2010.09.023
4. Yuste R (2015) From the neuron doctrine to neural networks. Nat Rev Neurosci 16:487–497. https://doi.org/10.1038/nrn3962
5. Miehl C, Onasch S, Festa D, Gjorgjieva J (2023) Formation and computational

implications of assemblies in neural circuits. J Physiol 601:3071–3090. https://doi.org/10.1113/JP282750

6. Mölter J, Avitan L, Goodhill GJ (2018) Detecting neural assemblies in calcium imaging data. BMC Biol 16:143. https://doi.org/10.1186/s12915-018-0606-4
7. Scanziani M, Häusser M (2009) Electrophysiology in the age of light. Nature 461:930–939. https://doi.org/10.1038/nature08540
8. Nakai J, Ohkura M, Imoto K (2001) A high signal-to-noise Ca2+ probe composed of a single green fluorescent protein. Nat Biotechnol 19:137–141. https://doi.org/10.1038/84397
9. Badura A, Sun XR, Giovannucci A, Lynch LA, Wang SSH (2014) Fast calcium sensor proteins for monitoring neural activity. Neurophotonics 1:1–12. https://doi.org/10.1117/1.NPh.1.2.025008
10. Lin MZ, Schnitzer MJ (2016) Genetically encoded indicators of neuronal activity. Nat Neurosci 19:1142–1153. https://doi.org/10.1038/nn.4359
11. Avitan L, Pujic Z, Mölter J, Van De Poll M, Sun B, Teng H, Amor R, Scott EK, Goodhill GJ (2017) Spontaneous activity in the zebrafish tectum reorganizes over development and is influenced by visual experience. Curr Biol 27:2407–2419. https://doi.org/10.1016/j.cub.2017.06.056
12. Mölter J (2023) GitHub: janmoelter/sgc-assembly-detection. Commit d9a2d82. https://github.com/janmoelter/sgc-assembly-detection
13. Mölter J (2018) GitHub: GoodhillLab/neural-assembly-detection. Commit f541531. https://github.com/GoodhillLab/neural-assembly-detection
14. Newman MEJ, Reinert G (2016) Estimating the number of communities in a network. Phys Rev Lett 117:078301. https://doi.org/10.1103/PhysRevLett.117.078301
15. Abeles M, Bergman H, Margalit E, Vaadia E (1993) Spatiotemporal firing patterns in the frontal cortex of behaving monkeys. J Neurophysiol 70:1629–1638. https://doi.org/10.1152/jn.1993.70.4.1629
16. Seidemann E, Meilijson I, Abeles M, Bergman H, Vaadia E (1996) Simultaneously recorded single units in the frontal cortex go through sequences of discrete and stable states in monkeys performing a delayed localization task. J Neurosci 16:752–768. https://doi.org/10.1523/JNEUROSCI.16-02-00752.1996

Chapter 8

Identifying Neuronal Ensembles: A Graph Theoretical Approach

Miguel Serrano-Reyes and Jose Bargas

Abstract

This chapter presents a comprehensive method for analyzing complex neuronal data using graph theory and UMAP (Uniform Manifold Approximation and Projection), emphasizing the identification of neuronal ensembles: groups of neurons with correlated activity seen as the key elements of neuronal circuits correlated with sensorimotor, working memory, cognitive, and behavioral functions. The process is detailed from the initial data acquisition in calcium imaging experiments through the stages of data preparation, including signal binarization and the construction of a graph-based representation of the data. The core of our analysis lies in the application of community detection algorithms, such as the Louvain algorithm, to uncover the modular structure within neuronal networks, revealing how neurons group into functionally significant ensembles.

Our approach aims to improve our understanding of the structural and functional connectivity of neuronal circuits that process cognitive functions like memory and decision-making. It also allows comparing control and pathological circuits, as well as pharmacological bioassays to evaluate drugs' effects. The methodologies employed are designed to be adaptable across various neuronal datasets, providing researchers with the tools needed to explore these techniques further. This work contributes to the broader neuroscience community by offering new insights into the organizational principles of the brain and potential strategies for addressing neurological disorders. Accompanying this chapter, a GitHub repository includes all relevant codes and examples to support reproducibility and further exploration.

Key words Neuronal ensembles, Graph theory, UMAP (Uniform Manifold Approximation and Projection), Community detection, Calcium imaging, Network analysis, Modularity, Data visualization, Louvain algorithm

1 Introduction

Understanding the dynamics of neuronal activity in brain networks is crucial for advancing our knowledge of brain function and treating neurological disorders [1]. This chapter explores the use of graph theory to analyze complex neuronal data, providing insights into how neurons interact within networks and contribute to cognitive processes.

Luis Carrillo-Reid (ed.), *Identification, Characterization, and Manipulation of Neuronal Ensembles*, Neuromethods, vol. 215,
https://doi.org/10.1007/978-1-0716-4208-5_8,

Advancements in neuroscience underscore the importance of studying neuronal populations. Neurons function collaboratively in groups, forming intricate networks that underpin sensory information processing, memory storage, and decision-making [2, 3]. Graph theory offers a strategic framework for the representation and analysis of these networks, modeling neurons as nodes and their interactions as edges to uncover architectural complexities and identify functional groups that may play key roles in cognitive processes [4, 5].

This chapter outlines the workflow involved in analyzing neuronal activity from calcium imaging experiments. It covers the initial data capture, the binarization of signals to simplify neuronal activity into a manageable format, constructing a graph-based representation of the experimental data, detecting communities within the graph, and visualizing these interactions to illustrate the functional architecture of a brain circuit.

Integrating calcium imaging data with graph-based modeling provides a comprehensive overview of how neurons form functional ensembles and interact within larger networks [1, 6]. This integration is crucial for deciphering the processing of information by neuronal circuits and for understanding their responses to various stimuli [7]. Furthermore, exploring neuronal ensembles using these techniques offers insights into the organizational principles of the brain, which are essential for understanding how the brain functions during health and disease [4, 5, 8]. These insights could lead to new therapeutic approaches and enhance our understanding of the brain's complex functions. The methodologies described are adaptable to various neuronal datasets, facilitating the exploration of new research avenues in neural network analysis.

2 Materials

To complement the theoretical and methodological discussion presented in this chapter, a set of practical materials is available through a dedicated GitHub repository [9]. This repository contains example codes, custom functions, and sample datasets to enable readers to perform their own simulations and analyses. Detailed instructions are also provided for setting up the work environment, installing necessary dependencies, and step-by-step guides for running simulations and analyzing the results.

These materials are designed with accessibility and reproducibility in mind, ensuring that researchers, students, and professionals can practically explore the concepts and techniques discussed in this work. Readers are encouraged to modify and adapt these examples to their own research projects, exploring new questions and expanding the frontiers of knowledge in neuroscience. All interested parties are invited to clone the repository, engage in active experimentation, and contribute their ideas and improvements via pull requests. For any questions or suggestions, opening an issue in the repository or contacting the team directly is encouraged.

3 Methods

3.1 General Considerations

Neuronal activity, complex and multifaceted, can be represented and stored in multiple ways, depending on the temporal resolution and the recording period of the experimental data. In the field of calcium imaging, where experiments may extend over several minutes, it becomes crucial to develop an efficient data structure that not only captures the essence of this activity but also facilitates subsequent analysis. In response to this need, the concept of the "neuronal activity raster matrix" emerges, known technically and commonly in the scientific literature as a raster matrix [10, 11].

It is essential to emphasize that it is assumed the reader already has a solid foundation regarding how data is obtained and prepared from calcium imaging techniques, including signal capture, selection of regions of interest (ROIs), and extraction and calculation of calcium signals from these ROIs (Fig. 1).

3.1.1 Binarization Process and Raster Plot Formation

Once calcium signals are acquired and prepared, the next step is their binarization. This process converts the continuous calcium signal data into a binary format, where a '1' represents the detection of neuronal activity (such as action potentials that generate calcium entrance, and thus fluorescence), and a '0' indicates no activity (Fig. 2). This binary transformation is crucial as it simplifies the complex data into a format that can be more easily analyzed

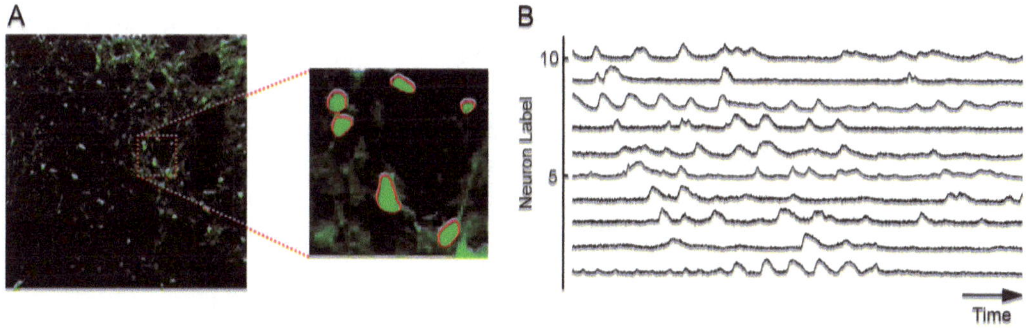

Fig. 1 Illustrative image of data preparation techniques in calcium imaging experiments on mouse brain. (**a**) Shows a slice of mouse brain treated with GCaMP6f, highlighting the neuron somas in green. A close-up emphasizes the regions of interest (ROIs) selected for analysis, which can be identified using either manual or automated methods. (**b**) Displays an example of calcium signals extracted from the ROIs. The fluorescence signal is calculated by comparing the estimated activity with the baseline fluorescence, using various techniques currently available in research. (Images used and modified under permission from [8, 12])

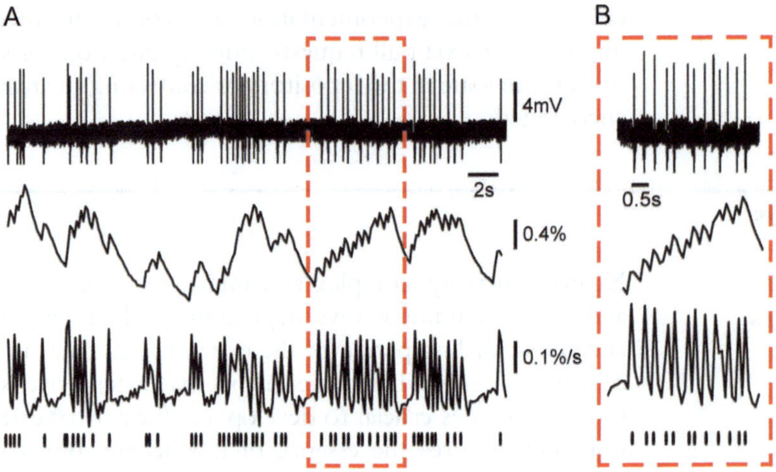

Fig. 2 Detailed visualization of the binarization process in neuronal activity analysis. (**a**) The display shows four stages of data processing. The top row is an extracellular recording of neuronal activity, that is, action potentials. The second row shows the calcium signal (ΔF/F0) associated with such activity, the third row shows the time derivative of this signal (d(ΔF/F0)/dt), highlighting that rapid upward changes indicate neuronal firing while downward descents do not. The fourth row shows the binary representation of activity: black bars represent detected neuronal activity ('1') and white spaces indicate no activity ('0'). (**b**) This panel provides a zoomed view of the red dashed region from Panel A, illustrating the transition from raw data to binary format. The example here is a guideline for binarization processes. Multiple models and methods can be used; however, the approach shown offers a clear method of translating complex neuronal signals into a binary format that supports statistical analysis and replication in neuroscience research. (Images are used and modified under permission from [12])

[13]. Furthermore, calcium signals descents are not associated with action potentials and thus yield spurious information if the main interest is neuronal communication. To ensure precise implementation, the process involves a series of methodical steps:

- *Data Preprocessing*: The first stage involves preparing the raw data. This may include correcting for artifacts, adjusting the baseline, and smoothing the signal to ensure that the calcium signal truly reflects neuronal activity [4].
- *Threshold Selection*: Next, determining the appropriate threshold for binarization is essential. This involves analyzing the calcium signal's characteristics, such as the intensity and duration of activity peaks. The threshold is typically set using statistical methods or dynamically adjusted automated techniques, depending on the signal's distribution [14].
- *Threshold Application*: After establishing the threshold, it is applied to the calcium signals of each neuron. This converts the signal values into a binary sequence, with '1' indicating activity above threshold (firing of action potentials) and '0' indicating no activity.

- *Verification and Correction*: The last step involves reviewing the binarized data. This is crucial to ensure accuracy, particularly to identify and correct any errors, such as the misclassification of artifacts as activity or missing real events of low intensity [13].

The raster matrix built as shown in Fig. 2 standardizes the experimental setup. Once it is confirmed that the fluorescence intensity of the fluorophore corresponds to neuronal activity, this threshold is used to automate the extraction of binary activity data from an entire calcium imaging video. This involves selecting all regions of interest (ROIs), calculating their calcium signals, and binarizing these signals. The result is a raster matrix, a binary (N × F) matrix where 'N' represents the total number of neurons monitored and 'F' represents the total number of frames captured (Fig. 3). This matrix provides a clear temporal representation of each neuron's activity, facilitating the identification of patterns and dynamics in neuronal coactivity over time, essential for converting complex, high-dimensional data into a manageable and analyzable format [15, 16].

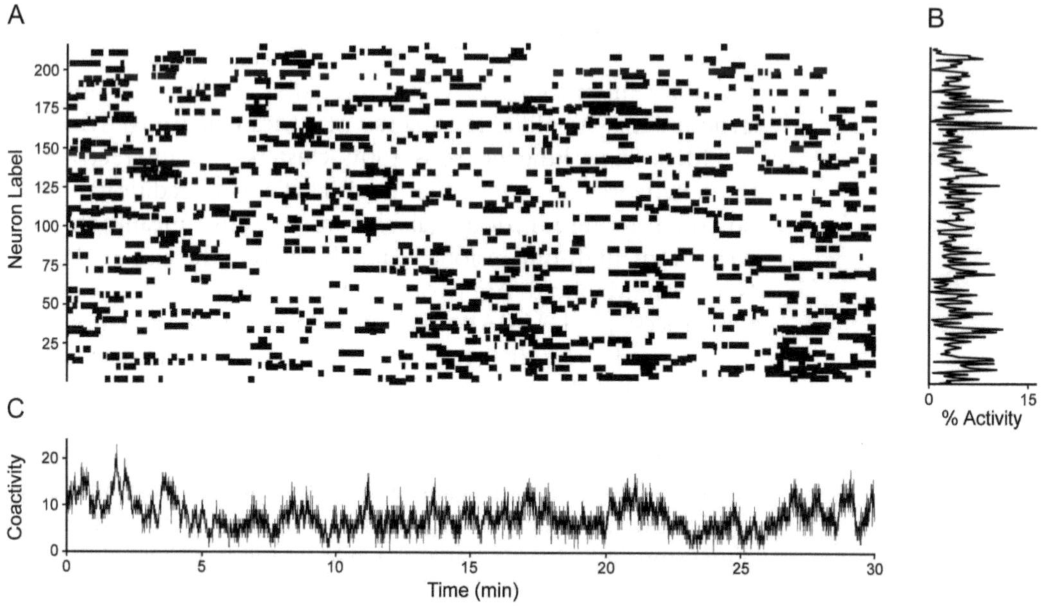

Fig. 3 Raster plot representation of neuronal activity from a calcium imaging experiment. (**a**) Displays the binary activity of over 200 neurons recorded during a 30-min calcium imaging session in the striatum of a mouse under control conditions. Each row represents the activity of a neuron, marked with a neuron label, where black bars indicate activity ('1') and white spaces represent inactivity ('0'). (**b**) Shows the percentage of activity for each neuron throughout the experiment. A line graph illustrates the number of frames each neuron was active, providing a quantitative overview of individual neuronal engagement. (**c**) Depicts the coactivity among neurons, with a line trace indicating the number of coactive neurons in each movie frame throughout the experiment. This panel highlights the temporal dynamics of neuronal coactivity, reflecting concurrent neuronal interactions. This figure provides a comprehensive view of neuronal activity, essential for detailed examination and computational analysis of neural dynamics. To enhance basal striatal activity, 3 μM NMDA were added to the bath saline

3.2 Modeling Neuronal Activity

To understand the complexity captured in raster plots, it is necessary to examine how neurons interact and organize themselves. This interest is deeply rooted in the history of neuroscience. Since the early days of brain research, scientists have endeavored to understand how neurons in the brain work together to produce behaviors, perceptions, and emotions [17]. A significant insight into the collective activity of neurons was provided by Donald Hebb, who introduced the concept that neurons firing together tend to strengthen their mutual connections [18]. This idea evolved into what is now recognized as the theory of neuronal ensembles, suggesting that networks of neurons with similar activity patterns constitute the basic functional units of brain circuits [19, 20].

The exploration of neuronal networks has advanced significantly since Hebb's era, driven by innovations in imaging technology and data analysis [20, 21]. These advances have opened a vast array of techniques ranging from statistical analysis to machine learning applications, revolutionizing the tools available and enabling the exploration of neuronal activity with unprecedented resolution and scale. Among these techniques, graph-based modeling has been adopted due to its effectiveness in representing complex networks [22, 23]. This method visualizes neurons as nodes and their interactions as edges, providing a clear and intuitive structure that simplifies the analysis of intricate neuronal dynamics.

Identifying how neurons form clusters and interact within these networks presents a significant challenge. Traditional methods often require intricate calculations to derive meaningful patterns from high-dimensional data. However, the use of UMAP (Uniform Manifold Approximation and Projection) streamlines this process [24]. UMAP excels in reducing dimensionality while preserving the essential structures within complex datasets, effortlessly revealing how neuronal groups form and function together [25]. This capability makes UMAP an invaluable tool for constructing graphical representations of neuronal activity, transforming complex data into manageable and analyzable visualizations [26].

UMAP is widely recognized for its ability to reduce the dimensionality of large datasets while preserving their global and local structures, making it ideal for simplifying the complexity of raster plots [8, 27]. However, the feature of UMAP that is particularly relevant here is its ability to create a graphical representation of the data. UMAP builds a weighted graph that effectively models the relationships between data points—in this case, neurons and their activity. This graph-based approach offers a structured way to analyze complex interactions within the data, facilitating a deeper understanding of neuronal connectivity [25].

The decision to employ graph modeling for analyzing neuronal activity stems from the comprehensive analytical strategies provided by graph theory. By adopting this model, researchers gain access to advanced techniques for examining the network structure, assessing the relative importance of specific neurons (centrality), identifying highly interconnected subnetworks (communities or modules), and tracing communication pathways within the neuronal network. These techniques allow for a detailed exploration of how neurons are interconnected and how these connections influence overall brain function [28, 29].

3.2.1 Using UMAP to Obtain a Representative Graph of Neuronal Activity

After establishing the need to model neuronal activity and highlighting UMAP as an innovative tool for this purpose, it is essential to delve into how exactly UMAP facilitates the acquisition of a representative graph depicting the complex relationships between neurons.

UMAP begins its process by assuming that the data, such as the neuronal activity patterns captured in our raster plots, are uniformly distributed across a high-dimensional topological space. This space is merely a mathematical representation where each point (or neuron, in our case) has its place, defined by its activity over time. The initial and crucial step in the UMAP process is the construction of a high-dimensional graph. Simply put, this involves creating a map of connections between neurons based on their similarity or proximity in this complex space. To achieve this, UMAP employs what are known as "fuzzy simplicial complexes," which practically means connecting each data point to its nearest neighbors (not in terms of anatomical vicinity but in terms of correlated activity implicit in neuronal ensembles definition). The strength or weight of each connection reflects how similar two neurons are in terms of their activation patterns [25].

This process starts by extending "radii" from each data point, connecting them with other points where these radii intersect. The key feature of UMAP is its adjustment of the size of these radii individually for each point, ensuring that each neuron is connected at least to its nearest neighbor. This guarantees the preservation of local structure—how each neuron relates to its immediate surroundings—in the final graph [25].

In practice, common implementations of the algorithm utilize a range of hyperparameters that can significantly impact the overall process [30]. There are two particularly critical for extracting the weighted graph:

- *n_neighbors (Number of Nearest Neighbors)*: This parameter defines the approximate number of nearest neighbors that will be considered for each data point when constructing the initial graph. A smaller number focuses the analysis on local structure, capturing fine details of neuronal activity, whereas a higher value tends to highlight the global structure, connecting data points that belong to broader groupings [30].

- *metric (Metric)*: This parameter determines the distance calculation method in the multidimensional space of the data. UMAP supports a variety of metrics such as Euclidean, Jaccard, cosine, and correlation, each suitable for diverse types of data and analysis needs. This flexibility allows UMAP to adapt to various data characteristics, enhancing its ability to capture meaningful patterns in the data [30].

By focusing on connecting each neuron with its nearest neighbors, UMAP not only highlights local clusters of similar activity but also preserves the global structure of the dataset. This means that the resulting graph not only displays groups of neurons that activate together (local structure) but also illustrates how these groups interact within the broader context of circuit activity (global structure) [8, 26].

Figure 4 presents a visual analysis demonstrating the influence of the 'n_neighbors' parameter across various metrics including the cosine of the angle between the row vectors that display the activity of each neuron, correlation of neuronal activity, and Jaccard similarity. Each graphical representation illustrates the distribution of connection strengths, where 'connection strength' in a weighted graph analogously corresponds to the degree in unweighted graphs, quantifying the intensity and significance of connections between nodes. These distributions clearly show a transition from a power-law-like distribution to a more normal distribution as the 'n_neighbors' parameter increases. Notably, structural patterns in the graphs remain consistent across different metrics for the same settings, indicating that while the 'metric' choice can be aligned to data-specific traits, it does not fundamentally alter the overarching interpretations of network structure. These findings underscore the importance of parameter selection in UMAP and suggest that adjusting 'n_neighbors' can provide valuable insights into the underlying neuronal activity, encouraging a nuanced approach to analyzing neuronal network dynamics.

With the establishment of a framework using UMAP to analyze neuronal networks, and the detailed exploration of the 'number of nearest neighbors' settings across various metrics, a foundation has been laid for constructing the weighted graph, G. This graph encapsulates the complex interactions and connectivity patterns among neurons as revealed by UMAP [8]. Researchers are encouraged to utilize these parameters to experiment further and refine their understanding of the underlying dynamics in neuronal activity.

3.2.2 Identification of Neuronal Ensembles Using Graph Theory

With the weighted graph G constructed through UMAP, the focus shifts to identifying neuronal ensembles—groups of neurons that exhibit similar activation patterns and are more functionally connected to each other than to the rest of the network. Graph

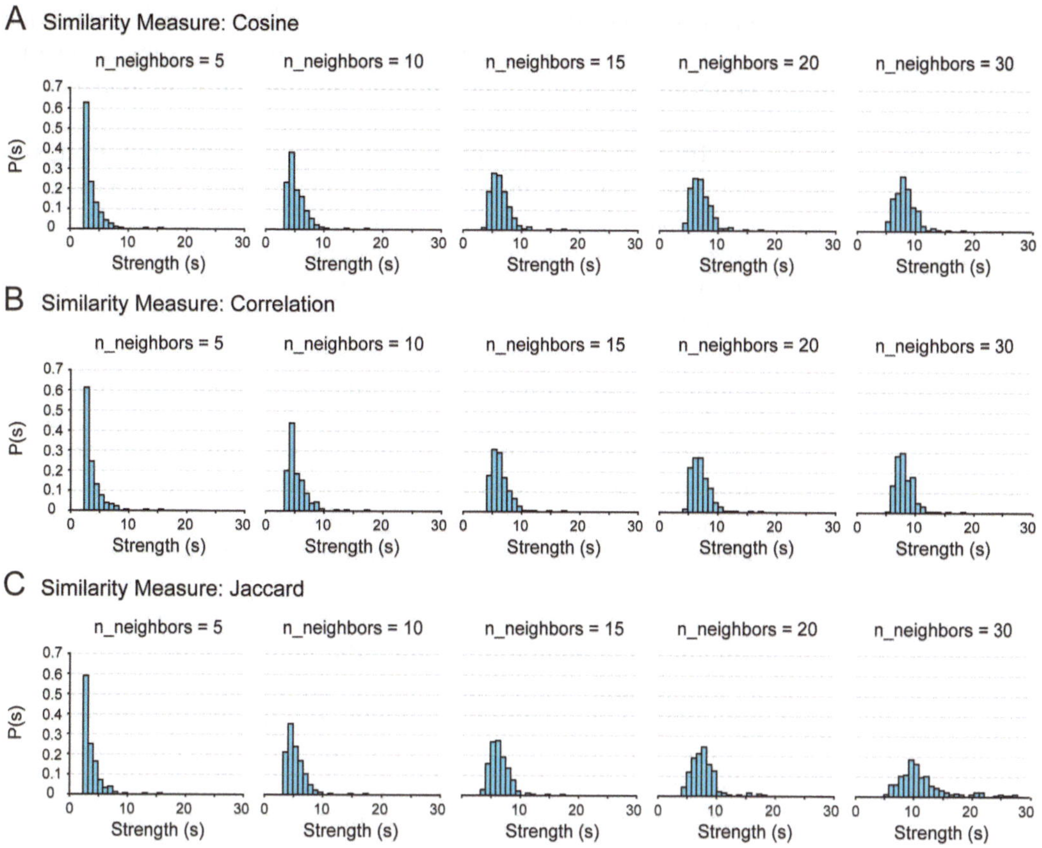

Fig. 4 Comparative analysis of neuronal connection strengths across different metrics and 'number of nearest neighbors' settings in graphs generated by UMAP. (**a**) Utilizing the cosine similarity measure between the row vectors that display the activity of each neuron, this panel presents histograms that detail the distribution of connection strengths 's' as the 'number of nearest neighbors' (n_neighbors) varies from 5 to 30. P (s) represents the probability density of observing a given connection strength, highlighting a shift from a power-law-like (skewed) to a more normal-like distribution. This shift indicates significant changes in local connectivity patterns as neighborhood size increases. (**b**) The correlation metric (e.g., Pearson coefficient) shows how connection strengths evolve with increasing n neighbors, aligning with the transitions observed with the cosine metric. (**c**) Using the Jaccard metric, appropriate for binary data, the histograms demonstrate how the distribution of connection strengths adapts under different 'n_neighbors' settings. Across all metrics, the consistency in the patterns of connection strength distributions suggests a robust methodological framework provided by UMAP, allowing researchers to select a 'metric' that best suits their data characteristics without losing analytical value. This visual comparison encourages researchers to experiment with various 'number of nearest neighbors' to explore distinct aspects of neuronal activity and network dynamics

theory provides a robust methodological arsenal for this task, utilizing concepts such as modularity and various community detection algorithms [29, 31].

Modularity and Community Detection

Modularity is a key indicator used to evaluate the strength of divisions within a network into distinct modules or communities. It measures the density of links inside communities compared to links between communities. A high modularity score indicates that the network has a strong division into modules, suggesting that connections within each module are denser than connections between different modules [31].

In practical terms, modularity helps in identifying groups of neurons that may function as coherent ensembles. These ensembles are characterized by dense intra-group connections which are indicative of strong functional relationships. The identification of such ensembles is crucial for understanding the organizational principles underlying neuronal networks [5, 32].

The computation of modularity involves comparing the actual density of edges within a community to a model that represents randomness. If the density within a group exceeds what would be expected by chance (given the network's overall connectivity), the arrangement is considered to have good modularity [31]. This parameter allows algorithms to assess the quality of different community divisions without a predefined number of clusters. By iteratively optimizing modularity, algorithms like Louvain and Leiden guide the clustering process, effectively determining the most natural division of the network into communities. These algorithms continually adjust clusters to maximize modularity, refining the network's division until no significant improvements can be achieved. This iterative process allows the algorithms to evaluate various potential divisions and select the one that best captures the dense interconnections of real communities. This ensures that the detected communities reflect genuine underlying patterns of connectivity and are not merely artifacts of the algorithm's initial conditions [8, 33].

Louvain Algorithm for Community Detection

The Louvain algorithm is recognized for its efficiency in handling large networks and its capability to deliver accurate community detection results. Known for its simplicity and widespread use, the Louvain method is selected for in-depth discussion due to its effective balance of computational simplicity and detailed community analysis. This algorithm is ideal for examining neuronal networks, where it can reveal complex functional structures across varied organizational levels [34, 35].

The Louvain algorithm uses an iterative process of optimization and aggregation, which is essential for both delineating clear community structures and adapting these structures across different scales of analysis. By optimizing community divisions based on modularity, the algorithm significantly aids in the practical detection and interpretation of neuronal ensembles [36].

Key Processes of the Louvain Algorithm

- *Initialization*: Each neuron starts as its own community, establishing a baseline for modularity optimization.

- *Local Community Assignment*: Neurons are evaluated and reassigned to optimize modularity, with each neuron moving to the community that enhances overall modularity. This process repeats until no further improvements can be made.
- *Community Graph Construction*: A new graph is created using the identified communities as nodes, with edges representing the connections between original community members.
- *Iteration*: The community assignment is reapplied to the new graph, iterating until there is no significant increase in modularity.
- *Finalization*: The process concludes when further iterations no longer enhance modularity, resulting in a network division that accurately reflects the coherent activation patterns of neuronal ensembles.

This streamlined explanation of the Louvain algorithm highlights its utility in identifying and analyzing community structures within complex networks, especially in neuronal studies. The method's ability to iteratively refine community boundaries based on modularity makes it particularly effective for revealing detailed and functionally significant patterns in large datasets [36].

Exploring Community Detection Dynamics with the Louvain Algorithm

Exploring how parameters like 'n_neighbors' and the choice of similarity metric—cosine, correlation, and Jaccard—affect community detection is a key part of using the Louvain algorithm. This analysis not only demonstrates the algorithm's flexibility and robustness but also its adaptability to different network structures and settings. By adjusting these parameters, researchers can understand their impact on modularity and the formation of clusters. This is crucial for fine-tuning the algorithm to meet specific research needs, allowing for an optimal balance between detailed community structures and high modularity.

Figure 5 presents the results of 1000 simulations that analyze the impact of these parameter variations on modularity and the distribution of cluster sizes across different metrics. Each panel in the figure offers a quantitative and visual representation of community structure variations, providing a clear basis for comparing the effectiveness of each setting in revealing meaningful patterns in neuronal data.

3.3 Visual Synthesis of Neuronal Ensembles Analysis

Once optimal parameters are selected using the Louvain algorithm and the graph of the experimental data is obtained via UMAP, this section presents a visual synthesis where the methodologies discussed are applied to neuronal data, illustrating the classification of neurons into distinct ensembles. These visualizations underscore the effectiveness of graph-based modeling techniques in

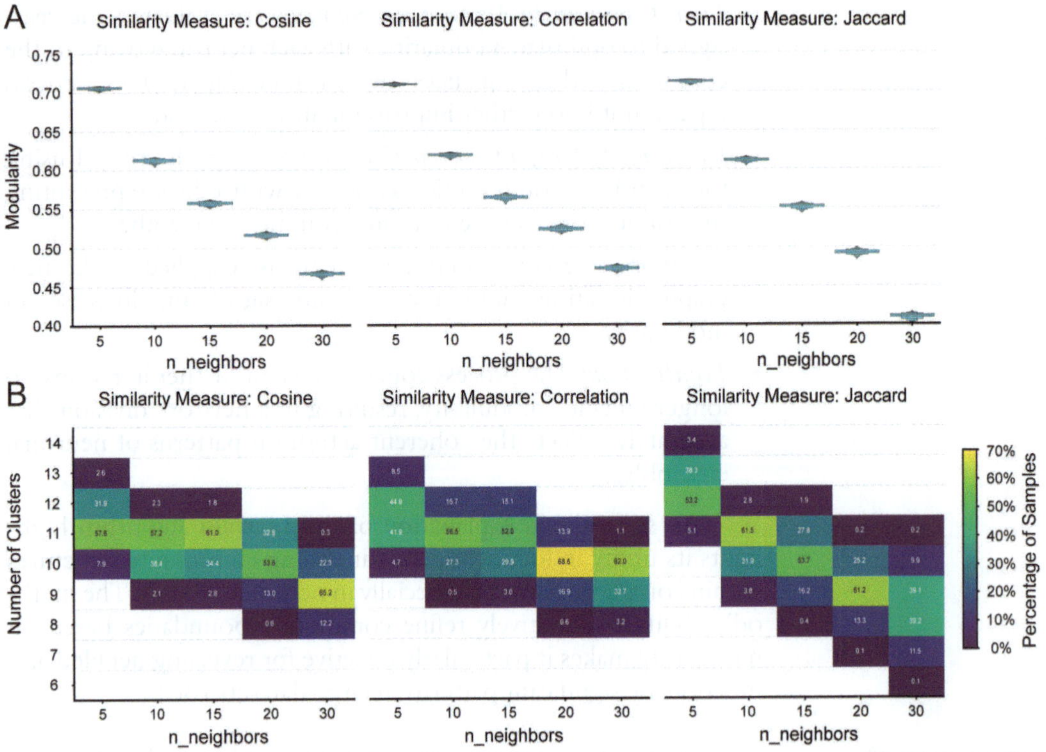

Fig. 5 Analysis of modularity and cluster distribution in graphs generated by UMAP across different nearest neighbors settings. (**a**) Displays modularity results from 1000 simulations at various 'n_neighbors' settings, evaluated using cosine, correlation, and Jaccard similarity measures (see above). The enhanced boxplots, chosen for their robust depiction of distribution traits, consistently indicate the highest modularity at 'n_neighbors' = 5 across all similarity measurements, suggesting this as one optimal setting for clustering in this specific dataset. (**b**) The heatmaps in this panel display the frequency distribution of the number of clusters identified in the simulations for each setting of 'n_neighbors' and similarity measure. A gradient from deep blue to vivid yellow illustrates the increasing percentage of simulations that resulted in each cluster count, offering insights into how changes in 'n_neighbors' and the choice of similarity measure impact the structural composition of the clusters. This color-coded approach highlights the variance in cluster frequencies, providing essential data for fine-tuning graph-based clustering parameters, facilitating the selection of the most effective settings for comprehensive network analysis

deciphering complex patterns of neuronal interactions, effectively demonstrating how neurons are grouped into functional ensembles and interact within larger networks.

Integration of Neuronal Data into a Raster Plot

The original raster plot of neuronal activity has been sorted to better illustrate the identified neuronal ensembles. Neurons belonging to the same cluster are positioned adjacently and color-coded, which not only clarifies the grouping by visualizing together the row vectors of neurons belonging to the same ensemble but also provides insights into the biological functions of these ensembles, that is, their sequence of activation that can in principle be associated with a behavioral, memory, or cognitive task. By organizing the data in this way, the plot offers a vivid depiction of how certain groups of neurons may coordinate their activity, potentially playing crucial roles in specific cognitive or behavioral processes.

Graphical Representation in Circular and Force-Directed Layouts

Two distinct graphical layouts provide additional perspectives on the neuronal network:

- Circular Layout*: Neurons are arranged in a circle and color-coded according to their identified neuronal ensemble affiliations in the raster plot. This visual consistency enhances the understanding of their interconnectivity and the distinct boundaries between groups, highlighting how different neuronal ensembles are linked.*
- Force-Directed Layout*: Implemented using a force-directed algorithm, this layout arranges neurons in a way that allows the strength and distribution of connections to dictate their placement. It effectively illustrates the density and centrality of connections within each cluster, offering insights into the structural integrity and functional significance of neuronal ensembles [37].*

Figure 6 encapsulates these concepts by showing the modified raster plot alongside the network visualizations in both circular and force-directed formats. The figure acts as a comprehensive visual summary, demonstrating the connectivity and functional dynamics of neuronal clusters derived from the analysis.

4 Conclusions

This chapter described a detailed framework using graph theory and UMAP to analyze complex neuronal data, with a specific focus on identifying neuronal ensembles. These ensembles, which represent groups of neurons with similar activation patterns, are crucial for understanding the brain's circuits functional architecture and how neurons connect and interact within larger networks.

The application of community detection algorithms, such as the Louvain algorithm, combined with advanced dimensionality reduction techniques like UMAP, has enabled the revelation of complex patterns of neuronal activity that are vital for advancing our understanding of important cognitive functions such as memory and decision-making. This approach not only enhances our ability to analyze complex neural networks but also opens new avenues for exploring therapeutic strategies targeted at brain disorders.

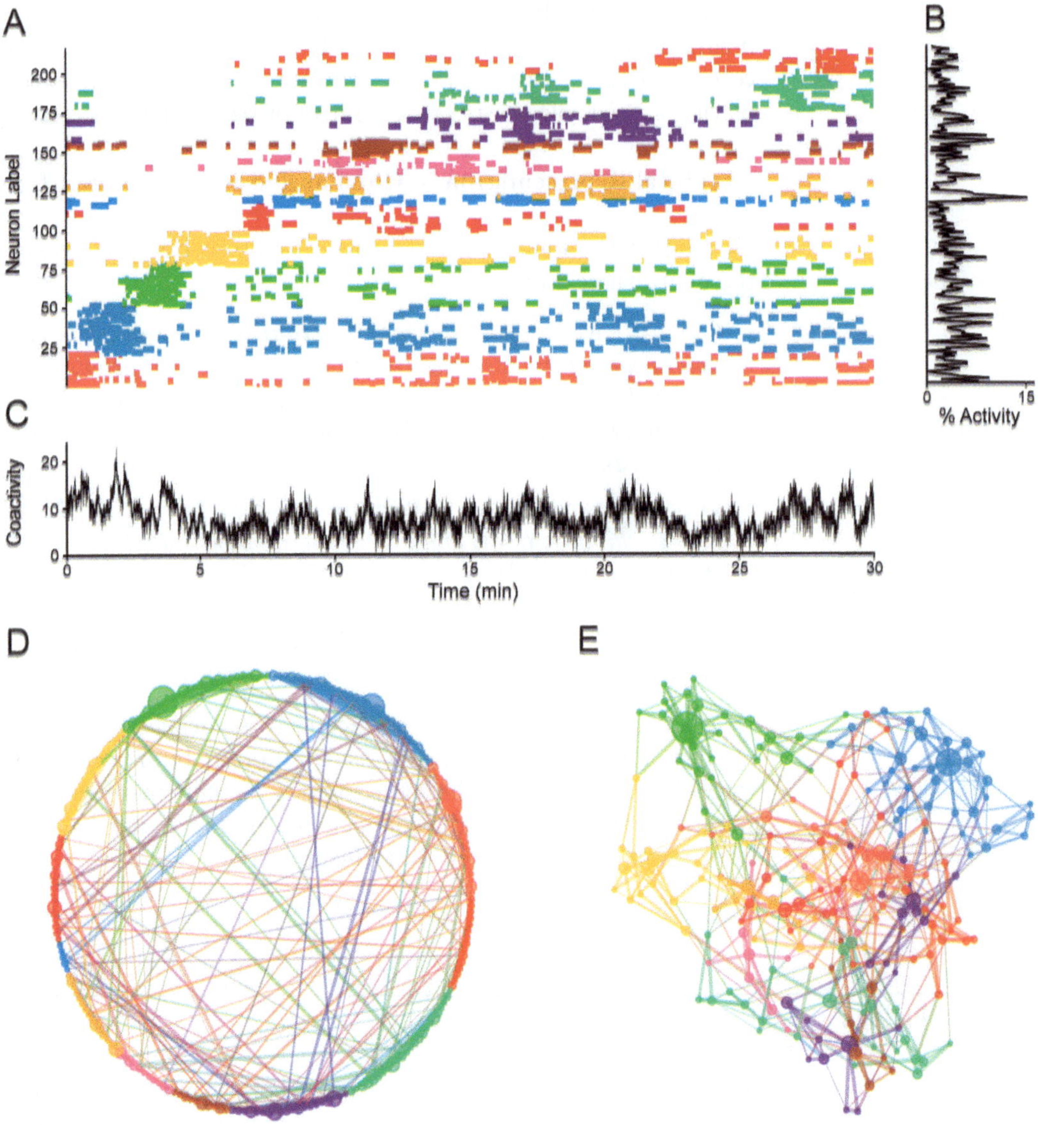

Fig. 6 Visualization of neuronal ensembles and interaction dynamics. (**a**) Raster plot displays neurons rearranged by clustering (color-coded), enhancing visual clarity of groupings and their functional implications, e.g., sequence of activity along time. (**b**) Vertical bar graph shows the percentage of activity of each neuron over time, highlighting neuronal engagement. (**c**) Line graph of coactivity over time illustrates the temporal dynamics of neuronal interactions. (**d**) Circular layout depicts neurons arranged by clusters with colored interconnections, emphasizing ensemble boundaries and functional connectivity. (**e**) Force-directed layout shows structural and connection density within and among neuronal clusters, visualizing how interaction strength influences neuron positioning. This figure encapsulates the connectivity and dynamics of neuronal clusters, highlighting graph-based modeling techniques

As these methodologies continue to evolve, they promise to deepen our understanding of the brain's network interactions further. This progress is essential for improving our ability to unravel complex cognitive processes and effectively address challenges in neuroscience. With this framework, researchers have robust tools to explore new research questions and expand the frontiers of knowledge in neural sciences.

5 Notes

All figures presented in this chapter and the details of their generation can be found in the accompanying GitHub repository [9]. This repository includes example codes, scripts, and data used in creating these visualizations, allowing readers to explore the methodologies in more detail and apply them to their own datasets. This access ensures that all materials are transparent and reproducible, supporting further exploration and understanding of the concepts discussed.

References

1. Bullmore E, Sporns O (2009) Complex brain networks: graph theoretical analysis of structural and functional systems. Nat Rev Neurosci 10(3):186–198. https://doi.org/10.1038/nrn2575
2. Jazayeri M, Movshon AJ (2006) Optimal representation of sensory information by neural populations. Nat Neurosci 9:690–696. https://doi.org/10.1038/nn1691
3. Hölscher C, Munk M (2008) Inf Process Neuronal Popul. https://doi.org/10.1017/CBO9780511541650
4. Pérez-Ortega J, Duhne M, Lara-González E, Plata V, Gasca D, Galarraga E, Hernández-Cruz A, Bargas J (2016) Pathophysiological signatures of functional connectomics in parkinsonian and dyskinetic striatal microcircuits. Neurobiol Dis 91:347–361. https://doi.org/10.1016/j.nbd.2016.02.023
5. Serrano-Reyes M, García-Vilchis B, Reyes-Chapero R, Cáceres-Chávez VA, Tapia D, Galarraga E, Bargas J (2020) Spontaneous activity of neuronal ensembles in mouse motor cortex: changes after GABAergic blockade. Neuroscience 446:304–322. https://doi.org/10.1016/j.neuroscience.2020.08.025
6. Fries P (2005) A mechanism for cognitive dynamics: neuronal communication through neuronal coherence. Trends Cogn Sci 9(10):474–480. https://doi.org/10.1016/j.tics.2005.08.011
7. Dechery JB, MacLean JN (2017) Functional triplet motifs underlie accurate predictions of single-trial responses in populations of tuned and untuned V1 neurons. PLoS Comput Biol 14. https://doi.org/10.1371/journal.pcbi.1006153
8. Serrano-Reyes M, Pérez-Ortega JE, García-Vilchis B, Laville A, Ortega A, Galarraga E, Bargas J (2022) Dimensionality reduction and recurrence analysis reveal hidden structures of striatal pathological states. Front Syst Neurosci 16. https://doi.org/10.3389/fnsys.2022.975989
9. Serrano-Reyes, M. (2024). Explore neuronal data with 'NeuralEnsembleGraphTools'. https://github.com/MiguelSerranoReyes/NeuralEnsembleGraphTools. Accessed 10 May 2024
10. Carrillo-Reid L, Tecuapetla F, Tapia D, Hernández-Cruz A, Galarraga E, Drucker-Colin R, Bargas J (2008) Encoding network states by striatal cell assemblies. J Neurophysiol 99:1435–1450. https://doi.org/10.1152/jn.01131.2007
11. Grienberger C, Konnerth A (2012) Imaging calcium in neurons. Neuron 73(4):862–885. https://doi.org/10.1016/j.neuron.2012.02.011
12. Padilla-Orozco M, Duhne M, Fuentes-Serrano A, Ortega A, Galarraga E, Bargas J, Lara-González E (2022) Synaptic

determinants of cholinergic interneurons hyperactivity during parkinsonism. Front Synaptic Neurosci 14. https://doi.org/10.3389/fnsyn.2022.945816

13. Ranganathan G, Koester H (2010) Optical recording of neuronal spiking activity from unbiased populations of neurons with high spike detection efficiency and high temporal precision. J Neurophysiol 104(3):1812–1824. https://doi.org/10.1152/jn.00197.2010
14. Theis L, Berens P, Froudarakis E, Euler T, Tolias AS, Bethge M (2016) Benchmarking spike rate inference in population calcium imaging. Neuron 90:471–482. https://doi.org/10.1016/j.neuron.2016.04.014
15. Cossart R, Aronov D, Yuste R (2003) Attractor dynamics of network UP states in the neocortex. Nature 423:283–288. https://doi.org/10.1038/nature01614
16. Ikegaya Y, Aaron G, Cossart R, Aronov D, Lampl I, Ferster D, Yuste R (2004) Synfire chains and cortical songs: temporal modules of cortical activity. Science 304(5670):559–564. https://doi.org/10.1126/science.1093173
17. Fentress JC (1999) The organization of behaviour revisited. Can J Exp Psychol 53:8–19. https://doi.org/10.1037/H0087296
18. Hebb DO (1949) The organization of behavior: A neuropsychological theory. New York, NY: Wiley, pp. 62, 229.
19. Lopes-dos-Santos V, Conde-Ocazionez S, Nicolelis M, Ribeiro S, Tort A (2011) Neuronal assembly detection and cell membership specification by principal component analysis. PLoS One 6. https://doi.org/10.1371/journal.pone.0020996
20. Carrillo-Reid L, Yang W, Bando Y, Peterka D, Yuste R (2016) Imprinting and recalling cortical ensembles. Science 353(6300):691–694. https://doi.org/10.1126/science.aaf7560
21. Pérez-Ortega J, Akrouh A, Yuste R (2024) Stimulus encoding by specific inactivation of cortical neurons. Nat Commun 15:3192. https://doi.org/10.1038/s41467-024-47515-x
22. Aviyente S, Villafañe-Delgado M (2018) Graph signal processing on neuronal. Networks. https://doi.org/10.1016/B978-0-12-813677-5.00031-6
23. Wang PY-P, Sapra S, George V, Silva G (2020) Generalizable machine learning in neuroscience using graph neural networks. Front Artif Intell 4. https://doi.org/10.3389/frai.2021.618372
24. Allaoui M, Kherfi ML, Cheriet A (2020) Considerably improving clustering algorithms using UMAP dimensionality reduction technique: a comparative study. In: El Moataz A, Mammass D, Mansouri A, Nouboud F (eds) Image and signal processing. ICISP 2020, Lecture notes in computer science, vol 12119. Springer, Cham. https://doi.org/10.1007/978-3-030-51935-3_34
25. McInnes, L., Healy, J., & Melville, J. (2018). UMAP: uniform manifold approximation and projection for dimension reduction. arXiv preprint arXiv:1802.03426. 10.48550/arXiv.1802.03426
26. Becht E, McInnes L, Healy J, Dutertre C, Kwok I, Ng L, Ginhoux F, Newell EW (2018) Dimensionality reduction for visualizing single-cell data using UMAP. Nat Biotechnol 37:38–44. https://doi.org/10.1038/nbt.4314
27. Sainburg T, McInnes L, Gentner T (2020) Parametric UMAP embeddings for representation and semisupervised learning. Neural Comput 33:2881–2907. https://doi.org/10.1162/neco_a_01434
28. Fornito A, Zalesky A, Bullmore ET (2016) Fundamentals of brain network analysis. Academic, London
29. Sporns O (2018) Graph theory methods: applications in brain networks. Dialogues Clin Neurosci 20(2):111–121. https://doi.org/10.31887/DCNS.2018.20.2/osporns
30. McInnes L (2024) Uniform Manifold Approximation and Projection. https://github.com/lmcinnes/umap. Accessed 10 May 2024
31. Newman MEJ (2006) Modularity and community structure in networks. Proc Natl Acad Sci 103(23):8577–8582. https://doi.org/10.1073/pnas.0601602103
32. Bruno AM, Frost WN, Humphries MD (2015) Modular deconstruction reveals the dynamical and physical building blocks of a locomotion motor program. Neuron 86:304–318. https://doi.org/10.1016/j.neuron.2015.03.005
33. Sadovsky AJ, MacLean JN (2013) Scaling of topologically similar functional modules defines mouse primary auditory and somatosensory microcircuitry. J Neurosci 33(35):14048–14060. https://doi.org/10.1523/JNEUROSCI.1977-13.2013
34. Waltman L, van Eck NJ (2013) A smart local moving algorithm for large-scale modularity-based community detection. Eur Phys J B 86(11):471. https://doi.org/10.1140/epjb/e2013-40829-0

35. Que X, Checconi F, Petrini F, Gunnels J (2015) Scalable community detection with the Louvain algorithm. In: In 2015 IEEE international parallel and distributed processing symposium (IPDPS), Hyderabad, pp 28–37. https://doi.org/10.1109/IPDPS.2015.59
36. Blondel VD, Guillaume J-L, Lambiotte R, Lefebvre E (2008) Fast unfolding of communities in large networks. J Stat Mech Theory Exp 2008(10):P10008. https://doi.org/10.1088/1742-5468/2008/10/P10008
37. Fruchterman TMJ, Reingold EM (1991) Graph drawing by force-directed placement. Softw Pract Experience 21(11):1129–1164. https://doi.org/10.1002/spe.4380211102

Chapter 9

Dissecting Neural Population Dynamics: Elastic Net and Conditional Random Fields for Deciphering Functional Networks and Information Processing in Learning and Memory

Masakazu Agetsuma, Issei Sato, Yasuhiro R. Tanaka, and Luis Carrillo-Reid

Abstract

Recent technical advancements in molecular, genetic, and optogenetic methods for model animals, as well as whole brain activity recordings, have allowed investigators to elucidate key brain regions regulating learning and memory. Neurons are proposed to work together to encode and regulate memories; therefore, the next crucial step is to develop methods to identify the exact neural network structure underlying memory encoding and investigate how these neural populations process information. Here, we describe an established methodological pipeline for analyzing longitudinal two-photon imaging data from the prefrontal cortex using a regularized regression method and graphical modeling, named elastic net and conditional random field modeling, respectively. This pipeline allowed the detection of learning-dependent changes in neural network topologies and information processing in the specific neural population encoding associative fear memory.

Key words Associative learning, Graphical modeling, Longitudinal two-photon imaging, Neural network topology, Regularized regression

1 Introduction

Learning and memory are crucial brain functions that allow animals to adapt to changes in the environment, thereby enhancing their survival in the wild. Associative learning, such as classical conditioning, is one of the simplest types of learning that has been intensively studied over the past century [1, 2]. It is based on repeated pairings of a neutral conditioned stimulus (CS) such as a tone, and an unconditioned stimulus (US) such as foot shock, that eventually elicit a conditioned response (CR) to the CS alone, e.g., a freezing response in an associative fear learning paradigm.

Luis Carrillo-Reid (ed.), *Identification, Characterization, and Manipulation of Neuronal Ensembles*, Neuromethods, vol. 215, https://doi.org/10.1007/978-1-0716-4208-5_9,

Technical developments in molecular, genetic, and optogenetic methods over the last 2 decades have inspired deeper investigation into the brain regions important for regulating associative learning and memory and, more recently, have enabled the tagging of populations of neurons in the brain whose specific manipulation modulates associative memory [3]. Findings from these studies indicate that information processing by specific neuronal populations likely underlies associative memory [3]. However, it remains unclear how information is stored and processed by neural populations to encode and retrieve associative memories. Moreover, although neurons are thought to form connections with each other to build a neuronal network and construct an information-processing system as a functional group, the detailed mechanisms of the neural networks underlying most brain functions are still elusive.

To tackle these basic questions, we developed a novel pipeline to dissect the computational architecture underlying associative memory. This pipeline, based on longitudinal two-photon imaging, regularized regression methods, and graphical modeling, enabled the characterization of learning-dependent changes in the internal neural network topology and computation of the prefrontal cortex upon memory acquisition [4].

In this chapter, we focus mainly on describing the mathematical analysis methods, which are theoretically applicable not only to prefrontal neural activity data recorded during fear-conditioning procedures but also to other brain regions and other types of memory. Experimental methods for obtaining neural activity data from the prefrontal region are briefly described in Subheading 2 and are provided in more detail in our recent publication [4].

2 Materials

We focused our analysis on the prefrontal neural network. The prefrontal cortex (PFC) is a brain region that regulates associative fear memory, which is evolutionarily conserved in mammals from humans to rodents [5–10]. Dysfunction of the PFC may lead to various psychiatric diseases, including post-traumatic stress disorder [11], and the associative fear learning paradigm has been used as a research model to investigate the mechanisms underlying this disorder. The dorsal part of the rodent medial prefrontal cortex (dmPFC) is an important brain region for retrieving associative fear memory [12–17]. During fear memory retrieval and evoked freezing responses (i.e., CR), activated individual neurons [18] or an enhanced synchrony of neural populations [15] in the dmPFC is observed, while pharmacological or optogenetic silencing of the dmPFC and its projections to specific downstream targets suppresses fear-memory retrieval [12, 13], indicating that associative fear memory is normally stored in the dmPFC. Recent studies uncovered how the dmPFC works together with other brain regions [13, 17, 19], including the basolateral amygdala, hippocampus, and paraventricular nucleus of the thalamus, each of which has also been intensively studied in learning and defensive behaviors as well as in human psychiatric disorders [3, 11, 20–23]. Therefore, the dmPFC is an interesting target for addressing the fundamental question of what structural and computational alterations in prefrontal networks are required to organize novel associative memories. Studies of the dmPFC may also contribute to our understanding of how novel associative memory is stored in the dmPFC together with preexisting networks, such as those regulating sensory and motor information.

2.1 Longitudinal Recordings of Neural Population Activity During Fear Conditioning in Head-Fixed Mice

The mice were head-fixed under a microscope objective and placed on a running disk (Fig. 1a). Disk rotation was recorded to assess the mouse locomotor state. Tones and foot shocks were delivered as the CS and US, respectively. Two different tones were used; one was associated with the US (CS+) and the other was explicitly not associated with the US (CS−). Behavioral analyses revealed that the mice learned to exhibit a freezing-like response, i.e., they decreased their locomotion as a CR, specifically during the CS+ presentation, only after the fear conditioning (FC) [4]. We refer to this expression of the CS+-evoked CRs during the early post-FC session as memory retrieval [13]. On the other hand, the CR observed during the early phase was extinguished after repeated exposure to the CS+ only (we refer to this progressive decrease in the CR observed after the repeated CS+-only presentation during the late post-FC session as extinction [15, 16, 24]). Thus, this experimental setup could be used to study brain computations during fear memory retrieval

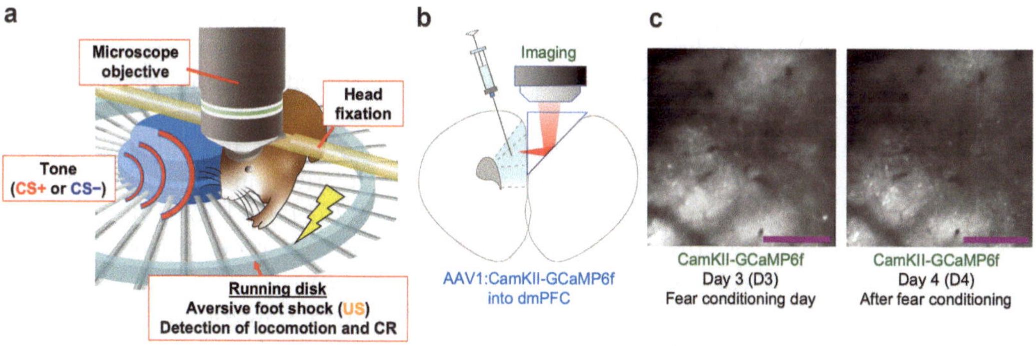

Fig. 1 Cued fear conditioning during two-photon microscopy. (**a**) Developed system for cued-fear conditioning under a two-photon microscope. (**b**) Microprism implantation along the midline for optical access to the dmPFC without cutting nerves. GCaMP6f was expressed in the dmPFC excitatory neurons by the AAV under the CamKII promoter regulation. (**c**) In vivo two-photon microscopy to detect activities at the single-cellular resolution visualized by GCaMP6f, chronically (day [D] 3 and D4) from the same set of dmPFC neurons. Scale bar, 250μm

and following extinction [4]. Locomotor speed negatively and significantly correlated with the freezing-like response; mice with less locomotion showed more freezing-like responses and vice versa [4]. On the other hand, heart-rate monitoring and dmPFC silencing experiments showed that the CR is accompanied by a physiological change in the heart rate and the expression of this change is under regulation by the dmPFC, confirming that our head-fixed system and the fear-conditioning protocol are useful for observing changes in prefrontal neural population activity contributing to associative fear learning [4].

To monitor the neural activities in the dmPFC by two-photon microscopy, a 2-mm microprism along the rostral midline of the brain to optically access the dmPFC region could be used (Fig. 1b). Using a genetically encoded Ca2+ indicator, GCaMP6f, expressed by an adeno-associated virus, neuronal activity from a wide region of the prefrontal area could be chronically visualized. The GCaMP6f was expressed under regulation of the CaMKII promoter to specifically record the activity of the excitatory neurons [4].

We focused on analyzing the activities of the surface layer neurons (150–200μm below the pial surface along the midline) in the dmPFC area [4]. In most of the data analyses, the neural representations during the first 3 trials of the fear conditioning on D3 (D3-early) were compared with those during the first 3 trials on D4 (D4-early) to assess the changes occurring as a result of the fear conditioning (Fig. 1c). The data obtained during the last 3 trials on D3 (D3-late) were used to assess the late conditioning phase, and the data obtained during the last 3 trials on D4 (D4-late) were used to assess the extinction phase.

To decipher network structural changes evoked by learning, single-neuron responses to the CS+ and CS− before and after acquisition of the fear memory were used in combination with conditional random fields (CRFs) [4] to characterize the possible signal flow and its experience-dependent dynamic change within the dmPFC [4].

For data analyses, commercial software and publicly available codes could be used (details are described in Subheading 2). For the statistical analysis, MATLAB R2014a and R2019b were used (MathWorks, Natick, MA). Graphs were produced by MATLAB R2014a/R2019b (MathWorks) and Excel (Microsoft). When plotting the entire large number of samples, CategoricalScatterplot was used (https://github.com/AbstractGeek/CategoricalScatterplot). Raw images of the GcaMP6f signals in the dmPFC were processed to correct for brain motion artifacts using the publicly available code for the enhanced correlation coefficient image alignment algorithm [25] (https://jp.mathworks.com/matlabcentral/fileexchange/27253-ecc-image-alignment-algorithm-image-registration). The regions of interest for the detection of neural activity were automatically selected using a constrained non-negative matrix factorization algorithm in MATLAB R2014a/R2019b with a publicly available code [26] (https://github.com/flatironinstitute/CaImAn-MATLAB). To infer the spike probabilities from the ΔF/F as an alternative estimate of neuronal activation, we used the publicly available code "constrained_foopsi.m" [26] (https://github.com/flatironinstitute/CaImAn-MATLAB). For the elastic net, we used the "lassoglm" function of MATLAB R2019b. For the CRF modeling, we used the publicly available code [27, 28] (https://github.com/hanshuting/graph_ensemble; https://github.com/darikoneil/Identification-of-Pattern-Completion-Neurons).

3 Methods

Based on the neuronal population activity data longitudinally recorded throughout the fear conditioning procedure, we first extracted the neural population encoding the associative memory and then detected learning-dependent changes in the dmPFC network topology especially within the memory-encoding population.

Unsupervised algorithms, such as conventional Principal Component Analysis and other dimensional reduction algorithms, are widely used to reveal neural representations embedded in neural population activities [15, 29]. These algorithms, however, do not directly indicate the neurons corresponding to respective states or animal behaviors. Also, some recent studies introduced unsupervised methods to extract subgroups of neurons by detecting the neurons within a population that are repeatedly coactivated [27, 30], but in our study of the dmPFC, this procedure did not work well to extract a repeated pattern of neural population activity that accurately explained all of the CR events from our data. Moreover, we attempted to confirm the specificity of the neural population for the CR by evaluating whether the CR-coding neural population and the regular motor-coding neural population overlapped or were distinct from each other. To address this goal, we eventually selected the elastic net, which allows independent identification of neural populations encoding distinct behavioral states.

We also introduced a probabilistic graphical model method, the CRF model [27, 28]. This method evaluates the contribution of specific neurons to overall network activity by modeling the conditional probability distribution of a given neuronal population firing together or of a suppressive relationship. We generated a graphical model in which each node represents a neuron in a given neural population and the edges represent the dependencies between neurons, which enabled us to estimate the functional connectivity between dmPFC neurons that were simultaneously recorded. Among the various mathematical algorithms used to evaluate the possible functional connectivity within the neural population, the CRF model is one of the most reliable methods because the results of the calculation (functional connectivity) have been carefully evaluated by two-photon holographic optogenetics and consequential behavioral modulation [27, 28]. In the present method, we calculated the ratio of these relevant connections (both coactive and suppressive) for all possible connections of each node as a "functional connectivity score" for each neuron. The procedures are described below.

3.1 Extraction of Neuronal Ensembles by the Elastic Net

The elastic net is a regularization and variable selection algorithm based on a regression model [31]. We used the elastic net to systematically extract a group of neurons (ensemble) encoding the freezing-like conditioned response (named the CR ensemble, CRE; neurons encoding suppressed locomotion triggered by CS+ as a result of the memory retrieval), or that encoding the memory-unrelated regular stationary state (named regular stationary state [RS] ensemble, RSE; neurons encoding the stationary state during no CS presentation) (Figs. 2, 3 and 4). Models were fitted on neural

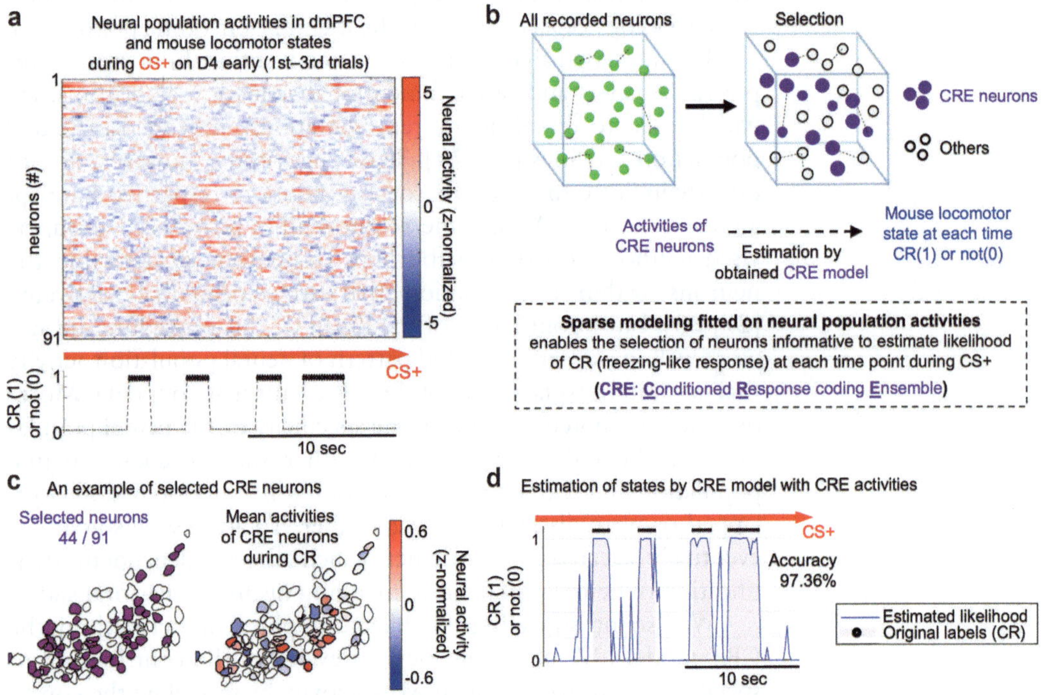

Fig. 2 Extraction of the CR ensemble using elastic net and in vivo imaging data. (**a**) Normalized neuronal population activity during CS+ (top) and locomotor state (bottom; freezing-like conditioned response [1] or not [0]). (**b**) Schematic diagram showing how we extracted the CR ensemble (CRE). (**c**) An example of the extracted CRE. (left) Selected neurons. (right) Mean neural activity during CR (freezing-like response) is shown in color. (**d**) Time-course changes of neural representation encoded by the CRE (of the one shown in panel **c**) under the CS+ presentation on D4 (memory retrieval phase). Black dots on the top of the graph and pink color in the graph indicate the timing of the actual CR, while the blue line shows information estimated by this CRE. The plots show a part of the whole length of the data (during the CS+ presentation). Overall estimation accuracy was 97.36% in this example

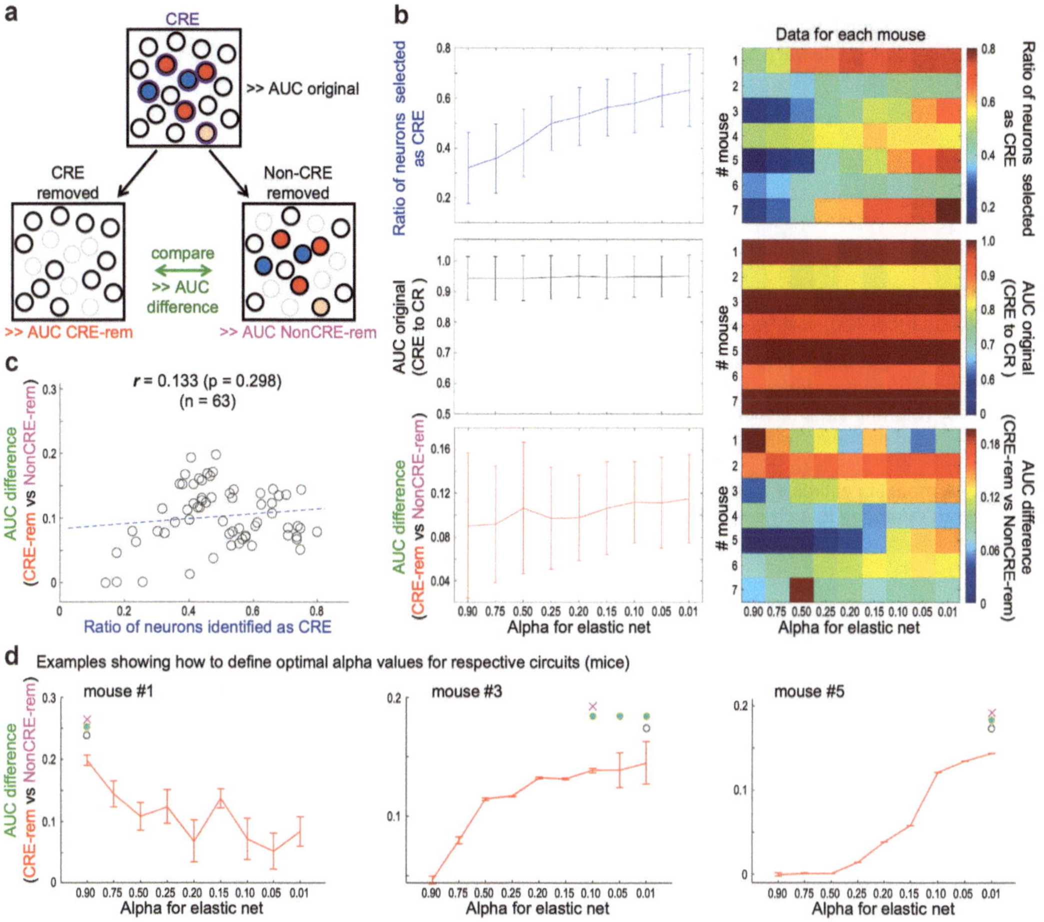

Fig. 3 Optimization process of the CR ensemble size by adjusting the alpha value for elastic net, which also revealed a redundant coding feature. (**a**) We systematically tested a wide range of alpha values and evaluated whether informative neurons were left as unselected neurons at each alpha by measuring the information (fitting performance of a model) of the remaining and unselected neurons. The AUC of the ROC was calculated to accurately evaluate the fitting performance of a model built at each alpha for each mouse. The fitting performance of the obtained model based on the whole original data was calculated as "AUC original." We further compared the difference in fitting performances between "AUC CRE-rem" and "AUC NonCRE-rem." AUC CRE-rem is the AUC value calculated by an elastic net model built with the neurons excluding original CR ensemble neurons. AUC NonCRE-rem is the AUC value calculated by the neurons excluding neurons other than original CR ensemble neurons. The "AUC difference" between those two values was further calculated, and in principle, we defined the best alpha based on the maximum AUC difference for each mouse independently. (**b**) Summary (Mean and S.D., left) and raw data (data for each mouse, right) for ratio of neurons identified as CR ensemble (per whole neurons of each mouse) (top), AUC original (middle), and AUC difference (bottom), at each alpha. (**c**) Pearson's correlation was used to calculate the *r* and p values, revealing that the ratio of neurons identified as CR ensembles (per whole neurons) and the AUC difference were not significantly correlated (n = 63 samples [7 mice × 9 alphas] were analyzed to determine the possible relationship). The blue dotted line corresponds to the result of the line fitting (linear regression). (**d**) Examples showing how to determine the optimal alpha values for respective circuits (mice). In principle, we defined the best alpha based on the maximum AUC difference for each mouse independently, but in some examples as in mouse #3, several alphas revealed statistically insignificant results among the AUC differences. In this case, the largest alpha among those with the same AUC difference was selected. Error bars show s.e.m.

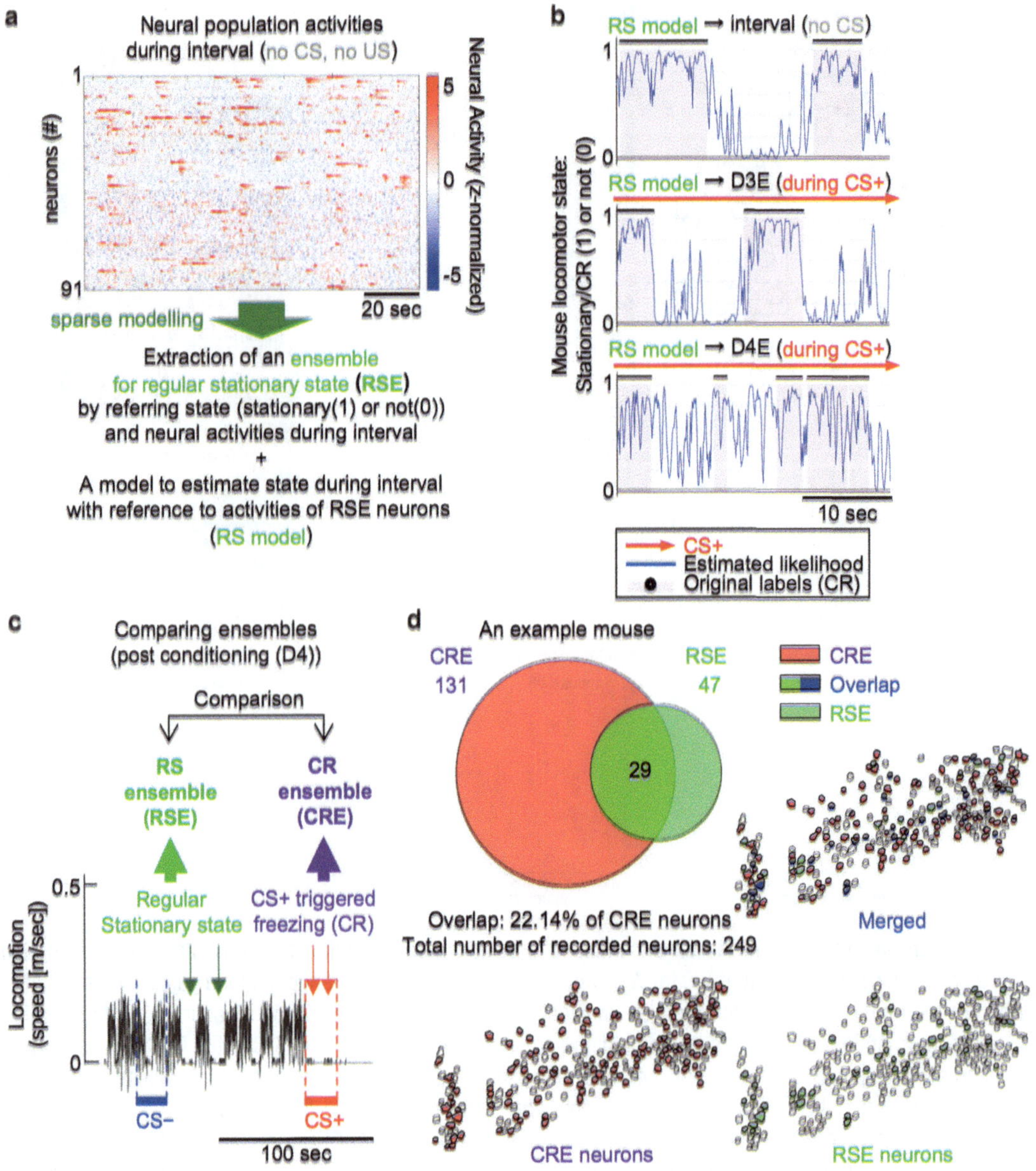

Fig. 4 Comparison of the CR and RS ensembles. (**a**) Schematic diagram for extracting the RS ensemble (RSE) with building the RS model. Normalized neural activity is shown. (**b**) Estimating and decoding locomotion states by the RS model. In an example mouse, the RS model possessed high performance for estimating RS (day[D]4-interval; top) and decoding locomotion states during CS+ at D3-early (D3E; middle). However, the performance decreased for the locomotion states during CS+ at D4-early (D4E; bottom). (**c**) Schematic diagram for comparing the overlap between the CR ensemble (CRE) and the RSE. (**d**) A Venn diagram and a spatial map of an example mouse showing the limited overlap between the CRE and RSE

population activities to calculate the likelihood of each locomotor state at each time point with high accuracy (Fig. 2) [4]. Sparse models that rely on the activities of a limited number of neurons with nonzero coefficients were produced, which enabled the selection of neurons informative for estimating the corresponding locomotor states (the CR or the RS). Because this method allowed us to independently identify different ensembles (CR ensemble and RS ensemble) from the same group of neurons, it could be used to verify whether neurons in the CR ensemble were unique or mostly overlapped with the RS ensemble. For the calculation, we used the "lassoglm" function of MATLAB R2019b.

When extracting the CR ensemble, we used neural activity data only during the CS+ presentation of D4-early (i.e., when mice showed the CR as a result of memory retrieval), and identified neurons informative for distinguishing whether animals exhibited the CR (freezing-like response) or were locomoting during the CS+ (Fig. 2). Therefore, our method produced no bias toward the CS+-responsive neurons in selecting neurons for the CR ensemble. While mice exhibited the CR as suppressed locomotion during D4-early, they locomoted intermittently during the CS+ presentation. Both labels (CR and locomotion) are required to perform the regression based on the elastic net; only mice for which the data contained at least 10% of each label (CR and locomotion) during the CS+ presentation were used to extract the CR ensemble. On the other hand, to extract the RS ensemble, we used data only during the no-CS period (whole D3 or whole D4 data, respectively), including the initial term (before the first CS presentation on each day) and the intertrial interval (which includes both the pre-CS+ and pre-CS− terms without any bias), excluding the 30-s of data after each CS+/CS− trial (when the neural representation might still be affected by the preceding CS).

Learning for the elastic net was formulated as follows:

$$\min_{\beta_0,\boldsymbol{\beta}}\left(\frac{1}{N}\sum\nolimits_{i=1}^{N}\left(-y_i\, log\, \tilde{y}_i-(1-y_i)\log(1-\tilde{y}_i)\right)+\gamma P_\alpha(\boldsymbol{\beta})\right),\text{where} \tag{1}$$

$$\tilde{y}_i=\frac{1}{1+\exp\left(-\left(\beta_0+\boldsymbol{x}_i^T\boldsymbol{\beta}\right)\right)}\quad\text{and} \tag{2}$$

$$P_\alpha(\boldsymbol{\beta})=\frac{(1-\alpha)}{2}\|\boldsymbol{\beta}\|_2^2+\alpha\|\boldsymbol{\beta}\|_1=\sum_{j=1}^{p}\left(\frac{(1-\alpha)}{2}\beta_j^2+\alpha\left|\beta_j\right|\right), \tag{3}$$

where N is the number of observations (time points); y_i is the state (CR/stationary $y_i = 1$ or locomotive $y_i = 0$) at observation i ($i = 1,\ldots,N$); $\mathbf{x}_i$ is input data (neuronal activity), a p-dimensional vector (neural activities of p neurons) at observation i; T denotes the transpose of a vector; γ is a positive regularization parameter;

parameters β_0 and $\boldsymbol{\beta}$ are a scalar variable and a p-dimensional vector, respectively. β_j is the coefficient for the corresponding neuron j estimated from training data. Because this method is designed to sparsely leave the coefficients β_j for the respective neurons, we could identify neurons with nonzero coefficients as those providing substantial information (i.e., CR-ensemble neurons or RS-ensemble neurons).

The elastic net is a hybrid of ridge regression and LASSO (Least Absolute Shrinkage and Selection Operator) regularization: when alpha (α) = 1, elastic net is the same as LASSO, while as α approaches 0, the elastic net approaches the ridge regression. The parameter α indicates the L^1 ratio as formulated as the penalty term $P_\alpha(\boldsymbol{\beta})$; it interpolates between the L^1 norm of $\boldsymbol{\beta}$ and the squared L^2 norm of $\boldsymbol{\beta}$. Parameter α directly controls the sparseness of $\boldsymbol{\beta}$. Tuning α is effective for two well-known problems in LASSO [31]. 1) LASSO is sensitive to correlations between variables and can choose only one if there are two highly correlated and useful variables, whereas by tuning α, the elastic net is more likely to select both useful variables, which leads to more stable variable selection. In other words, in LASSO, highly correlated variables have "zero" coefficients except for one variable, whereas in the elastic net, the highly correlated variables can take nonzero values together by tuning α. 2) The number of variables that can be selected is limited in LASSO when the amount of training data is small. For example, in LASSO, the maximum number of variables that can be selected is N when $N < p$. In contrast, the elastic net can increase the number of variables up to p by tuning α. Therefore, in the elastic net, tuning α can include (select) variables that may be missed in LASSO. In fact, it appeared that our recording data actually included strongly correlated pairs of variables (neuronal activities), suggesting that elastic net is suitable for our neural population activity data; hyper-parameter α enabled adjustment of the size of the selected neurons depending on the data, and lowering α value tended to increase their inclusion [4].

The tuning parameter γ controls the overall strength of the penalty. We varied γ with seven values spaced logarithmically evenly between $10^{-2.5}$ and $10^{-3.3}$, and the γ value with minimum expected deviance, which was calculated by sevenfold cross-validation, was systematically selected for each dataset, at each α (for the α tuning, a wide range of α values were tested as explained below).

To prepare the template neural activity dataset for modeling while avoiding an imbalance in the number of labels for respective states (e.g., CR or not [locomotive] when extracting the CR ensemble), the same number of neural activity vectors (neural population activities at randomly selected time points) from respective states were resampled (chosen randomly with replacement from the original vector set). A total of 900 samples for each state were used to produce each model. The produced models and nonzero-coefficient neurons varied slightly trial by trial, even with the same template dataset. To accurately define each ensemble (and the nonzero-coefficient neurons), we performed this procedure (random resampling and modeling) 100 times and obtained the distribution of the coefficient values for each neuron (we performed this 20 times, instead, when evaluating the performances of "CR ensemble removed," "non-CR ensemble removed," "RS ensemble removed," and "non-RS ensemble removed," which are explained later, because we additionally repeated the whole modeling procedure for them 10 times). Gaussian fitting was performed to define the centroid and 95% confidence interval of each distribution for each coefficient. Then, to build the model, the 95% confidence interval was used to determine whether or not each coefficient was significantly different from zero (enabling us to maintain sparsity), and the centroid was used to define the final coefficients of nonzero coefficient neurons.

The goodness of a model was considered "fitting performance" when we evaluated the performance to estimate locomotor states from neural activity data containing all or part of the training dataset. On the other hand, "decoding performance (or predictability)" was defined as the score obtained by the prediction of the states from data excluding the training dataset.

The obtained model was evaluated by calculating the accuracy or the area under the curve (AUC) of the receiver operating characteristic (ROC) curve. Accuracy denotes the ratio of correct estimations per whole observations (time points). The AUC of the ROC curve, which is another parameter for validating the performance of the model and is suggested to be more accurate when labels are imbalanced, was also calculated. In our previous study, we observed that those scores were very similar to each other overall [4].

Based on the above-described procedure, a wide range of α values were systematically tested to find the optimal value. In some previous studies, the α value was arbitrarily fixed for the analysis [32–34]. Because the dmPFC neurons in our data [4], as well as cortical neurons in general [35–38], include strongly correlated pairs of variables (neuronal activities), however, it appeared important to carefully optimize α when attempting to maximally select informative neurons and further compare them with unselected neurons (CR ensemble vs non-CR ensemble neurons) or an independently selected group (vs RS ensemble) [4].

Ideally, if all the informative neurons can be selected into the CR ensemble, the remaining neurons should have poor information and show poor fitting performance. According to this idea, to optimize the α value and the number of selected neurons, we built a model at each α for each mouse (Fig. 3) and compared the difference in fitting performance between "CR ensemble removed (CRE removed; when all CRE neurons were removed)" and "non-CR ensemble removed (non-CRE removed; when non-CR ensemble, i.e., neurons that were not extracted as the CR ensemble, were removed)" by calculating the difference between their AUCs ("AUC CRE-rem" and "AUC non-CRE-rem" in Fig. 3a): "AUC CRE-rem" is the AUC value calculated by an elastic net model built with the neurons excluding all the original CR ensemble neurons; "AUC nonCRE-rem" is the AUC value calculated by the neurons excluding neurons other than original CR ensemble neurons, randomly selected, and the number of excluded neurons was the same as the number of original CR ensemble neurons (so that the number of neurons used to calculate "AUC nonCRE-rem" is set to be the same as that used for calculating "AUC CRE-rem"); when the number of the neurons unselected for the CR ensemble is less than the number of the original CR ensemble neurons and if the difference is k neurons, "AUC nonCRE-rem" is calculated by removing both randomly selected CR ensemble neurons ($n = k$ neurons) and all the non-CR ensemble neurons. The "AUC difference" (the difference between "AUC CRE-rem" and "AUC nonCRE-rem") was calculated to estimate the degree of remaining information, and in principle, we defined the best α based on the maximum AUC difference for each mouse independently. In some cases, further statistical evaluation was required to define the optimal α as explained below. In those cases, we repeated these procedures 10 times for both "AUC CRE-rem" and "AUC nonCRE-rem" for the following statistical evaluation.

We applied these methods to the dmPFC neural activity data for extracting the CR ensemble [4]. Although the fitting performance of the original CR ensembles (i.e., AUC original in Fig. 3a, b) was not drastically affected by α (Fig. 3b, middle), the size of the CR ensemble was affected, and a smaller α generally resulted in a larger number of selected neurons for each mouse (Fig. 3b, top), suggesting that the CR information might be redundantly encoded in the dmPFC (this possibility is discussed below). Also, the influence of α on the AUC difference was more complicated. As explained above, we defined the best α based on the maximum AUC difference for each mouse independently. There was no clear relationship between the AUC difference and size of the extracted CRE neurons (Fig. 3c). In some cases, as shown in Fig. 3d (mouse #3), the other α(s) had an AUC difference(s) that was not significantly different from the maximum AUC difference. In such cases, the α of the smallest ensemble size among those αs, i.e., the largest α among them, was selected to avoid unnecessarily including additional neurons that did not improve the AUC difference (e.g., in mouse #3, $\alpha = 0.1, 0.05, 0.01$ showed similar AUC differences and there was no statistically significant difference among them [Wilcoxon rank sum test, α of maximum AUC difference vs the other αs, $n = 10$ estimates for each calculated as explained above]. Therefore, in this case, the largest $\alpha = 0.1$ among those three was selected to define the CR ensemble for this mouse).

These results revealed two important points. First, searching around the α values is important in some cases (at least it was crucial for extracting the CR ensemble). Considering this, we also screened several αs for RS ensembles. However, we found no difference among the various αs for RS ensembles even if we tested an additional number of reference frames. While we eventually fixed the α at 0.75 in the case of the RS ensemble [4], the difference between the cases of RS ensemble and CR ensemble confirms the importance of searching around the α values, especially when starting new analyses with new data.

Second, fear memory triggering the CR might be redundantly encoded in the dmPFC at the level of population coding. By adjusting parameter α, we could not only optimize the number of selected neurons but also search for a possible overlap in information coding. Note that this method analyzes the information coding not as individual neurons one-by-one but as a population. Interestingly, at some α value, even without (i.e., after removing) the minimum number of neurons required to estimate the CR information at ~100% accuracy, the remaining neurons still possessed the CR information at high accuracy, similar to the pre-removed neuron set (in Fig. 3b, at the right bottom panel, zero [dark blue] for some α in the mouse #5, for example, means no difference between pre-removed and post-removed). We needed to adjust α and the size of the selected neurons to maximally extract the neural population encoding the CR information, depending on the mouse. This observation itself suggested that the CR information was redundantly encoded by multiple neural populations in the dmPFC (i.e., both minimally extracted population A and the remaining population B encoded the CR information).

In terms of the population coding, this tendency was not observed when we extracted the regular motor-coding population, i.e., the RS ensemble (Fig. 4a), suggesting that the redundant population coding of the CR information in the dmPFC is unique [4]. The observation at each single cell level demonstrated that many neurons show similar response types to tones or foot shocks [4], but our analyses based on the CRF model (described below) also revealed that they were functionally connected to each other in some cases and seemed to work together in the CR ensemble during the memory retrieval phase. Further investigation of how and why these individual neurons contribute to the redundant (or separate) groups of neurons could provide interesting information.

Eventually, to evaluate the specificity of the CR ensemble as a memory-encoding neural population, we carefully compared it with the RS ensemble (Fig. 4). First, we applied the RS model, a model for estimating the RS based on the neural population activities of the RS ensemble (Fig. 4a), to predict the locomotor states during CS+ before and after fear conditioning (Figs. 4b). The RS model showed high performance not only for estimating the locomotor state during the no-CS period (RS model to interval [no CS] in Fig. 4b, top) but also for decoding the locomotor state during the CS+ presentation on D3, i.e., during the early phase of the fear conditioning (RS model to D3-early [during CS+] in Fig. 4b, middle). This result suggests that the RS model was applicable to the data obtained when CS+-related activities were also observed before acquisition of the fear memory, and thus, like cross-validation by the data with the additional noise, confirmed the reliability of the RS model obtained using our elastic net-based method.

The fitting performance for the RS estimation by the RS model was also similar between D3 and D4 (i.e., the day of the fear conditioning vs the day of the post-FC session) [4]. The decoding performance of the RS model to the locomotion state during the CS+ on D4-early, i.e., during memory retrieval, was significantly reduced, however, compared with that of D3-early (RS model to D4-early [during CS+] in Fig. 4b, bottom). These results suggest that the locomotion states during the memory retrieval, or the CR, could not be explained by the RS model [4].

In addition, the cellular overlap between the different ensembles in each mouse can be investigated (Fig. 4c, d). We found that most of the neurons selected as the CR ensemble were unique and did not overlap with the RS ensemble (an example is shown in Fig. 4d; see [4] for details). Furthermore, the reduced decoding performance at D4-early (i.e., memory retrieval phase) was substantially recovered at D4-late (i.e., extinction phase) [4].

These results suggest that the CR ensemble extracted by our method based on the elastic net was unique, not the simple motor-coding population, and dominantly and exclusively explained the locomotor states of the mice during the CR (i.e., CS+-evoked memory retrieval) as an encoder of the acquired associative memory.

3.2 Conditional Random-Field Models to Evaluate Functional Connectivity and Cellular Decoding Performance

Once the memory-encoding ensemble is extracted, various further analyses can be performed to investigate how it emerges and what specific features it has [4]. For example, we evaluated coactivity within the CR ensemble by calculating pairwise correlation coefficients, which are widely used to evaluate neural population activity and are reportedly related to improved network computation [35–38]. We observed that the emergence of the CR ensemble after fear conditioning was accompanied by an enhancement of the coactivity within the CR ensemble [4]. These results led us to hypothesize that the enhanced functional connectivity within the CR ensemble is a major contributor to enhancing the coactivity.

To evaluate changes in the functional connectivity at the population level, we used CRFs [27, 28], which model the conditional

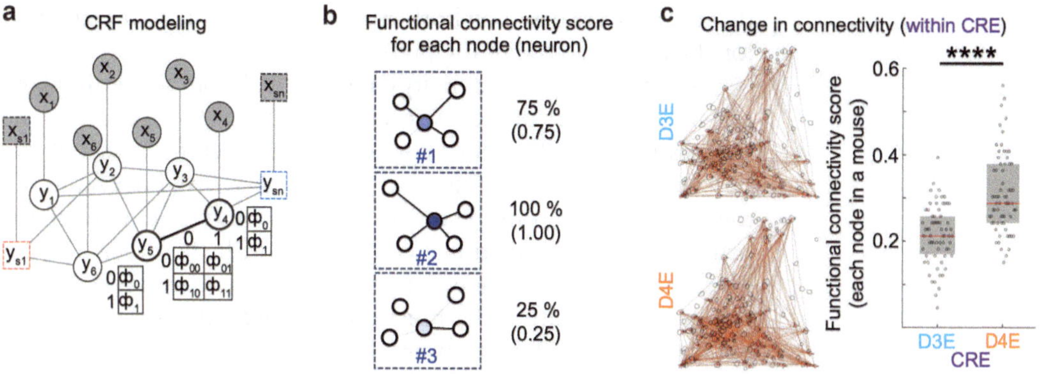

Fig. 5 Conditional random fields modeling. (**a**) Graphical representation of CRFs. Circles represent neurons and squares represent added nodes depicting different experimental conditions. Gray nodes represent observations (x) and white nodes represent modeled neurons (y) connected by edges indicating their mutual dependencies (excitatory or inhibitory). Node potentials are defined as two possible states (active or not) whereas edge potentials are defined over four possible states of each edge depending on the state of adjacent nodes. (**b**) Graphical representation of the functional connectivity score for three example neurons within a group of the same total number of neurons. Circles represent neurons. Black lines between them depict functional connections, while pale gray lines indicate no functional connectivity. (**c**) Enhanced functional connectivity within the CRE in an example mouse after fear conditioning ($n = 67$ CRE-neurons marked by red ellipses [left, spatial maps] or black dots [right, functional connectivity scores]). The functional connectivities within the CRE during the first three trials of the fear conditioning on D3 (D3E) were compared with those during the first three trials on D4 (D4E) to assess the changes occurring as a result of the fear conditioning. **** $p < 0.0001$ by two-tailed, Wilcoxon signed-rank test. Red bars, median; gray boxes in (**c**) indicate 25th–75th percentiles

probability distribution of the interaction among neuronal ensemble members as described below. The CRFs enable capturing the contribution of specific neurons to the overall network activity defined by population vectors belonging to a given neuronal population (Fig. 5a). We generated a graphical model in which each node represents a neuron in a given neural population and edges represent the dependencies between neurons, which enabled us to estimate the functional connectivity between dmPFC neurons that were simultaneously recorded (~300 neurons) [4].

For this analysis, the spike probabilities were inferred from the $\Delta F/F$ as an alternative estimate of neuronal activation using a constrained sparse non-negative calcium deconvolution method [26]. We used the code "constrained_foopsi.m" [26], and the parameters used in the calculation were not manually selected but estimated from the data by the code. After inference of the spike probability and further thresholding by 2 standard deviations, the obtained binary data were further binned (bin size: 1 s) for the CRF modeling. Importantly, the results obtained by CRF modeling were consistent with the results of the coactivity analyses based on the $\Delta F/F$ (and z-normalized $\Delta F/F$) [4], providing substantial support that the analyses based on both estimates complemented each other for the data analyzed in the present study.

Based on the neural activity data recorded from the dmPFC, we calculated the ratio of these relevant connections (both coactive and suppressive) per all possible connections for each node as a "functional connectivity score" for each neuron (Fig. 5b). For example, if there are 75 neurons recorded simultaneously and one of the neurons possesses functional connections with the other 37 neurons, the functional connectivity score for this neuron is 0.5.

Before comparing the functional connectivity between different categories (different ensembles [e.g., within-CR-ensemble vs within-non-CR-ensemble] or different cell types [e.g., US-responsive neurons vs non-US-responsive neurons] [4]), we first calculated the whole network connectivity using all simultaneously recorded neurons (i.e., without stratifying by ensemble or cell type) and further separated them into different categories.

The CRF modeling also allowed us to evaluate the information encoded by individual neurons [27, 28]. In the present study, to measure the information for a given stimulus (CS+ or CS−), we computed the standard ROC, taking as ground truth the timing of a particular CS. The AUC from the ROC curve that represents the performance of each neuron was calculated and used to compare the encoded information in different ensembles, different neuron types, and different days (e.g., D3 fear conditioning session vs D4 post-FC session). As recently demonstrated [27], high ranks for this value in the neural population indicate high potential to recall the neural and cognitive representation of a given stimulus.

3.3 Details for Constructing the CRF Model

The following is a detailed description of how the CRF model was constructed [28, 41]. We generated a graphical model in which each node represents a neuron in a given neural population and the edges represent the dependencies between neurons (Fig. 5a). Variables X^N and Y^N describe an N-dimensional space where the dimensionality describes the total number of active neurons. We used indicator feature vectors $x = [x^1, x^2, \ldots, x^M]$, where $x^m \in X^N$, for each edge and node, and target binary population activity vectors $y = [y^1, y^2, \ldots, y^M]$, where $y^m \in Y^N$ for N neurons and M samples (time points). To consider specific experimental conditions, we used binary indicator feature vectors (x_{s1} to x_{sn} squares in Fig. 5a) and target binary population activity vectors (y_{s1} to y_{sn} squares in Fig. 5a). For each time point, the conditional probability is expressed as follows:

$$p(y^m|x^m;\theta) = \frac{\exp(\langle \phi(x^m, y^m), \theta \rangle)}{Z(x^m;\theta)}, \tag{4}$$

where ϕ is a vector of the distribution expressed in log-linear form, θ is a vector of parameters for the log-linear distribution, and Z is the partition function as follows:

$$Z(x^m;\theta) = \sum_{y\in\mathcal{Y}} exp(\langle\phi(x^m, y), \theta\rangle). \tag{5}$$

The conditional probability can be factored over a graph structure $G = (V, \mathcal{A})$, where V is the collection of nodes representing observation variables and target variables and $\mathcal{A}$ is the collection of subsets of V. Based on the graph structure, the model parameters can be written separately for nodes and edges as $\phi = \{\phi_V, \phi_A\}$ and $\theta = \{\theta_V, \theta_A\} \in \mathfrak{R}^N$, respectively. Given binary x and y, node parameters ϕ_V and θ_V include two sets of distributions and the parameters corresponding to the node state 0 and 1, whereas edge parameters ϕ_A and θ_A include four sets of distributions and the parameters, corresponding to the edge states 00, 01, 10, and 11. The conditional dependencies can then be written as follows:

$$p(Y|X;\theta) = \frac{\exp\left(\sum_{i\in V}\theta_i\phi_i(X, Y_i) + \sum_{\alpha\in\mathcal{A}}\theta_\alpha\phi_\alpha(X, Y_\alpha)\right)}{Z(X;\theta)}. \tag{6}$$

This model is a generalized version of Ising models, which are applied previously to model neuronal networks [39]. The log likelihood of each observation can then be written as follows:

$$\ell(\theta; X^m, Y^m) = \langle\phi(X^m, Y^m), \theta\rangle - \log Z(X^m). \tag{7}$$

To reduce the complexity of the model, we first learned a sparse graph structure that represents variable dependencies and then constructed a CRF on the learned structure. Given the inferred binary spikes from the raw imaging data, we constructed CRF models in two steps: (1) structure learning and (2) parameter learning.

For structure learning, we learned a sparse graph structure $G = (V, \mathcal{A})$ using L^1-regularized neighborhood-based logistic regression for each node r [40] as follows:

$$\min_{\theta_{\backslash r}} \left\{\ell(\theta^s; x) + \lambda_s \left\|\theta^s_{\backslash r}\right\|_1\right\}, \tag{8}$$

where

$$\ell(\theta^s; x) = -\frac{1}{n}\sum_{i=1}^{n} log \frac{exp\left(2x_r\sum_{t\in V\backslash r}\theta^s_{rt}x_t\right)}{exp\left(2x_r\sum_{t\in V\backslash r}\theta^s_{rt}x_t\right) + 1} \tag{9}$$

and

$$\theta^s_{\backslash r} = \{\theta^s_{ru}, u \in V\backslash r\}, \tag{10}$$

where $\backslash r$ denotes except r and θ^s is a vector of regression parameters for structure learning. Here, λ_s is a regularization parameter that controls the sparsity (or conversely, the density) of the constructed graph structure, leaving only relevant functional connectivity, including both coactive and suppressive relationships. This is essentially a logistic regression of variable X_r on the other variables $X_{\backslash r}$, with L^1-regularization. The regression coefficients thus represent the neighborhood structure and the sign pattern. L^1-regularization is implemented as described previously [27, 28]. We constructed a large number of structures (100+) across a broad range of λ_s, from which we stochastically selected a number of structures for subsequent parameter learning. We ensured proper bracketing by including the largest and smallest λ_s structures in this priming set. In further iterations of parameter learning, the sets of the obtained structures were systematically tested, and the best model was selected using the log likelihood of the learned models as feedback.

For parameter learning, we aimed to learn the potential node and edge parameters (ϕ and θ). Based on the learned structure, we used the Bethe approximation to estimate the partition function and iterative Frank-Wolfe methods for parameter estimation by maximizing the log likelihood of observations with a quadratic regularizer [41] as follows:

$$\ell(\theta; X, \Upsilon) = \sum_{m=1}^{M} \ell(\theta; X^m, \Upsilon^m) - \frac{\lambda_p}{2} \|\theta\|^2. \quad (11)$$

λ_p is a regularization parameter that controls the learned parameters and helps prevent overfitting.

Cross-validation was performed to find the best combination of λ_s and λ_p via held-out model likelihood. We varied λ_s with 100 values (between 0.00001 and 0.5, sampled uniformly) and λ_p with 4 values (1, 10, 100, and 10,000). When evaluating the model parameters, we used 80% of the time points of the activity data (of all simultaneously recorded neurons), which were randomly selected from all time points, and the remaining 20% was used for cross-validation to evaluate the fitting of the constructed model. Cross-validation was performed by using all possible combinations of the above parameters and calculating the likelihood of the data. The best model parameters were determined by selecting the parameter set with a locally maximum likelihood in the parameter space, which determine the best combination of λ_s and λ_p, and eventually allow us to estimate what pairs of nodes are functionally connective within the neural population and calculate the functional connectivity score.

Using this model, we can also evaluate the cellular contribution in the population for predicting the stimulus conditions (CS+ and CS−) [27, 28]. For the ith neuron in the population, we compared its activity or inactivity in all M imaging frames (time points) for each model. With the two resulting population vectors in the mth frame among all samples, we calculated the log probability of them coming from the trained CRF model as follows:

$$p_{i,1}^m = p\left(y^m | x_{\setminus i}^m, x_i^m = 1; \theta\right) \quad (12)$$

$$p_{i,0}^m = p\left(y^m | x_{\setminus i}^m, x_i^m = 0; \theta\right). \quad (13)$$

Then, we computed the log likelihood ratio as follows:

$$\ell_{i,1-0} = \left\{ \log p_{i,1}^{m} - \log p_{i,0}^{m} \right\}, m = 1, \ldots, M \quad (14)$$

and calculated the standard ROC curve with the ground truth as the timing of each tone (CS+, CS−). The prediction ability of all nodes (neurons) for all presented stimuli (tones) is then represented by an AUC matrix A, where $A_{i,\ d}$ represents the AUC value of node i predicting tone d. Additionally, we can calculate the node strength $S = \{s_i\}$ of each neuron in the CRF model [27, 28]. We then compute the predicted AUC A^r and node strength S^r of each node r from 100 CRF models trained on shuffled data. The final ensemble for tone d is defined as follows:

$$\left\{ i | A_{i,d} > \text{mean}\left(A_d^r\right) + \text{std}\left(A_d^r\right), S_i > \text{mean}(S^r) + \text{std}(S^r) \right\}. \quad (15)$$

Note that, in our recent study, the functional connectivity score is used to evaluate the change in functional connectivity as a result of the fear conditioning in the study [4], instead of the node strength value; the node strength value is designed for evaluating the relative strength of each node (neuron) among the entire nodes [27, 28] but not for comparing functional connectivity between multiple experimental days.

3.4 Analyses Based on a Combination of the Elastic Net and CRF Modeling

Once the neural population encoding associative fear memory has been extracted by the elastic net and the functional connectivity was determined by CRF modeling, different questions could be answered by combining metrics of both approaches: we found that the functional connectivity and the information coding for the CS+ was specifically enhanced within the CR ensemble, as a result of the fear conditioning (a part of the results is shown in Fig. 5c; see [4] for details). In contrast, the changes in information coding for the CS− were not significantly different between the CR ensemble and the non-CR ensemble [4]. Importantly, our method did not produce any bias to the CS+ in selecting CR ensemble neurons, as explained above. Therefore, these results indicate that the fear conditioning drove a reorganization of functional connectivity in the dmPFC and that the neural population encoding the CR was dominantly associated with the CS+ information in the post-conditioning dmPFC network. These changes may lead to the formation of an information-processing neural network to trigger the CR by the CS+ (i.e., a neural network for the CS+-to-CR transformation).

Furthermore, taking the advantage of the metrics obtained by our approach, we investigated the mechanism underlying the experience-dependent functional reorganization of the dmPFC neural network upon the associative fear learning, which might occur via activity-dependent modulation during the repeated CS+-US pairing. We found that there are neurons responsive to the US during fear conditioning. Interestingly, they were predominantly and significantly more involved in the CR ensemble after the fear conditioning. Similarly, we evaluated the CS responsiveness of the neurons becoming the CR ensemble and found that neurons responding to the CS+ during D3-late (i.e., after the repeated CS+-US pairing) were marginally but significantly more included into the CRE on D4, though the difference between the CRE and the non-CRE was smaller than in the case of the US responsiveness [4]. These results suggest that the US-responsive neurons were preferably integrated into the CR ensemble, in which functional connectivity might also be modulated and strengthened by US-evoked activity, perhaps together with the paired CS+-responsive neurons. To test this possibility, we performed further analyses based on the CRF modeling to assess the functional connectivity of these various types of neurons [4]. We found that the US-responsive neurons became functionally more connected within the CR ensemble than the Non-US-responsive neurons (i.e., neurons that were not defined as US-responsive neurons). This higher connectivity was not innate but a result of the fear conditioning. The information coding for the CS+ was also significantly higher in the US-responsive neurons, specifically in the CR ensemble. As expected from this enhanced information coding for the CS+, we observed that the functional connectivity between the US-responsive neurons and the neurons activated by the CS+ was significantly enhanced within the CR ensemble as a result of the fear conditioning. These results suggest that the US-responsive neurons were dominantly associated with the CS+-activated neurons when they became integrated into the newly emerged CR ensemble as hubs of this novel associative dmPFC network [4].

In summary, using the methods described in this chapter, we observed that the functional connectivity between the CS+-activated neurons and the US-responsive (activated) neurons was enhanced as a result of fear conditioning [4]. The information coding for the CS+ in such US-responsive neurons was also enhanced in the CR ensemble but not in the outside network (non-CR ensemble). We also observed that the US-responsive neurons possessed significantly higher functional connectivity within the CR ensemble than the other neurons, for which enhancement was experience-dependent. The CR ensemble encoded the CR information and was distinct from the regular motor-coding neurons. These results visualized the possible signal flow from the CS+ to the CS+-activated neurons in the CR ensembles, then to the US-responsive neurons, and further to the entire CR ensembles, in which most of the association or responsiveness was enhanced as a result of the associative fear learning paradigm. In other words, the US-responsive neurons in the dmPFC gained features of pattern completion cells [27] and became a hub of the

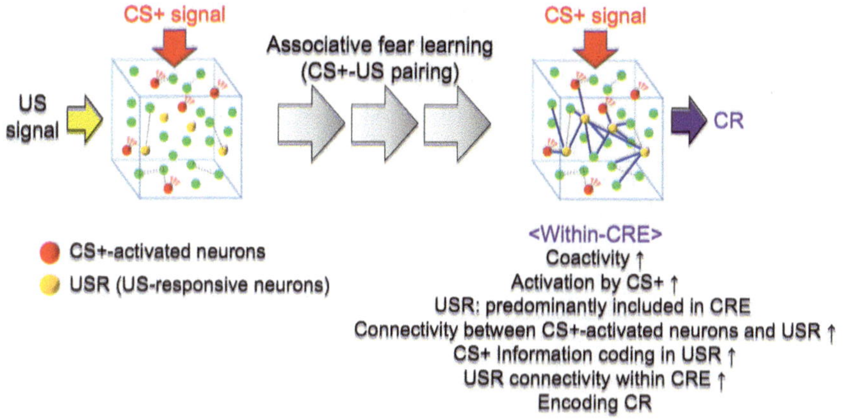

Fig. 6 Our pipeline uncovered learning-dependent dynamic changes in the dmPFC network. Our methodological pipeline for longitudinal two-photon imaging and computational dissection of the neural population allowed us to investigate learning-dependent dynamic modulation of functional connectivity and population coding for associative fear learning in the dmPFC network

novel neural ensemble linking the CS+ to the CR, a memory-evoked response, after the repeated CS+ and US pairings (Fig. 6).

4 Conclusion

This chapter illustrates a pipeline for longitudinal two-photon imaging and computational dissection of the neural population, which allowed the investigation of learning-dependent dynamic modulation of population coding for associative fear learning, structured on an activity-dependent hub-network formation within the dmPFC. These mathematical analysis methods could be applicable not only to prefrontal neural activity data recorded during fear-conditioning procedures but also to other brain regions and other types of memory described by sparse data.

5 Notes

Prior to investigating population coding, we assessed dmPFC single-neuron responses to the CS+ and CS− before and after the acquisition of the fear memory ($n = 1165$ neurons from $N = 7$ chronically recorded mice) [4]. We found that approximately 60% of neurons exhibited a change in neural activity following exposure to the CS+ and/or CS−, and approximately 20% of neurons showed responses to both the CS+ and CS−. The distributions of these types of neurons were consistent throughout the learning process. This type of responsiveness of individual neurons to variable task-relevant aspects has also been reported in the primate PFC

[42] and is proposed to enhance the number of tasks that each neural circuit containing a limited number of neurons can handle in a high-dimensional space implemented by a population of networked neurons [42, 43]. This encouraged us to further analyze the population coding for associative fear memory, which was followed by comparison with the single cellular responsiveness. The results that we obtained as described above [4] indicate the usefulness of the combination of the conventional measures based on the single cellular responsiveness and the new CRF graphical modeling technique.

Acknowledgments

This work was supported by the Japan Science and Technology Agency, PRESTO (to M.A.), JSPS KAKENHI Grant (grant number JP18K06536, JP18H05144, JP20H05076, JP21H02801, JP22H05081, JP22H05519 to M.A.; JP20H03357, JP22H05162 to Y.R.T.), JSPS Bilateral Program (JPJSBP1-20199901 to M.A.), AMED (grant number JP19dm0207086 to M.A.), the grant of Joint Research by the National Institutes of Natural Sciences (NINS program No 01112008 and 01112106 to M.A.), grants from Brain Science Foundation and Shimadzu Foundation to M.A., and UNAM-DGAPA-PAPIIT (IN213923) and CONAHCYT (CF6653) to L.C-R.

References

1. Pavlov PI (2010) Conditioned reflexes: an investigation of the physiological activity of the cerebral cortex. Ann Neurosci 17(3): 136–141. https://doi.org/10.5214/ans.0972-7531.1017309
2. Watson JB, Rayner R (1920) Conditioned emotional reactions. J Exp Psychol 3(1):1–14. https://doi.org/10.1037/h0069608
3. Josselyn SA, Tonegawa S (2020) Memory engrams: recalling the past and imagining the future. Science 367(6473):eaaw4325. https://doi.org/10.1126/science.aaw4325
4. Agetsuma M, Sato I, Tanaka YR, Carrillo-Reid L, Kasai A, Noritake A, Arai Y, Yoshitomo M, Inagaki T, Yukawa H, Hashimoto H, Nabekura J, Nagai T (2023) Activity-dependent organization of prefrontal hub-networks for associative learning and signal transformation. Nat Commun 14(1):5996. https://doi.org/10.1038/s41467-023-41547-5
5. Miller EK, Cohen JD (2001) An integrative theory of prefrontal cortex function. Annu Rev Neurosci 24:167–202. https://doi.org/10.1146/annurev.neuro.24.1.167
6. Calhoon GG, Tye KM (2015) Resolving the neural circuits of anxiety. Nat Neurosci 18(10): 1394–1404. https://doi.org/10.1038/nn.4101
7. Le Merre P, Ahrlund-Richter S, Carlen M (2021) The mouse prefrontal cortex: Unity in diversity. Neuron 109(12):1925–1944. https://doi.org/10.1016/j.neuron.2021.03.035
8. Shin LM, Liberzon I (2009) The neurocircuitry of fear , stress , and anxiety disorders. Neuropsychopharmacology 35(1):169–191. https://doi.org/10.1038/npp.2009.83
9. Jezzini A, Bromberg-Martin ES, Trambaiolli LR, Haber SN, Monosov IE (2021) A prefrontal network integrates preferences for advance information about uncertain rewards and punishments. Neuron 109(14):2339–2352 e2335. https://doi.org/10.1016/j.neuron.2021.05.013
10. Burgos-Robles A, Kimchi EY, Izadmehr EM, Porzenheim MJ, Ramos-Guasp WA, Nieh EH, Felix-Ortiz AC, Namburi P, Leppla CA, Presbrey KN, Anandalingam KK, Pagan-Rivera PA, Anahtar M, Beyeler A, Tye KM (2017)

Amygdala inputs to prefrontal cortex guide behavior amid conflicting cues of reward and punishment. Nat Neurosci 20(6):824–835. https://doi.org/10.1038/nn.4553

11. Fenster RJ, Lebois LAM, Ressler KJ, Suh J (2018) Brain circuit dysfunction in post-traumatic stress disorder: from mouse to man. Nat Rev Neurosci 19(9):535–551. https://doi.org/10.1038/s41583-018-0039-7
12. Corcoran KA, Quirk GJ (2007) Activity in prelimbic cortex is necessary for the expression of learned, but not innate, fears. J Neurosci 27(4): 840–844. https://doi.org/10.1523/JNEUROSCI.5327-06.2007
13. Do-Monte FH, Quinones-Laracuente K, Quirk GJ (2015) A temporal shift in the circuits mediating retrieval of fear memory. Nature 519(7544):460–463. https://doi.org/10.1038/nature14030
14. Courtin J, Chaudun F, Rozeske RR, Karalis N, Gonzalez-Campo C, Wurtz H, Abdi A, Baufreton J, Bienvenu TC, Herry C (2014) Prefrontal parvalbumin interneurons shape neuronal activity to drive fear expression. Nature 505(7481):92–96. https://doi.org/10.1038/nature12755
15. Dejean C, Courtin J, Karalis N, Chaudun F, Wurtz H, Bienvenu TC, Herry C (2016) Prefrontal neuronal assemblies temporally control fear behaviour. Nature 535(7612):420–424. https://doi.org/10.1038/nature18630
16. Klavir O, Prigge M, Sarel A, Paz R, Yizhar O (2017) Manipulating fear associations via optogenetic modulation of amygdala inputs to prefrontal cortex. Nat Neurosci 20(6):836–844. https://doi.org/10.1038/nn.4523
17. Jercog D, Winke N, Sung K, Fernandez MM, Francioni C, Rajot D, Courtin J, Chaudun F, Jercog PE, Valerio S, Herry C (2021) Dynamical prefrontal population coding during defensive behaviours. Nature 595(7869):690–694. https://doi.org/10.1038/s41586-021-03726-6
18. Burgos-Robles A, Vidal-Gonzalez I, Quirk GJ (2009) Sustained conditioned responses in prelimbic prefrontal neurons are correlated with fear expression and extinction failure. J Neurosci 29(26):8474–8482. https://doi.org/10.1523/JNEUROSCI.0378-09.2009
19. Kitamura T, Ogawa SK, Roy DS, Okuyama T, Morrissey MD, Smith LM, Redondo RL, Tonegawa S (2017) Engrams and circuits crucial for systems consolidation of a memory. Science 356(6333):73–78. https://doi.org/10.1126/science.aam6808
20. Bienvenu TCM, Dejean C, Jercog D, Aouizerate B, Lemoine M, Herry C (2021) The advent of fear conditioning as an animal model of post-traumatic stress disorder: learning from the past to shape the future of PTSD research. Neuron 109(15):2380–2397. https://doi.org/10.1016/j.neuron.2021.05.017
21. Sotres-bayon F, Quirk GJ (2010) Prefrontal control of fear : more than just extinction. Curr Opin Neurobiol 20(2):231–235. https://doi.org/10.1016/j.conb.2010.02.005
22. Tovote P, Fadok JP, Luthi A (2015) Neuronal circuits for fear and anxiety. Nat Rev Neurosci 16(6):317–331. https://doi.org/10.1038/nrn3945
23. Herry C, Johansen JP (2014) Encoding of fear learning and memory in distributed neuronal circuits. Nat Neurosci 17(12):1644–1654. https://doi.org/10.1038/nn.3869
24. Herry C, Ciocchi S, Senn V, Demmou L, Muller C, Luthi A (2008) Switching on and off fear by distinct neuronal circuits. Nature 454(July):600–606. https://doi.org/10.1038/nature07166
25. Evangelidis GD, Psarakis EZ (2008) Parametric image alignment using enhanced correlation coefficient maximization. IEEE Trans Pattern Anal Mach Intell 30(10):1858–1865. https://doi.org/10.1109/TPAMI.2008.113
26. Pnevmatikakis EA, Soudry D, Gao Y, Machado TA, Merel J, Pfau D, Reardon T, Mu Y, Lacefield C, Yang W, Ahrens M, Bruno R, Jessell TM, Peterka DS, Yuste R, Paninski L (2016) Simultaneous denoising, deconvolution, and demixing of calcium imaging data. Neuron 89(2):285–299. https://doi.org/10.1016/j.neuron.2015.11.037
27. Carrillo-Reid L, Han S, Yang W, Akrouh A, Yuste R (2019) Controlling visually guided behavior by holographic recalling of cortical ensembles. Cell 178(2):447–457 e445. https://doi.org/10.1016/j.cell.2019.05.045
28. Carrillo-Reid L, Han S, O'Neil DA, Taralova E, Jebara T, Yuste R (2021) Identification of pattern completion neurons in neuronal ensembles using probabilistic graphical models. J Neurosci 41:8577. https://doi.org/10.1523/jneurosci.0051-21.2021
29. Rozeske RR, Jercog D, Karalis N, Chaudun F, Khoder S, Girard D, Winke N, Herry C (2018) Prefrontal-periaqueductal gray-projecting neurons mediate context fear discrimination. Neuron 97(4):898–910 e896. https://doi.org/10.1016/j.neuron.2017.12.044
30. Ghandour K, Ohkawa N, Fung CCA, Asai H, Saitoh Y, Takekawa T, Okubo-Suzuki R, Soya S, Nishizono H, Matsuo M, Osanai M,

Sato M, Ohkura M, Nakai J, Hayashi Y, Sakurai T, Kitamura T, Fukai T, Inokuchi K (2019) Orchestrated ensemble activities constitute a hippocampal memory engram. Nat Commun 10(1):2637. https://doi.org/10.1038/s41467-019-10683-2

31. Zou H, Hastie T (2005) Regularization and variable selection via the elastic net. J R Stat Soc Ser B (Stat Methodol) 67(2):301–320. https://doi.org/10.1111/j.1467-9868.2005.00503.x
32. Runyan CA, Piasini E, Panzeri S, Harvey CD (2017) Distinct timescales of population coding across cortex. Nature 548(7665):92–96. https://doi.org/10.1038/nature23020
33. Moradi E, Khundrakpam B, Lewis JD, Evans AC, Tohka J (2017) Predicting symptom severity in autism spectrum disorder based on cortical thickness measures in agglomerative data. NeuroImage 144(Pt A):128–141. https://doi.org/10.1016/j.neuroimage.2016.09.049
34. Banerjee A, Parente G, Teutsch J, Lewis C, Voigt FF, Helmchen F (2020) Value-guided remapping of sensory cortex by lateral orbitofrontal cortex. Nature 585(7824):245–250. https://doi.org/10.1038/s41586-020-2704-z
35. Agetsuma M, Hamm JP, Tao K, Fujisawa S, Yuste R (2018) Parvalbumin-positive interneurons regulate neuronal ensembles in visual cortex. Cereb Cortex 28(5):1831–1845. https://doi.org/10.1093/cercor/bhx169
36. Harris KD, Mrsic-Flogel TD (2013) Cortical connectivity and sensory coding. Nature 503(7474):51–58. https://doi.org/10.1038/nature12654
37. Hofer SB, Ko H, Pichler B, Vogelstein J, Ros H, Zeng H, Lein E, Lesica NA, Mrsic-Flogel TD (2011) Differential connectivity and response dynamics of excitatory and inhibitory neurons in visual cortex. Nat Neurosci 14(8):1045–1052. https://doi.org/10.1038/nn.2876
38. Cossell L, Iacaruso MF, Muir DR, Houlton R, Sader EN, Ko H, Hofer SB, Mrsic-Flogel TD (2015) Functional organization of excitatory synaptic strength in primary visual cortex. Nature 518(7539):399–403. https://doi.org/10.1038/nature14182
39. Schaub MT, Schultz SR (2012) The Ising decoder: reading out the activity of large neural ensembles. J Comput Neurosci 32(1):101–118. https://doi.org/10.1007/s10827-011-0342-z
40. Ravikumar P, Wainwright MJ, Lafferty JD (2010) High-dimensional Ising model selection using ℓ1-regularized logistic regression. Ann Stat 38:3. https://doi.org/10.1214/09-aos691
41. Tang K, Ruozzi N, Belanger D, Jebara T (2016) Bethe learning of graphical models via MAP decoding. JMLR Workshop Conf Proc 51:1096–1104
42. Rigotti M, Barak O, Warden MR, Wang XJ, Daw ND, Miller EK, Fusi S (2013) The importance of mixed selectivity in complex cognitive tasks. Nature 497(7451):585–590. https://doi.org/10.1038/nature12160
43. Fusi S, Miller EK, Rigotti M (2016) Why neurons mix: high dimensionality for higher cognition. Curr Opin Neurobiol 37:66–74. https://doi.org/10.1016/j.conb.2016.01.010

Chapter 10

Analyzing Functional Neuronal Ensembles in a Between-Subjects Paradigm

Jordan P. Hamm

Abstract

The hypothesis that certain psychiatric or neurological diseases can be best understood—and therefore treated—at a systems level is an attractive one. For testing such a hypothesis, animal models are useful, and the ability to meaningfully compare multicellular activity *between* brains is needed. Identifying when population-level functional dynamics are typical or healthy, and when they are aberrant or pathological, is not straightforward, especially when there exists no within-animal baseline state to which to compare. This chapter focuses on practical considerations for applying popular ensemble-identification analyses (such as tSNE, SVM, k-means) to comparisons between groups, such as one might carry out between a wild-type animal and an animal with a genetic mutation affecting a disease-relevant pathway. While the methods are many, the principles are few(er). We focus first and foremost on the choice of metric: precisely what is thought to differ between groups? What aspect of multineuronal activity—stability, number, size, diversity, etc.—might be depleted or augmented in the disease state of interest? And could other behavioral or neural properties, such as arousal, motor output, or baseline neural firing rates, better account for change in your chosen metric, rather than a specific loss or disorganization of neural ensembles? We provide an example analysis pipeline wherein these points are considered.

Key words Neuronal ensembles, Dimensionality reduction, Functional connectivity, Pathological conditions, Neurological disorders

1 Introduction

A popular hypothesis posits that some psychiatric and neurological diseases are fundamentally disorders of neuronal ensembles [1–5]. While myriad genetic variants and molecular alterations may present differently across individuals with the same diagnosis, investigation of brain function at a systems level may reveal a unified pathophysiology: loss, instability, or hyperstability of spatiotemporal patterns of brain activity, made up of recurrently connected, reliably co-activating neuronal ensembles that are thought to underlie stable memory, perception, behavior, and thought [6–11]. Current pharmacological approaches are ultimately

Luis Carrillo-Reid (ed.), *Identification, Characterization, and Manipulation of Neuronal Ensembles*, Neuromethods, vol. 215, https://doi.org/10.1007/978-1-0716-4208-5_10,

insufficient in treating all aspects of several major psychiatric disorders, such as negative or cognitive symptoms in schizophrenia [12], so this systems-level hypothesis offers an alternate and potentially optimistic vista. Perhaps, interventions and therapeutics aimed at restabilizing or destabilizing pathological neuronal ensembles could better address treatment-resistant aspects of these disease states [13].

If sufficiently qualified, operationalized, and detailed with respect to brain region, behavioral state, and diagnosis, this framework can provide provocative—if not potentially powerful—hypotheses with broad implications for medicine as well as basic neural science [14]. However, such hypotheses are also challenging to test directly. This is due in part to our methodological limits. Analyzing brain activity at a population level requires functional recordings with single cell resolution. To do this noninvasively in humans—ideally recording in some individuals with and some without a diagnosis—is simply outside the reach of current methods.

So, animal models must be utilized to test a neural ensemble hypothesis of, e.g., schizophrenia. A challenge for this type of translational research is that *wholistic models "of" a complex neuropsychiatric disease may not be attainable* [15], and major funding agencies now even discourage such approaches [16]. The notion of a mouse with schizophrenia does not only "offend our sensibilities" [17], but it is hardly verifiable to begin with, as the core symptomatology of schizophrenia involves delusions, hallucinations, and disorganized behavior and thought: all of which one cannot directly ascertain in animals with current methods (and one certainly cannot carry out a structured clinical interview with a mouse). And even in cases where it seems that a disease pathophysiology is known, such as a once widely accepted role for amyloid-beta plaques in Alzheimer's disease, translational research taking this biomarker for granted as modeling Alzheimer's in mice has led to failed clinical trials and lost resources [18], exposing our incomplete understanding of the underlying disease etiology [19].

Instead, it has been proposed that using animal models "for" disease is the more circumspect and directly effective approach for advancing the field of neuropsychiatry [16]. To clarify, one should focus animal research on understanding disease-relevant pathways and testing mechanistic hypotheses about interactions among levels of analysis and biological scales. For example, instead of attempting to create a "schizophrenic mouse" and then discovering the brain functions affected or using it to test a novel drug in a rescue experiment, neuropsychiatry would be better served by studying the molecular pathways affected by a known genetic risk variant, 22q11.2-microdeletion [2] (which predisposes for a variety of pathologies, including but not limited to schizophrenia, intellectual disability, and heart defects) and then testing whether and which of these pathways plays a direct role in, e.g., degrading ensemble activity in the cortex. Mapping simpler causal pathways—how certain genes lead to certain molecular pathologies, which cellular changes lead to which circuit aberrations, which network dysfunctions lead to which symptom clusters—is proposed as the way forward for studying complex diseases like schizophrenia [20]. Therefore, even if a given gene does not play a direct role in all individuals with a disease, it is still relevant to understand whether and how that gene impacts neural ensembles, and the molecular and cellular mechanisms connecting these two levels of analysis, as this may help refine, restrict, and/or seed hypotheses about the disease state in humans.

Optimal approaches for utilizing animal models for disease research not-withstanding, another hurdle for testing an "ensemble hypothesis of neuropsychiatric disease" is in *quantification of ensembles in brain recordings.* It is this hurdle that we seek to address in this chapter. Provided that one has an animal model with utility for understandings some aspect of a disease, one must arrive at a metric that captures "diseased neuronal ensembles." It becomes less important whether one can identify which neurons are connected to which other neurons [6], and which neuron groups correlate with or cause a given percept or behavior [13, 21], and more important whether such connectivity is present at all, and in the right balance. For this, the literature provides no consistent metric nor analytical standard [1, 2, 5, 22, 23].

A final important challenge to consider: studies of developmental diseases often *require between-subjects designs*, and most major neuropsychiatric diseases purported to involve ensemble pathophysiology are neurodevelopmental in nature. That is, they require comparison of ensemble activity in two separate brains, in two entirely different sets of neurons in two separate animals. While the reason for this is intuitive in the traditional translational paradigm (i.e., you have one mouse with a disease and one without... of course you need between-subjects comparisons), it is not entirely a given in the new translational paradigm discussed above (i.e., animal models *for* disease). If all we care about is the connection between molecules, cells, and circuits, do we need to carry out experiments in separate animals? If we take again the example of schizophrenia, which is a major disorder of perception and cognition that gives rise to, but is not purely or exclusively defined, psychotic symptoms, then we must consider a few points. Schizophrenia is neurodevelopmental in the sense that it is accepted to be neural *and* developmental. Therefore, many disease-relevant pathways involve protracted periods of brain development, such as dendritic pruning in adolescence [12] or white matter growth in early adulthood [24]. Further, risk for schizophrenia is also highly influenced by maternal infection, which may impact neuronal migration in very early perinatal development. Manipulations affecting these periods are not something which cannot be washed out or repeated in a given organism. For this point, between-subjects analyses are needed, even if one is not claiming to study or cure a "schizophrenic mouse." Another point: to test the impact of a genetic mutation, one must appreciate that most genes play distinct roles throughout development, and some have little, or even different, roles in adulthood. Thus, silencing a gene or gene product in an adult animal, recording ensemble activity before and after using a "within subjects" design, may have limited utility for understanding disease relevant pathways.

Therefore, the ability to test ensemble-hypotheses about neurological and psychiatric disease states rests on the use of animal models (which, given our current genetic and imaging tools, are typically mice), carefully crafted mechanistic hypotheses, precise metrics that capture the state of neural ensembles, and between-subjects designs. Of course, the latter three will be useful regardless of whether one's experimental goals directly involve disease.

2 Example

Here we present an example where neocortical activity is compared between a wild-type mouse and a mouse engineered to have changes in a given disease-relevant gene that affect its expression. For the sake of discussion, let us say, several *single nucleotide polymorphisms*, or SNPs, in and around gene X predispose toward developing disease Y in humans. Perhaps, these SNPs each reduce how effectively the gene product, protein Z, is transcribed, translated, or functions. Having any or all of the SNPs carries an overall odds ratio of 10 (meaning, having a SNP in gene X increases your chance of having disease Y by 10 times).

2.1 Formalizing the Hypothesis

The central hypothesis being tested in our example is *that disease Y is a disease of ensemble disorganization in neocortex*. Let us formalize this. This means (1) that in all cases of disease Y, ensemble disorganization is causative of disease Y (if we were to directly disorganize neocortical groups, through any means or by direct manipulation of cell-synchronies, we would recapitulate disease Y phenomenology), (2) that one should observe ensemble disorganization in every case of disease Y, and (3) that disease Y can be effectively treated by resolving ensemble disorganization (or, alternatively, that any treatment that resolves all aspects of disease Y should also affect ensembles in cortex).

Of course, we do not have this type of access in humans, and we do not have the ability to directly study disease Y in animals. This is because, in disease Y, we cannot purely ascertain whether a mouse has or even can have disease Y. Disease Y has symptoms that must be reported by the patient directly, and there are no diagnostic

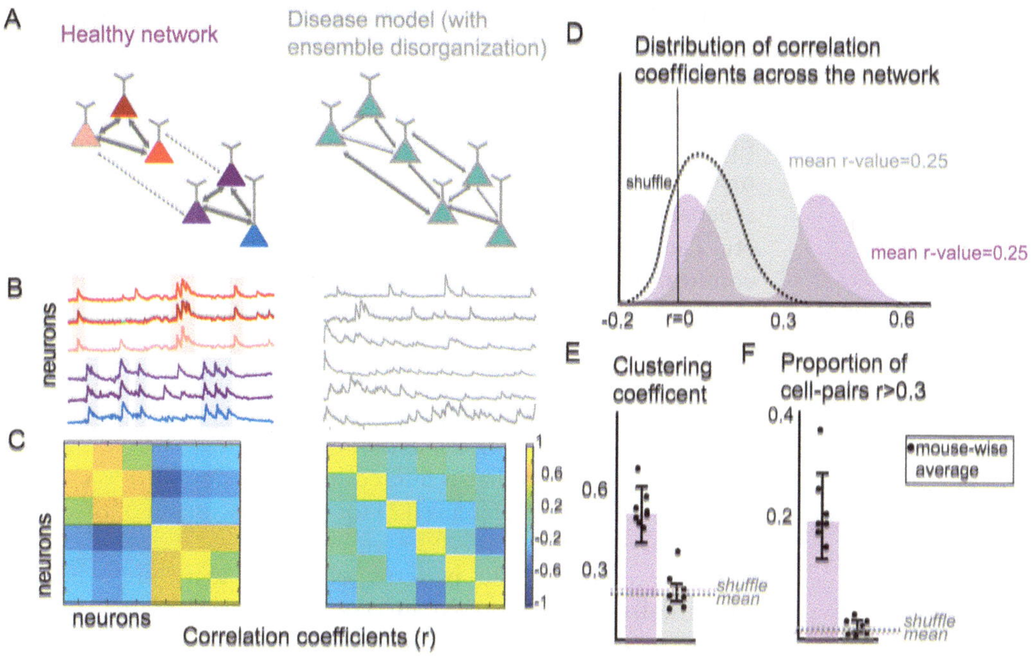

Fig. 1 Cell-wise analysis of ensemble organization between subjects. (**a**) Depiction of hypothesis. Schematic of healthy network on the left, with well-defined ensembles with strong connections within cell groups (red and blue) and weak connections between cell groups. On the right is a hypothesized disease network, with less defined ensemble groups, with more randomly distributed functional correlations/connections among neurons. (**b**) Example calcium imaging data from theoretical networks in A. (**c**) Correlation matrices from cells in A and B. (**d**) Distribution of cell-cell functional correlations, theoretically from networks in A and B, compared to null baseline distribution of correlations generated from spatiotemporally shuffled datasets. Note—overall average functional correlations are not different between the networks, but the healthy network has both more strong functional connections (cell-cell correlations) *and* more weak functional connections. (**e**) Theoretical "clustering coefficients" from networks in A. (**f**) Proportion of strongly correlated cell-cell pairs from D. (**e** and **f**) depict group-wide averages and distributions, deriving one value from each mouse/brain network (individual dots) for subsequent between-subjects statistical comparisons, such as a t-test or factorial ANOVA. *Note: all data in this figure was generated artificially, or semi-artificially by manipulating real calcium imaging datasets, and is meant for demonstration purposes only*

biomarkers of disease Y. What we can do is test a softer corollary of the hypothesis. We can test that *targeted mutations in gene X disorganize ensembles in the neocortex*. Because variants affecting gene X influence the development of disease Y, this is an extension of point (1) in the above paragraph.

Next, we need to define what is meant by *ensemble disorganization* in neocortex. Do we mean a loss in synaptic connections? Decrease in functional connectivity (effective or correlational)? A loss of stable neuronal groups? A change in the frequency of ensemble activations? A change in the number of neurons per ensemble? A change in the number of ensembles per population? An increase in the flexibility or rigidity of ensemble group membership? Not only will this definition determine your metric (below), but it will also determine your methods. If you equate "ensemble disorganization" with a loss in synaptic connections, we would suggest that quantifying synapses in fixed tissue is your best option. However, in what is formally meant by ensembles, we would argue that the raw number/density of synaptic connections is not what is meant by "ensemble organization." Sure, ensembles imply synapses, but changes to ensemble number, discreteness, stability, etc. can occur in the absence of sheer increases or decreases in connection probability. See Fig. 1a, for example. For the same reason, decreases in overall *functional* connectivity—shared activity or coactivity between all neurons and all other neurons—are also something that can occur semi-independently of actual changes to ensemble organization. What is implied by our example hypothesis in Fig. 1a is that the *pattern of connections*, rather than the magnitude or number, is what is altered in the disease of interest. The reader is referred to reference [2] for more discussion of this topic.

2.2 Choosing a Metric

Cell-wise analyses: The patterns of connectivity are what are hypothesized to be altered in our disease-relevant manipulated mouse model. How is that measured exactly? Patterns of *synaptic* connectivity in a given sample of brain tissue can be assessed directly with a number of approaches, such as paired patch clamp experiments with >3 or 4 cells per recording [25, 26] or electron microscopy with network reconstruction and imputation [27]. However, these methods are highly resource intensive and (a) often do not provide a large-scale picture of connectivity beyond a few cells at a time (e.g., with patching) or (b) they require huge efforts and cannot, currently, be carried out in multiple mice at a time in any reasonable timeframe.

On the other hand, two-photon calcium imaging or dense-array extracellular electrophysiology with, e.g., Neuropixels probes [28] can be carried out in individual labs on multiple animals with few personnel. These techniques do not provide a picture of monosynaptic connectivity, but they can be used to ascertain "functional connectivity"; here (in our example in Fig. 1) we simply mean shared or similar activity across time as expressed by Pearson's correlation coefficients. Although this, again, does not reveal monosynaptic connectivity directly, it may not matter for the purposes of our hypothesis: ensembles may be Hebbian or non-Hebbian, or they may co-activate because of polysynaptic motifs or shared bottom-up drive or top-down modulation. In fact, our hypothesis may posit that any number of anatomical changes may lead independently to ensemble disorganization, different across patients in the genetic causes and molecular/structural pathways but converging on a similar functional disorganization of ensembles! And if we find differences in ensemble activations (functional connectivity or coactivity) in a given, e.g., genetic model, our next question may be to work backward

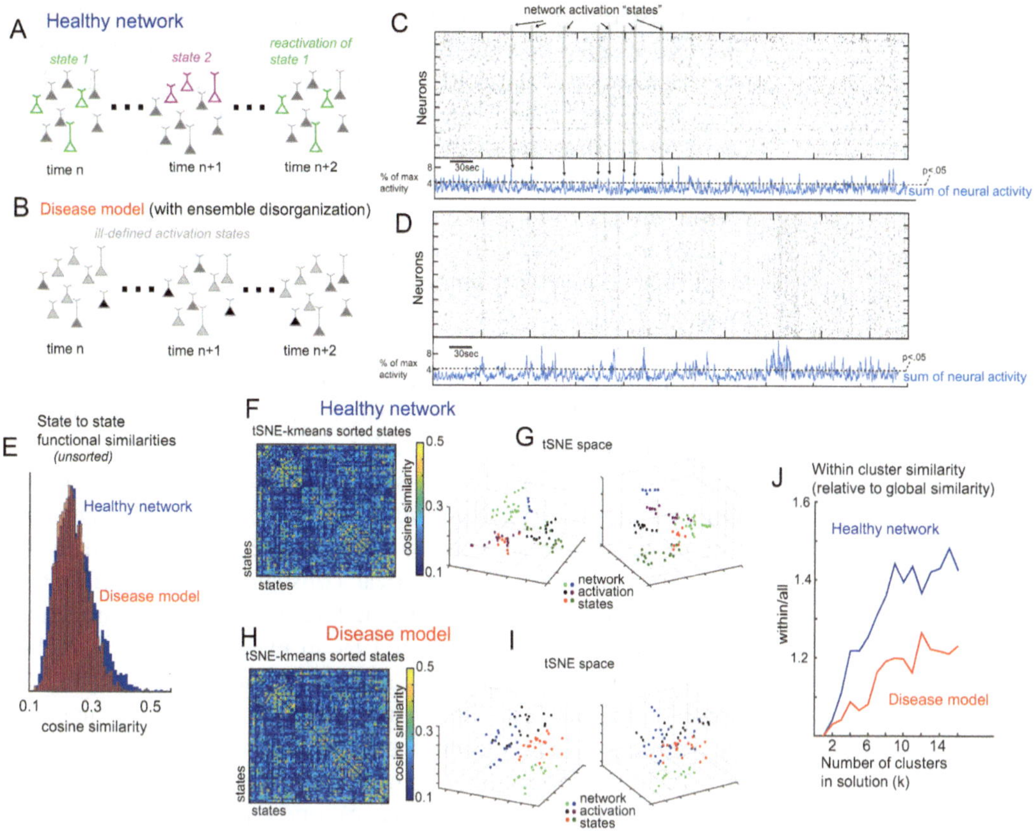

Fig. 2 Time-wise or frame-wise analysis of ensemble organization between subjects. (**a**) Depiction of hypothesis. Schematic of healthy network vs (**b**) a hypothesized disease network. (**c**) Example calcium imaging data raster plot. Bottom: data summed across neurons from each timepoint, relative to the boot-strapped $p < 0.05$ significance line. This was done to identify network activations, i.e., "states" (**d**) same as C, but for a disease model (artificially generated data from C for the purposes of demonstration). (**e**) Histograms of cosine similarities, computed between all network activation states in C and D for each condition. (**f**) Matrix of cosine similarity values between network activation states from C after sorting into clusters via a t-SNE—k-means approach. Similarities were calculated in native space (200 neurons), not t-SNE or PCA space. (**g**) Clusters from analysis in F, viewed in t-SNE space. (**h** and **i**) same as F and G, but for the disease model. (**j**) t-SNE-k-means analysis was carried out with varying numbers of clusters in the solution (i.e., "k") for each mouse. Y-axis reflects cosine similarities calculated between all network activations within-clusters and then averaged and then divided by the average global similarity (among all network activations, regardless of cluster membership). *Note: all data in this figure was generated artificially, or semi-artificially by manipulating real calcium imaging datasets, and is meant for demonstration purposes only*

and figure out why: monosynaptic connections, reduced top-down modulation, and reduced local inhibitory tone. We could then link these structural and/or molecular determinants to various risk genotypes, supporting our convergent ensemble hypothesis.

For quantifying functional connectivity, one obvious metric is Pearson-correlation (Fig. 1c). It is important to eliminate artifactual influences on estimation of correlation coefficients, such as shared instrument noise and the decay times of calcium indicators. The use of spike deconvolution algorithms [29] or conversion to positive first derivatives [30] can circumvent these problems. However, even very clean activity estimates may not be normally distributed, so measures other than Pearson correlation are better suited for estimating shared activity between neurons, such as cosine similarity [11]. The interested reader is referred to a recent review for a thorough discussion of the best practices for analyzing functional connectivity among neuronal networks with fluorescent indictors [31].

To be sure, loss of all connectivity, or maximization of all connectivity, will undoubtedly disrupt ensemble groups (just as an image on a television screen will be lost if all pixels suddenly unrelated to one another [static] or completely linked to one another [all black or all white screen]), but it is important to note *that changes in functional ensembles can occur in the absence of changes to average connectivity.* See example in Fig. 1d. More formally speaking, graph theoretical constructs like *modularity* or *clustering coefficients* get at these tertiary levels of connectivity. The interested reader is pointed to an excellent review on these algorithms as they are applied to neural or brain networks [32]. For example, one may estimate clustering coefficient, which quantifies the degree to which a given network contains clustered connectivity, which would resemble neuronal "ensembles." Starting with a connectivity matrix (e.g., Fig. 1a, c), one estimates the probability of A connecting to C if (1) A connects to B and (2) C connects to B. In our example, the "healthy" networks would contain very strong clustering (Fig. 1e) while the disease network would not, despite both having the same average amount of connectivity (Fig. 1c, d). An alternative approach, based on reference [2], is simply to estimate what value of correlation/similarity reflects "functionally connected" vs not connected and then quantify the proportion of cell pairs in a given mouse that exceed this number (Fig. 1f).

In either case, the reader is cautioned that parametric estimates of correlation significance may not apply to all datasets. That is, it is difficult to know whether correlations of $r = 0.25$ or $r = 0.30$ are strong, weak, or different from one another unless one knows the underlying null distribution of correlation coefficients for a given cell pair. Given variation in baseline activity, noise levels, sampling rate, or fluorescent indictor dynamics across cells, mice, or brain regions, what constitutes a significant correlation between cells may need to be ascertained on a case-by-case basis. One way of doing this is to shuffle cell activation time courses in time or in space (between neurons) within each recording and then to estimate the "null distribution" of functional connectivity for each pair, brain region, or experiment (Fig. 1d) [1, 2, 7, 11, 31, 33]. Another point to consider: if one is analyzing functional connectivity between neurons, it is known that functional connectivity drops off as a function of distance between cell bodies [1]. So, cell-cell functional connectivity may be best analyzed in that context as well.

Frame-wise analyses (activity patterns): Analysis of cell-cell functional connectivity, even with the use of graph theory metrics that estimate clustering and other forms of secondary and tertiary connectivity in networks, may nevertheless miss larger nested dynamics. It may be that ongoing background activity in dense, recurrently connected networks, like the mammalian neocortex, is sparse and only involves co-activations among a few local neurons while occasional population activity bursts—suggestive of underlying "attractor" states of the network—reveal ensemble patterns of activity [7]. In fact, this is exactly what the in vitro [7, 34] and in vivo [2, 6, 11] literature demonstrates. That is, the cortex exhibits bursts of *network wide activations*, separated by relatively quiescent periods, that basic cell-wise metrics (functional connectivity, clustering coefficient, etc.) will miss, as most timepoints/observation that go into such calculations involve the quiescent, sparsely active interstate intervals. The construct of interest (changes in ensemble activation patterns) will play a minimal role in influencing, e.g., the clustering coefficient, relative to the sparse background flux (which may make up over 90% of the recorded data).

Thus, another manner of testing our hypothesis—that a mutation in gene X disorganizes neural ensembles in the brain—is to focus on the activation patterns in these network activations, i.e., the "ensemble activations." To describe one metric/approach that gets at this hypothesis, we provide an example pipeline. Figure 2 depicts an example analysis of how to test such a hypothesis using calcium imaging data from a resting but awake animal. In Fig. 2a, we provide a visual depiction of our hypothesis, tweaked to fit a "frame-wise" paradigm. We see a schematic of a putative healthy network of neurons on the top with well-defined, recurring population-wide activity patterns made up of coactive neuronal ensembles (green and magenta). Compare this to Fig. 2b, where we depict a hypothesized disease network, with less defined ensemble groups, with more randomly distributed functional activity patterns across time. Notably, we do not propose that the disease effects more activity or less activity than in our wildtype, but we specifically propose that the network activation patterns—putative attractor states or ensemble activations—are not distinct from one another and also do not reliably reactivate across time. And again, because we cannot directly record from individuals with disease Y, we focus on testing whether a causal factor for disease Y—variations affecting gene X—alters brain networks by disorganizing ensemble activity.

We need to first identify the network activations in our neural timeseries datasets and then test whether the patterns of neural activity within these network activations—these sudden burst states—exhibit some structure over time. What constitutes as typical "structure" likely depends on the brain region, the size of the network, behavioral state, and timespan one is analyzing. Here we focus on layer 2/3 of visual cortex, where our past work has demonstrated that one typically finds between 3 and 7 spontaneously reactivating neural activity patterns in an awake mouse [1, 6]. The goal with this analysis will be to identify these activations systematically in each dataset/mouse and to determine their number, distinctness, and, most importantly, their reliability over time. To be sure, the biological significance of reduced number, distinctness, and reliability of ensemble activations is hypothetical. Whether and how it relates to the alterations in perception, cognition, and thought seen in neuropsychiatric disorders remains to be demonstrated.

In Fig. 2c, we present some example calcium imaging data in a raster plot, representative of the theoretical "healthy" networks in A. The dataset presented here was from a resting mouse with no explicit task or sensory stimulus, recorded at a 10 Hz frame rate with GCaMP6s. Each row represents the first derivative of the halo-subtracted fluorescence trace for each neuron (delta-F). Only the rises of calcium transients are used, and all are scaled within each neuron as a function of its maximum delta-F value [2]. To identify network wide activations (putative "ensemble activations"), activity is summed across neurons and "peak" frames are identified (see line plot below the raster plot [6, 11]). Then, for each mouse, the same dataset is used to determine which of these "peak" activation frames contain more activity than one would expect due to chance. The same dataset is shuffled over 1000 times, with each neuron activity trace moving a distinct random shift in time for each shuffle step, and activity peaks are calculated and saved into a null distribution. This approach maintains each neuron's average activity while shuffling its time course relative to the other neurons, determining what the expected coactivity probability would be if these neurons were independent, e.g., if they were all recorded in different mice. The 95th percentile of this null distribution is used to determine the $p < 0.05$ cutoff for subsequently analyzed "network activations." In Fig. 2c, this cutoff is depicted as the dotted line. The same analysis is carried out on the data in each mouse individually (as it has been done for the disease model mouse in Fig. 2d) to be sure that baseline activity rates of individual neurons do not bias the selection/identification of network-wide activations.

At this point, it is critical to more directly check two features: differences in frame-wise activity levels and differences in cell-wise activity levels. First, is the frequency of activation significantly different between groups? If so, this alone may be an important point to report. Perhaps, neurons simply are not coactivating in groups in the disease model. This would constitute an important finding! And it would not be unexpected in a brain system with reduced cortical synapses [35]. However, if there are very few actual ensemble activations present, this makes the subsequent steps in this analysis (i.e., k-means or t-SNE) challenging and potentially problematic. At the very least, it is recommended that one "yoke" the number of health activation frames to the number of disease activation frames, or vice versa. That is, if one identifies 100 network activations in the healthy dataset and only 20 in the disease dataset, identifying more than 2 or 3 clusters of network activations in the disease dataset may not be feasible. However, if one decides to still analyze the content of network activations (see below), it is advisable that the healthy dataset be limited to 20 frames (of the 100) that are size/distribution/activity-level matched to the disease dataset (i.e., match the peak sizes in the line plots below Fig. 2c, d).

Another point: if the average size of network activations (i.e., number of neurons or overall summed network activity) is noticeably different, then this could be interpretable as well. Perhaps, disynaptic inhibition or feed-forward inhibition in one cortical network is reduced [36, 37], leading to over-activation and poorly distinguished ensembles. Whether subsequent analysis of network activation similarity or k-means is needed or interpretable remains to be determined, as over-activation is a simpler explanation than "ensemble disorganization." Also, in cases of very little activity or high levels of activity (>50% of max) across the network, similarity metrics become saturated and difficult to interpret. A viable alternative, again, is to yoke one condition against the other: only analyze network activations greater than the threshold, but less than the max of the other group, e.g., 15% overall activation. This way, one could both observe (a) over-activation or under-activation in network activity and (b) also disorganized activity patterns during moderate network activations.

Next, one should also take care to ensure that the overall neural activity levels (i.e., within individual cells) are not drastically different between groups. For example, if one decides to compare ensembles between anesthetized and awake animal, one notices immediately that these brain states are not comparable. Studying ensemble activation patterns when "ensemble activations" (i.e. more than one neuron active at a time) do not exist is ill-posed. Again, a good strategy is to limit one dataset to the properties of another and, therefore, analyze apples to apples (i.e., remove high and/or low activity frames, or exclude highly active neurons to match distributions). And again, reduced or augmented spontaneous activity levels alone is an important feature to report for your disease model of choice, regardless of the status of neuronal ensembles.

For the sake of our example in Fig. 2, the frequency of network activations and overall neuronal activity levels are not significantly different between conditions. In Fig. 2e, we have computed functional similarities between all identified network activation frames from C to D (all the above threshold "peaks"), with neurons as the variables and timeframes as the observation using a modified cosine similarity index [1, 2, 31]. Notably, the mean and median between the distributions is not noticeably different, but the disease model has fewer frame pairs in the tails. Reporting this difference in the tails, or bootstrapping a statistically significant similarity value for each mouse and quantifying the proportion of network state "reactivations," is one strategy for establishing a disruption in neural ensemble patterns [2].

However, just comparing the overall distribution of frame-wise similarities may miss the fact that high and low similarities are only present among and between well-defined activation states. One approach [1] suggests first identifying, in each dataset, the putative unique activation patterns. To do this, we provide an example. One can carry out a three-dimensional t-distributed Stochastic Neighbor Embedding (t-SNE) conversion on the neurons by frames dataset to nonlinearly reduce the $\approx$200 neurons $\times$80 frames (i.e., 80 network activations, from Fig. 2c) to a three-dimensional t-SNE space (3 $\times$ 80) to optimally reveal "clustering" among the 80 frames. This aids in the identification of network activations with similar activity patterns across neurons. This step is followed by k-means clustering analysis with varying values for "k" to identify the optimal number of clusters. In Fig. 2g, we have visualized the results by clustering network-activation frames via k-means clustering with $k = 6$ in the t-SNE space. It is very important to note that the distances in this tSNE space are nonlinear and not easily interpreted. Euclidean distances within and among clusters in this space should not be used to compare between groups, as the parameters of the t-SNE algorithm are chosen to optimize cluster separation (e.g., perplexity =3), but not to allow for quantification. Once clusters are identified, we return to the full 200 neuron by 80 frame dataset and compute similarities within and between states with respect to the t-SNE/k-means identified clusters. In order to avoid bias in this step analysis, we can analyze within cluster similarities at a number of values of "k" (i.e., clusters in the solution) and then normalize these similarities as a function of the overall mean similarity across all states (i.e., in Fig. 2e). Then, from these curves, we can compare between groups using parametric statistics such as a mixed ANOVA with "k" as a within-subject factor, model (healthy vs disease) as the between subjects factor, and within-divided-by-global similarity as the dependent variable. Notably, the t-SNE step may produce different results with subsequent runs (as is the case with the matlab algorithm). In order to address this, it is advisable to repeat the steps in Fig. 2f–j 1000 times, creating a final distribution of curves for each mouse.

How does a reduction in this curve, as seen in Fig. 2j and also reference [1], relate to reduced number, distinctness, or reliability of ensembles in the cortex? A drop in within-cluster similarity suggests all three, as (a) if ensembles are not reliable (they do not repeat with fidelity over time) or (b) if ensembles are not distinct (different activation patterns are not sufficiently different from one another; e.g. a cloud like in Fig. 2i), then it is less warranted to call any of the network activations "ensembles" at all. The number of ensembles might be operationally defined as zero (or one: the cortex is one big, flexible ensemble). That said, it may be that the disease model has more ensemble activation patterns than the healthy comparison mouse—that each of the individual points in Fig. 2i would come to represent its own cluster as enough data is collected, and pattern repeats are observed. To test this, much longer recordings should reveal that, at values of k higher than 16 (see Fig. 2j), there is sudden a peak. We find this unlikely in our networks of 200 neurons given our past recordings, but if true, it is still a valuable finding. And, that said, exactly how *more ensembles* relate to a given disease pathway or phenomenology is not a simpler or more difficult problem to tackle than fewer/less distinct.

2.3 Controlling for Changes at Other Levels of Analysis

One way to think about challenge of testing this hypothesis is that because genes, molecules, synapses, cell functions, ensembles, circuits, networks, and behavior are all levels of analysis in the same overall system, any change fundamentally affecting one level (let us say, removing a molecule) will have impacts at all levels. Normalizing for neuronal activity levels is absolutely necessary. It could simply be the case that, in attempting to identify neuronal ensemble patterns, a given mouse simply is less or more "awake," or has weaker cholinergic tone, than another. One might conclude disordered neuronal ensembles, but if one were to focus on periods of equal activity levels (as described above, with a "yoking" approach), one would see that the patterned activity itself is not lost. In some cases, a decrease in neuronal activity could occur in the absence of a deficit in ensemble connectivity, and in others, it could co-occur with a deficit in ensemble activity. For example, a recent study of an Scn2a insufficiency mouse thoughtfully demonstrated both—a dearth of spontaneous activity in individual neurons as well as a reduction in paired activity in neuronal ensembles [5].

How else might one "normalize" for these variables? One way is to only select ensemble activations that have comparable numbers of neurons active. Another is to analyze ensemble complexity (i.e., numbers of ensembles, within/between similarity, etc.—as above) as a function of number active neurons, creating a function/curve for each group.

Another important variable to account for is locomotion and behavioral state. As has been well demonstrated in the mouse, small behavioral movements and body position (e.g., whisking, paw position, nose movements) not only affect but dominate cortical activity across time, trials, and states [28, 38, 39]. If one mouse group shows less or more motor flux, then this could account for differences in frequency, size, or consistency of ensemble activation. Again, recording activity for long enough and analyzing brain activity as a function of these variables (e.g., ensemble activation size and diversity at different locomotion speeds, or during vs between locomotion bouts) will help sort out whether these effects are correlated, and if ensemble disorganization occurs across bio-behavioral states in both the model and the comparison group.

3 Conclusion and Summary

The ability to compare multicellular activation patterns between two separate neural populations is paramount for testing systems-level hypotheses of major neuropsychiatric or neurological diseases. Although short-term manipulations, like pharmacological challenges or optogenetic lesions, can enable within-subjects designs that assess specific relationships between disease-relevant molecular (e.g., NMDA-receptors) or cellular functions and ensemble-level dynamics, many neuropsychiatric disorders are developmental in nature, and certain aspects of the disease can only be modeled sufficiently with manipulations that affect all or parts of the animals lifespan. The use of one wildtype group and one with a genetic mutation or physiological lesion enables such modeling, but it is important not to think of the lesion mouse as a "model of complex disease Y." Instead, it is a tool for understanding how, e.g., a particular disease-relevant gene impacts disease-relevant brain networks.

A challenge is how can one analyze ensembles in one group if the cells are not the same? The literature presents us with several solutions for this: counting the number of ensemble states, comparing within-ensemble similarities, testing for dimensionality, comparing correlation coefficients *en masse*, clustering coefficients, etc. We provide one example, which lands on quantifiable metric—within cluster similarity curves—that has proven relevant in the past. However, regardless of the choice of algorithm or metric, it remains important to analyze these metrics with respect to the animal's behavioral and motor state, as well as sex, age, and time of day.

References

1. Hamm JP, Shymkiv Y, Mukai J, Gogos JA, Yuste R (2020) Aberrant cortical ensembles and schizophrenia-like sensory phenotypes in Setd1a+/− mice. Biol Psychiatry 88:215. https://doi.org/10.1016/j.biopsych.2020.01.004
2. Hamm JP, Peterka DS, Gogos JA, Yuste R (2017) Altered cortical ensembles in mouse models of schizophrenia. Neuron 94(1): 153–167.e8. https://doi.org/10.1016/j.neuron.2017.03.019
3. Adams RA, Napier G, Roiser JP, Mathys C, Gilleen J (2018) Attractor-like dynamics in

belief updating in schizophrenia. J Neurosci 38(44):9471–9485. https://doi.org/10.1523/JNEUROSCI.3163-17.2018

4. Krystal JH, Anticevic A, Yang GJ et al (2017) Impaired tuning of neural ensembles and the pathophysiology of schizophrenia: a translational and computational neuroscience perspective. Biol Psychiatry 81(10):874–885. https://doi.org/10.1016/j.biopsych.2017.01.004
5. Li M, Eltabbal M, Tran HD, Kuhn B (2023) Scn2a insufficiency alters spontaneous neuronal Ca2+ activity in somatosensory cortex during wakefulness. iScience 26(11):108138. https://doi.org/10.1016/j.isci.2023.108138
6. Miller JEK, Ayzenshtat I, Carrillo-Reid L, Yuste R (2014) Visual stimuli recruit intrinsically generated cortical ensembles. Proc Natl Acad Sci 111(38):E4053–E4061. https://doi.org/10.1073/pnas.1406077111
7. Cossart R, Aronov D, Yuste R (2003) Attractor dynamics of network UP states in the neocortex. Nature 423(6937):283–288. https://doi.org/10.1038/nature01614
8. Rolls ET, Loh M, Deco G, Winterer G (2008) Computational models of schizophrenia and dopamine modulation in the prefrontal cortex. Nat Rev Neurosci 9(9):696–709. https://doi.org/10.1038/nrn2462
9. Yuste R (2015) From the neuron doctrine to neural networks. Nat Rev Neurosci 16(8):487–497. https://doi.org/10.1038/nrn3962
10. Carillo-Reid L, Yuste R, Carillo-Reid L, Yuste R (2020) What is a neuronal ensemble? In: Oxford research encyclopedia of neuroscience. Oxford University Press. https://doi.org/10.1093/acrefore/9780190264086.013.298
11. Carrillo-Reid L, Miller JEK, Hamm JP, Jackson J, Yuste R (2015) Endogenous sequential cortical activity evoked by visual stimuli. J Neurosci 35(23):8813–8828
12. Jauhar S, Johnstone M, McKenna PJ (2022) Schizophrenia. Lancet 399(10323):473–486. https://doi.org/10.1016/S0140-6736(21)01730-X
13. Carrillo-Reid L, Yang W, Kang Miller JE, Peterka DS, Yuste R (2017) Imaging and optically manipulating neuronal ensembles. Annu Rev Biophys 46(1):271–293. https://doi.org/10.1146/annurev-biophys-070816-033647
14. Wang XJ, Krystal JH (2014) Computational psychiatry. Neuron 84(3):638–654. https://doi.org/10.1016/j.neuron.2014.10.018
15. Morris SE, Sanislow CA, Pacheco J, Vaidyanathan U, Gordon JA, Cuthbert BN (2022) Revisiting the seven pillars of RDoC. BMC Med 20(1):220. https://doi.org/10.1186/s12916-022-02414-0
16. NOT-MH-19-053: Notice of NIMHs considerations regarding the use of animal neurobehavioral approaches in basic and pre-clinical studies. https://grants.nih.gov/grants/guide/notice-files/NOT-MH-19-053.html. Accessed 24 Oct 2023
17. Gordon JA (2010) Testing the glutamate hypothesis of schizophrenia. Nat Neurosci 13(1):2–4. https://doi.org/10.1038/nn0110-2
18. Selkoe DJ, Hardy J (2016) The amyloid hypothesis of Alzheimer's disease at 25 years. EMBO Mol Med 8(6):595–608. https://doi.org/10.15252/emmm.201606210
19. Herrup K (2023) How not to study a disease. The MIT Press
20. Gordon JA, Moore H (2012) Charting a course toward an understanding of schizophrenia. Neuron 76(3):465–467. https://doi.org/10.1016/j.neuron.2012.10.023
21. Carrillo-Reid L, Han S, Yang W, Akrouh A, Yuste R (2019) Controlling visually guided behavior by holographic recalling of cortical ensembles. Cell 178(2):447–457.e5. https://doi.org/10.1016/j.cell.2019.05.045
22. Seshadri S, Klaus A, Winkowski DE, Kanold PO, Plenz D (2018) Altered avalanche dynamics in a developmental NMDAR hypofunction model of cognitive impairment. Transl Psychiatry 8(1):3. https://doi.org/10.1038/s41398-017-0060-z
23. Liang B, Zhang L, Barbera G et al (2018) Distinct and dynamic ON and OFF neural ensembles in the prefrontal cortex code social exploration. Neuron 100:700. https://doi.org/10.1016/J.NEURON.2018.08.043
24. Whitford TJ, Grieve SM, Farrow TFD et al (2007) Volumetric white matter abnormalities in first-episode schizophrenia: a longitudinal, tensor-based morphometry study. Am J Psychiatry 164(7):1082–1089. https://doi.org/10.1176/appi.ajp.164.7.1082
25. Jiang X, Shen S, Cadwell CR et al (2015) Principles of connectivity among morphologically defined cell types in adult neocortex. Science 350(6264):aac9462-1-9. https://doi.org/10.1126/science.aac9462
26. Karnani MM, Jackson J, Ayzenshtat I et al (2016) Opening holes in the blanket of inhibition: localized lateral disinhibition by VIP interneurons. J Neurosci 36(12):3471–3480. https://doi.org/10.1523/JNEUROSCI.3646-15.2016
27. Gal E, London M, Globerson A et al (2017) Rich cell-type-specific network topology in

neocortical microcircuitry. Nat Neurosci 20(7):1004–1013. https://doi.org/10.1038/nn.4576

28. Stringer C, Pachitariu M, Steinmetz N, Reddy CB, Carandini M, Harris KD (2019) Spontaneous behaviors drive multidimensional, brainwide activity. Science 364(6437):255. https://doi.org/10.1126/science.aav7893
29. Pnevmatikakis EA, Soudry D, Gao Y et al (2016) Simultaneous Denoising, deconvolution, and demixing of calcium imaging data. Neuron 89(2):285. https://doi.org/10.1016/j.neuron.2015.11.037
30. Hamm JP, Yuste R (2016) Somatostatin interneurons control a key component of mismatch negativity in mouse visual cortex. Cell Rep 16(3):597–604. https://doi.org/10.1016/j.celrep.2016.06.037
31. Wenzel M, Hamm J (2020) Identification and quantification of neuronal ensembles in optical imaging experiments. J Neurosci Methods 351:109046. https://doi.org/10.1016/j.jneumeth.2020.109046
32. Bullmore E, Sporns O (2009) Complex brain networks: graph theoretical analysis of structural and functional systems. Nat Rev Neurosci 10(3):186–198. https://doi.org/10.1038/nrn2575
33. Ikegaya Y, Aaron G, Cossart R et al (2004) Synfire chains and cortical songs: temporal modules of cortical activity. Science (New York, NY) 304(5670):559–564. https://doi.org/10.1126/science.1093173
34. Luczak A, Barthó P, Harris KD (2009) Spontaneous events outline the realm of possible sensory responses in neocortical populations. Neuron 62(3):413–425. https://doi.org/10.1016/j.neuron.2009.03.014
35. Hall J, Trent S, Thomas KL, O'Donovan MC, Owen MJ (2015) Genetic risk for schizophrenia: convergence on synaptic pathways involved in plasticity. Biol Psychiatry 77(1):52–58. https://doi.org/10.1016/j.biopsych.2014.07.011
36. Lewis DA (2014) Inhibitory neurons in human cortical circuits: substrate for cognitive dysfunction in schizophrenia. Curr Opin Neurobiol 26:22–26. https://doi.org/10.1016/j.conb.2013.11.003
37. Van Derveer AB, Bastos G, Ferrell AD et al (2020) A role for somatostatin-positive interneurons in neuro-oscillatory and information processing deficits in schizophrenia. Schizophr Bull 47:1385. https://doi.org/10.1093/schbul/sbaa184
38. Musall S, Kaufman MT, Juavinett AL, Gluf S, Churchland AK (2019) Single-trial neural dynamics are dominated by richly varied movements. Nat Neurosci 22(10):1677–1686. https://doi.org/10.1038/s41593-019-0502-4
39. Bimbard C, Sit TPH, Lebedeva A, Reddy CB, Harris KD, Carandini M (2023) Behavioral origin of sound-evoked activity in mouse visual cortex. Nat Neurosci 26(2):251–258. https://doi.org/10.1038/s41593-022-01227-x

Chapter 11

Tagging and Manipulation of Fear Engrams

Marie-Theres Wittmann, Asim J. Rashid, Sheena A. Josselyn, and Paul W. Frankland

Abstract

The ability to manipulate neurons involved in memory formation and retrieval is a key factor to understanding neuronal circuits involved in cognition. In this chapter, we provide methods to tag and/or allocate neurons to a fear conditioning engram in the hippocampus and manipulate these neurons using optogenetic and chemogenetic approaches. For tagging, we describe IEG-based methods (TRAP2 mice or the viral RAM system) to express transgenes, such as designer receptors exclusively activated by designer drugs (DREADDs) or channelrhodopsin in fear engram cells. For allocation, the excitability of neurons in the CA1 region may be transiently and reversibly heightened using opsins, and heightened excitability increases the chance of the neurons being integrated into the fear conditioning engram. Allocated/tagged neurons then can be subsequently activated or silenced during the memory test using optogenetic or chemogenetic transgenes leading to heightened or decreased freezing, respectively.

Key words Memory engram, Tagging, Allocation, TRAP2, RAM system, Adeno-associated virus (AAV), Optogenetics, DREADDs, Hippocampus, CA1

1 Introduction

The physical representation of memory, i.e., neuronal ensembles, has been termed an engram by Richard Semon [1]. In recent years, considerable progress has been made to understand engrams by providing molecular methods to identify and manipulate engram cells during memory formation, consolidation, and retrieval [2].

Various methods, such as transgenic mouse lines and viral manipulations, are commonly used to tag engrams. Immediate early gene (IEG)-dependent techniques are prevalent. IEG expression is rapidly activated by learning, stress, sensory input, social interactions, seizures, and neurotrophic factors. Therefore, IEG promoters or their derivatives can express exogenous genes in an activity-dependent manner, e.g., the TRAP2 (targeted recombination in active neurons) mouse model [3, 4] or the viral-based RAM (robust activity marking“) system [5].

Luis Carrillo-Reid (ed.), *Identification, Characterization, and Manipulation of Neuronal Ensembles*, Neuromethods, vol. 215, https://doi.org/10.1007/978-1-0716-4208-5_11,

The TRAP2 mouse model uses an in-frame knock-in of a tamoxifen-dependent Cre (2A-iCreER^{T2}) into the c-fos locus, an immediate early gene [3]. Transgene expression in the RAM system is driven by an optimized synthetic activity-regulated promoter generated from elements of fos and Npas4 IEGs under a TET-system that is sensitive to doxycycline [5].

To manipulate engram neurons, typically chemogenetic elements (DREADD receptors) can be used such as hM3dq (excitatory) or hM4di (inhibitory) [6], or optogenetic elements such as channelrhodopsin2 (ChR2, excitatory) or halorhodopsin (NpHR, inhibitory) [7].

Another useful approach is to direct a memory to specific neurons via optogenetic allocation [8–10]. This method employs virus-mediated expression of a construct containing two genes. The first gene encodes an activating opsin, facilitating the selective allocation of opsin-expressing neurons in targeted brain regions through light stimulation. The second gene can either be a second opsin or a DREADD, which is used to manipulate the neurons that have been allocated. For instance, the NpACY construct encodes two opsin, the ChR2(H134R) variant (sensitive to blue light) and the NpHR3.0 variant (sensitive to red light). This construct enables the specific allocation of memories through blue light to opsin-expressing neurons. Subsequently, during a memory test, the same neurons can be inhibited using red light [8–10].

Certainly, combining the mentioned approaches offers a significant advantage. It allows for the tagging and differential manipulation, both activation and inhibition, of engrams for memory encoding or retrieval at different time points within the same mouse. This feature is key for studying how memory encoding and storage evolve over time. For instance, as memories undergo reorganization through systems consolidation and/or neuronal drift, there are changes in how engrams are organized. By differentially tagging and manipulating encoding and retrieval engrams, researchers can directly investigate these cellular-level changes, providing valuable insights into the dynamic nature of memory processes.

In this chapter, we describe the methods to (1) allocate and manipulate a fear memory in the CA1 region of the hippocampus using an adeno-associated virus (AAV)-mediated approach and (2) tag a fear memory using TRAP2 mice and/or the RAM system. Advantages and disadvantages of the different tagging systems and transgenes used are summarized in Tables 1 and 2, respectively.

As procedures for viral infusion into specific regions of the brain and insertion of optrodes for optogenetic manipulation [7] and details of the RAM system have been described in detail elsewhere [5], only a brief overview will be provided in the Methods section along with any critical factors to consider.

Table 1
Advantage and disadvantage of the different tagging systems

Engram tagging system	Advantages	Disadvantages
TRAP2 mice	Tagging of endogenous activity	CreERT2 expression may not be constant in all brain regions
	Tagging within a relatively short window (<6 h)	Cost of 4-hydroxytamoxifen
DeNardo et al. [4]	Flexibility in which Cre-dependent transgenes to express virally	
	Transgene expression is long lasting (>6 weeks)	
	No disruption of endogenous fos expression unlike original fosTRAP mice [11]	
RAM system	Tagging of endogenous activity	Possible background expression during longer dox-off condition
Sorensen et al. [5]	Can use any mouse line	Expression lasts for ~2 weeks
	Flexibility in which transgenes to virally express	
	Can complement other tagging systems (e.g., TRAP2)	
Optogenetic allocation	Can use different viral promoters for cell-specific expression	"Artificial" method
	Tagging can be done hours before memory encoding	
	Every tagged neuron can be subsequently excited or inhibited	
	Transgene expression is long lasting (>6 weeks)	

2 Materials

The catalog numbers and/or suppliers indicated for the equipment below are those which we have found to be reliable. However, in most cases equivalent items from other suppliers may be used and catalog numbers have not been included.

2.1 Mice

TRAP2 mice.

Fostm2.1(icre/ERT2)Luo/J Strain 030323 (Jackson Labs).

WT mice.

C57Bl/6 J Strain 000664 (Jackson Labs).

Table 2
Advantage and disadvantage of DREADDs vs. optogenetics

Transgene	Advantages	Disadvantages
DREADDS	Long-lasting neuronal activity modulation (several hours after injection, days for water administration)	Cost of administrated drug
	Broad tissue manipulation (all cells expressing DREADDS are targeted)	Administration of drug before experiment
		No temporal control
		Systemic administration of the drug
Optogenetics	Precise temporal control	Chronic fiber optic implant
	Precise regional control (axonal or dendritic excitation/ inhibition is possible)	Limited behavior due to need of light source cables
	Precise stimulation control	Short-term manipulations

2.2 Chemicals

z-4-hydroxytamoxifen (Toronto Research Chemicals, Inc. Cat# H954745),

Anhydrous ethanol.

Cremophor EL (Sigma-Aldrich Cat# C5135-500G).

Mouse diet containing doxycycline (40 mg/kg) (Bioserv cat#S4519).

C21 (DREADD Agonist 21) (Tocris Biochemicals, Cat# 5548).

2.3 Plasmids Used to Make Viruses

AAV-CaMKII-NpACY (available upon request from Dr. Karl Deisseroth, Stanford University).

AAV-nEF-C_{ON}/F_{ON}-iC++-EYFP (Addgene 137155).

pAAV-RAM-NpHR3.0-iRFP (Addgene 47631),

AAV-RAM-hM4di-mCherry (available upon request from the authors).

2.4 Fear Conditioning (Training and Testing)

NIR Fear Conditioning System for Mouse (Cat# MED-VFC-NIR-M, Med Associates, Inc.)

Video Fear Conditioning Interface Package (Cat# MED-SYST-VFC, Med Associates, Inc.)

2.5 Optogenetics

Blue (473 nm) laser (Laserglow Technologies, cat# LRS-0473series).

Red (655 nm) laser (Laserglow Technologies cat# LRD-0655 series).

Waveform generator (Agilent Technologies 33500B Series).

Mono Fiber-optic patch cords (Doric Lenses).

Mono Fiber-optic Cannula (Doric Lenses).

Digital Optical Power and Energy Meter (Thorlabs cat#PM100D).

3 Methods

3.1 Optogenetic Allocation and Manipulation of a Contextual Fear Memory Engram in the CA1 of WT Mice (Fig. 1)

As described in the introduction, optogenetic allocation uses a construct containing an activating opsin and a second opsin or DREADD. In this procedure, we use the NpACY construct that expresses activating ChR2(H134R) and inhibiting NpHR3.0-eYFP. First, A bilateral viral injections of AAV(DJ)-CaMKIIa-NpACY are performed in CA1 neurons of the mouse hippocampus. After 2–3 weeks to enable viral transgene expression, mice are exposed to blue light stimulation for 30 s just before subjecting them to fear conditioning (Fig. 1). This results in selective allocation to opsin-expressing neurons, which can be validated by performing a fear memory test with and without activation of NpHR by red light illumination.

1. Approximately 2–3 weeks after viral infusions (*see* **Notes 1–3**), habituate the mice to the optopatch cables with 2 min sessions, once a day, for 3 days.
2. On the day of allocation and fear conditioning, set up the optogenetic equipment and turn on the blue laser. Allow 15–20 min for the laser to warm up before setting the output to 10 mW, as measured using the optical power meter. Using the waveform generator, adjust the stimulation parameters as follows (*see* **Note 4**):

   473 nm square pulse

   Frequency: 4 Hz.

   Amplitude: 5 Vpp (pp: peak to peak).

   Duty cycle: 6%.
3. Attach optopatch cables to the mouse and place it in an empty cage. Stimulate with blue light for 30 s. Control mice have cables attached but with no laser turned on.
4. Place the mouse in the conditioning chamber to begin the fear conditioning protocol (Note—remove the cables or not). A typical fear conditioning session can be a 5 minute protocol with 3 × 0.5 mA foot-shocks at 120, 180, and 240 s (*see* **Note 5**).
5. Return the mouse to its home cage and clean the chamber before repeating the protocol with the next mouse.
6. For contextual fear testing the following day (*see* **Note 6**), attach the patch cables and place the mouse in a clean conditioning chamber which is in an identical state to when conditioning was performed. During a 5 min test session, red light

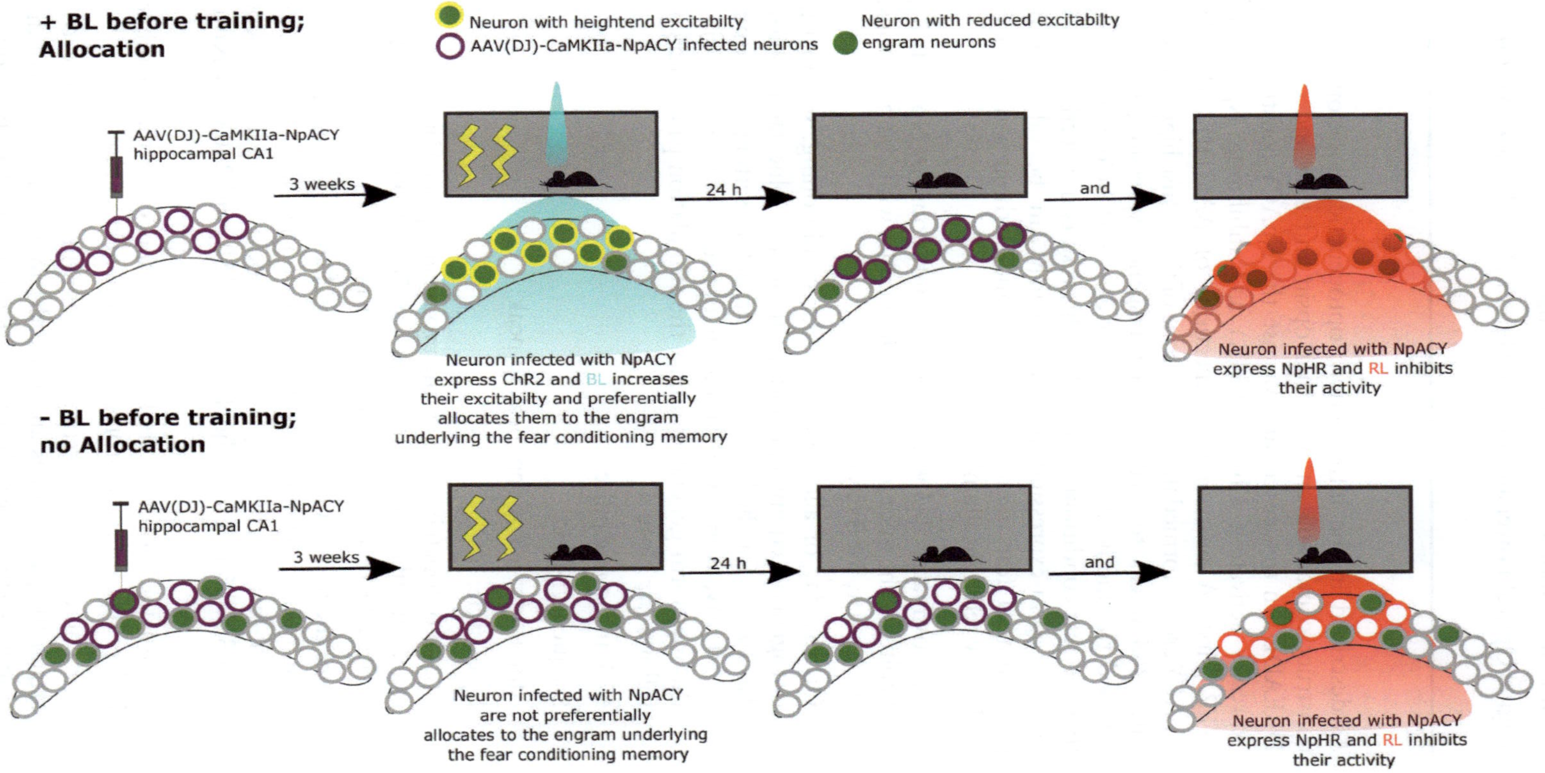

Fig. 1 Scheme of optogenetic allocation and manipulation of a contextual fear memory engram in the CA1 of WT mice. Mice are stereotactically injected with AAV (DJ)-CamKIIa-NpACY into the CA1 region of the hippocampus. After 3 weeks, experimental mice are subjected to blue light right before fear conditioning leading to an increased allocation of transduced cells into the fear memory engram. 24 h after mice are subjected to a fear conditioning test with red light being turned on in the middle of the test. Red light inhibits the activity of transduced cells and thereby the memory, mice freeze less. Control mice are not subjected to blue light before fear conditioning, and therefore, random cells with heightened excitability are allocated to the fear memory engram. Use of red light with these mice leads to no inhibition of the fear conditioning memory [see an example in [9] for the lateral amygdala]

illumination (633 nm) (7–10 mW square pulse with no frequency, amplitude, or duty cycle adjustments) can be activated during the third minute to inhibit opsin-expressing neurons. Success in allocation is manifested as decreased freezing during red light illumination when compared to freezing during the minute before and the minute after (*see* **Note 7**).

3.2 TRAPing and Manipulation of the Contextual Fear Engram (Fig. 2)

In the TRAP2 system, engram neurons are tagged using transgenes such as tdTomato, opsins, or DREADDs, which are expressed after Cre-dependent recombination. To achieve this, TRAP2 mice are crossbred with a reporter mouse model like tdTomato (Ai14 B6. Cg-Gt(ROSA)26Sortm14(CAG-tdTomato)Hze/j) or injected with a virus containing FLOX sites surrounding the transgene. Activity triggers Cre-ERT2 expression in TRAP2 mice, and Cre activity is induced through the injection of tamoxifen or 4-hydroxytamoxifen (4OHT). In this example, we utilized two AAV viruses, one containing NpACY and another with DIO-Cherry (AAV(DJ)-CaMKIIa-NpACY, AAV(DJ)-EF1a-DIO-mCherry). The protocol outlined below specifically pertains to the use of 4OHT (*see* **Note 8**).

3.2.1 Preparation of 4-Hydroxytamoxifen

1. To make a working solution, weigh out and resuspend 4OHT powder in 100% ethanol at a concentration of 40 mg/mL.
2. Heat the solution for 10 min at 50 °C and then vortex vigorously to dissolve the 4OHT.
3. Add an equal volume of cremophor and vortex the solution vigorously until homogeneous (see **Note 9**).
4. For injection, dilute the 4OHT solution 1:1 in saline, giving a 10 mg/mL concentration.

3.2.2 Fear Conditioning, Tagging, and Manipulation of the Encoding Engram

1. Mice are subjected to contextual fear conditioning as described in Subheading 3.1, including habituation to the optopatch cables. Depending on the fear conditioning system and number of chambers available, multiple animals can be conditioned at the same time.
2. After conditioning, place the mice back in their home cage and keep them in a quiet room until ready for injection.
3. Typically, 4OHT injection (50 mg/kg) is performed within 30 min of behavioral training or testing. As the solution is viscous, perform the injection slowly.
4. For optogenetic inhibition of engram activity during retrieval, attach the optopatch cords and subject mice to a contextual fear test 72 h after 4OHT injection as described in Subheading 3.1 (*see* **Note 10**). During the 5 min test session, turn on the red laser red light illumination (633 nm) (7–10 mW square pulse with no frequency, amplitude, or duty cycle adjustments) during the third minute to inhibit opsin-expressing neurons.

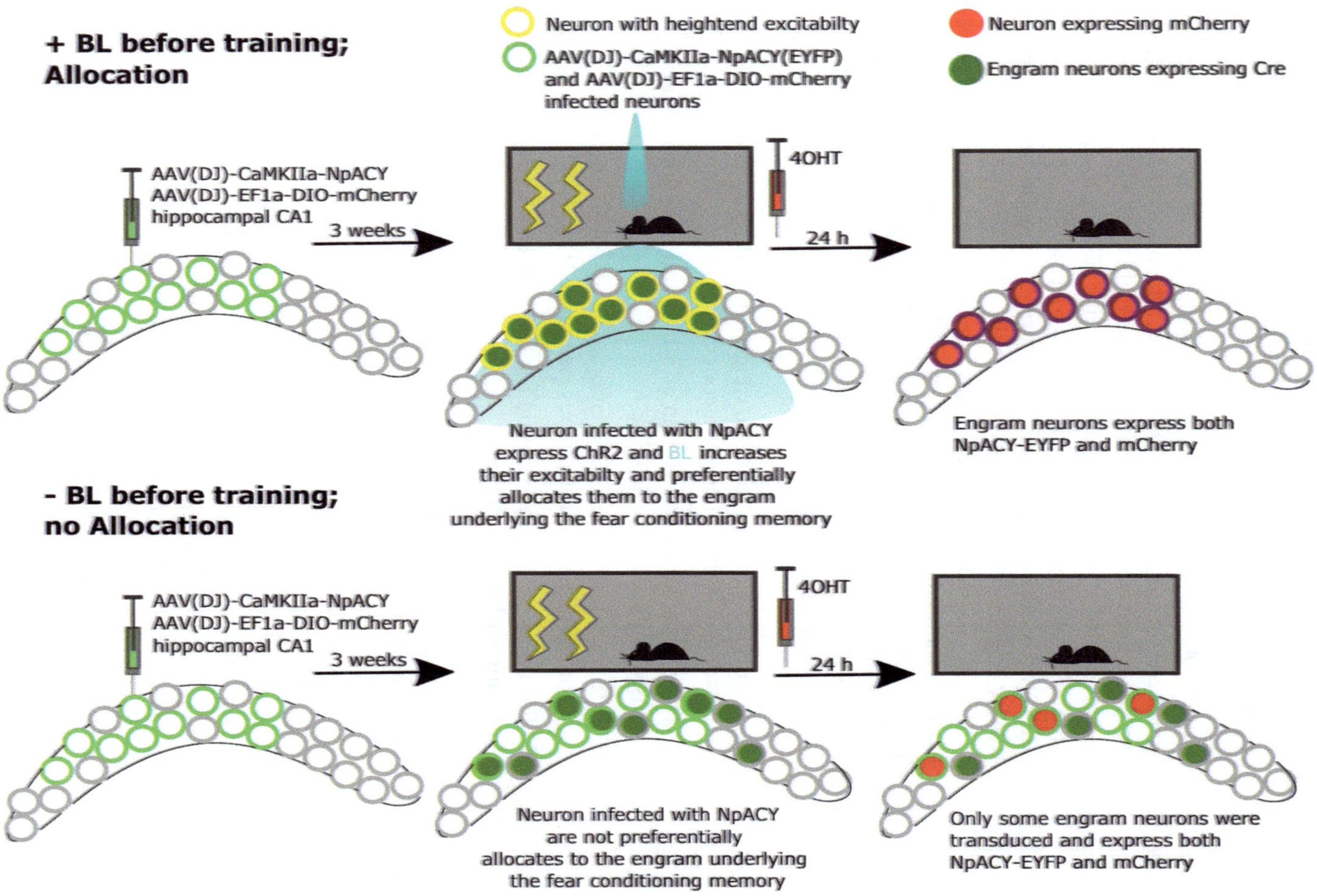

Fig. 2 Scheme of optogenetic allocation in the dorsal hippocampus combined with activity-dependent labeling in TRAP2 mice. The CA1 of mice is injected with AAV-viruses containing NpACY (NpHR3.0 and ChR2(H134R)-YFP) and a Cre-dependent mCherry. For 30 s just prior to a fear conditioning session, blue light (BL+) stimulation is performed to activate ChR2 and allocate the memory to opsin-expressing neurons. After the conditioning session, mice are injected with 4-hydroxytamoxifen to tag engram neurons with mCherry. A memory test can be performed 3d or more later to show that memory retrieval can be attenuated with red light (RL+) activation of NpHR, as would be expected with successful optogenetic allocation (see Fig. 1). Histological analysis will show that in the BL+ condition, there is greater overlap between ChR2+ and mCherry+ neurons than in mice that did not receive BL+ [see an example in [8] for the lateral amygdala and AAV(DJ)-DIO-

3.3 RAM-Mediated Labeling of the Contextual Fear Retrieval Engram After Engram Tagging in TRAP2 Mice (Fig. 3)

In the context of labeling retrieval engrams in mice through RAM-mediated processes following TRAP2 labeling during training, it is recommended to use an AAV-RAM virus containing a transgene that complements the labeling method employed during training. For example, if a blue-activated opsin was utilized for TRAP labeling, it is recommended to use a virus containing a DREADD or a red-activated opsin for best results. In our example, we use AAV-EF1a-DIO-mCherry and AAV-RAM-hM4di-EYFP to tag the encoding engram and the retrieval engram, respectively.

1. The day before viral infusions, mice are placed on a doxycycline diet (40 mg/kg).
2. Perform optogenetic allocation or tagging in TRAP2 mice as described above.
3. To tag the remote engram, 4 weeks after fear conditioning, place mice in a new cage with food that does not contain doxycycline and subject them to a contextual fear test 24 h later to induce expression from the RAM promoter.
4. After another 24 h, subject mice to another fear test (note—first test should not induce extinction). For mice expressing RAM-hM4di, inject mice with C21 (5 mg/kg) or saline 1 h before the fear test.
5. Collect tissue for histological analysis the same day by animal perfusion. Alternatively, place mice back on a doxycycline diet for further testing in the future. RAM expression typically lasts for 2 weeks without a measurable drop in expression.

4 Notes

1. While the DJ serotype of AAV works well to enable gene delivery and expression in the brain, other serotypes can be and have been used by other laboratories.
2. We perform bilateral virus injections (and bilateral optrode placement) with 1ul of virus injected per side from a viral stock with a titer of approximately 10^{11} mL.
3. Viral targeting and optrode placement is fundamentally important for these and following procedures in this chapter. Optimal coordinates for viral injection to CA1 or other brain regions should be empirically determined as they can vary slightly depending on equipment used and delineation of bregma on the mouse skull. Our coordinates for viral infusion into CA1 are AP −1.8, ML ± 1.5, and DV −1.5. For optrode placement (which can be done during the same session as viral infusion, or 2 weeks after), coordinates are the same except DV −1.2.

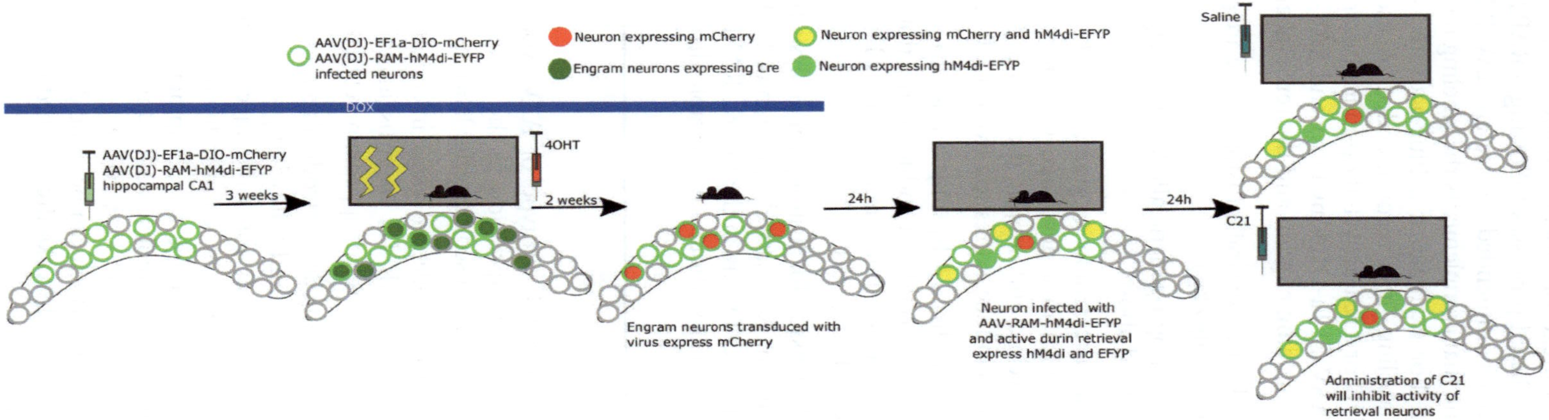

Fig. 3 Scheme of RAM-mediated labeling of the Contextual Fear Retrieval Engram after engram tagging in TRAP2 mice. The CA1 of mice is injected with viruses containing a Cre-dependent mCherry and a RAM-dependent expression of hM4di-EYFP. Mice are put on DOX 1 day prior to surgery. After a fear conditioning session, mice are injected with 4-hydroxytamoxifen to tag engram neurons with mCherry. A day before the memory test, mice are taken of DOX to allow for hM4di-EYFP expression. Mice are subjected to a fear test and neurons that are active during retrieval express hM4di. 30 min before a second fear test 24 h later, mice are either I.p. injected with C21 or saline. C21 inhibits the activity of retrieval neurons and mice should show a reduced freezing in comparison to saline injected mice. Histological analysis will show that there is a substantial overlap of mCherry and EYFP expressing neurons; however, the overlap is never 100%

4. Stimulation parameters will vary depending on the brain region used. For example, in the lateral amygdala, 20 Hz stimulation is used rather than 4 Hz for CA1.
5. Fear conditioning parameters can vary, often depending on the mouse strain used (e.g., single and double shock protocols). Ideally, conditioning parameters should allow for robust memory formation while minimizing generalization to different contexts at recent time points.
6. Fear testing can be performed the following day or at a later point that is within the time frame of virus expression (8 weeks or more).
7. Due to the presence of the optic cables, automated measuring of freezing cannot be performed by software associated with the commercial fear conditioning system (e.g., actimetrics etc.). Scoring of freezing can be done manually, or alternatively by using analysis tools such as DeepLabCut [11] or ezTrack [12].
8. While significantly more expensive, 4OHT provides a significant advantage over tamoxifen in that it enables tagging of active neurons within a shorter time window (<6 h) when compared to tamoxifen (<24–36 h) [13], thus increasing the temporal specificity of TRAPing (tagging in the TRAP2 system).
9. It is critical that 4OHT is fully dissolved. If the 4OHT has not gone into solution, the sample can be further solubilized by sonication for 5–10 min. In addition, 4OHT should be used right away or within 1–2 days of preparation (storage should be at −20 °C). Solubility has been shown to decrease with time [14].
10. The expression levels of the Cre-induced transgene increase with time. In our hands, we have observed high levels of expression of virally delivered transgenes up to 10 weeks after fear conditioning and 4OHT administration.

References

1. Semon R (1909) Die mnemischen Empfindungen in ihren Beziehungen zu den Originalempfindungen. Engelmann, Leipzig
2. Josselyn SA, Tonegawa S (2020) Memory engrams: recalling the past and imagining the future. Science 367(6473). https://doi.org/10.1126/science.aaw4325
3. Allen WE, DeNardo LA, Chen MZ, Liu CD, Loh KM, Fenno LE, Ramakrishnan C, Deisseroth K, Luo L (2017) Thirst-associated preoptic neurons encode an aversive motivational drive. Science 357(6356):1149–1155. https://doi.org/10.1126/science.aan6747
4. DeNardo LA, Liu CD, Allen WE, Adams EL, Friedmann D, Fu L, Guenthner CJ, Tessier-Lavigne M, Luo L (2019) Temporal evolution of cortical ensembles promoting remote memory retrieval. Nat Neurosci 22(3):460–469. https://doi.org/10.1038/s41593-018-0318-7

5. Sorensen AT, Cooper YA, Baratta MV, Weng FJ, Zhang Y, Ramamoorthi K, Fropf R, LaVerriere E, Xue J, Young A, Schneider C, Gotzsche CR, Hemberg M, Yin JC, Maier SF, Lin Y (2016) A robust activity marking system for exploring active neuronal ensembles. elife 5. https://doi.org/10.7554/eLife.13918
6. Roth BL (2016) DREADDs for neuroscientists. Neuron 89(4):683–694. https://doi.org/10.1016/j.neuron.2016.01.040
7. Vlasov K, Van Dort CJ, Solt K (2018) Optogenetics and chemogenetics. Methods Enzymol 603:181–196. https://doi.org/10.1016/bs.mie.2018.01.022
8. Rashid AJ, Yan C, Mercaldo V, Hsiang HL, Park S, Cole CJ, De Cristofaro A, Yu J, Ramakrishnan C, Lee SY, Deisseroth K, Frankland PW, Josselyn SA (2016) Competition between engrams influences fear memory formation and recall. Science 353(6297): 383–387. https://doi.org/10.1126/science.aaf0594
9. Lau JMH, Rashid AJ, Jacob AD, Frankland PW, Schacter DL, Josselyn SA (2020) The role of neuronal excitability, allocation to an engram and memory linking in the behavioral generation of a false memory in mice. Neurobiol Learn Mem 174:107284. https://doi.org/10.1016/j.nlm.2020.107284
10. Ramsaran AI, Wang Y, Golbabaei A, Aleshin S, de Snoo ML, Yeung BA, Rashid AJ, Awasthi A, Lau J, Tran LM, Ko SY, Abegg A, Duan LC, McKenzie C, Gallucci J, Ahmed M, Kaushik R, Dityatev A, Josselyn SA, Frankland PW (2023) A shift in the mechanisms controlling hippocampal engram formation during brain maturation. Science 380(6644):543–551. https://doi.org/10.1126/science.ade6530
11. Mathis A, Mamidanna P, Cury KM et al (2018) DeepLabCut: markerless pose estimation of user-defined body parts with deep learning. Nat Neurosci 21:1281–1289. https://doi.org/10.1038/s41593-018-0209-y
12. Pennington ZT, Dong Z, Feng Y et al (2019) ezTrack: An open-source video analysis pipeline for the investigation of animal behavior. Sci Rep 9:19979. https://doi.org/10.1038/s41598-019-56408-9
13. Guenthner CJ, Miyamichi K, Yang HH, Heller HC, Luo L (2013) Permanent genetic access to transiently active neurons via TRAP: targeted recombination in active populations. Neuron 78(5):773–784. https://doi.org/10.1016/j.neuron.2013.03.025
14. Felker A, Nieuwenhuize S, Dolbois A, Blazkova K, Hess C, Low LW, Burger S, Samson N, Carney TJ, Bartunek P, Nevado C, Mosimann C (2016) In vivo performance and properties of tamoxifen metabolites for CreERT2 control. PLoS One 11(4): e0152989. https://doi.org/10.1371/journal.pone.0152989

Chapter 12

Molecular Analysis of Neural Networks: Methods for Identifying and Characterizing Gene Expression Patterns in Neuronal Ensembles

Yinqing Li and Violeta Gisselle Lopez-Huerta

Abstract

Neuronal ensembles are functional units in the brain; ensembles are built by distinct neuron types that contribute differentially to information processing depending on the physiological range allowed by their gene expression and molecular properties. In this chapter, we describe single-nuclei RNA sequencing and biSNE analysis, a key method to characterize the molecular signatures and patterns of gene expression that define distinct subtypes of neurons. These subpopulations often underlie specific brain functions influencing behavioral states, both in health and disease.

Key words Neuronal ensemble, Transcriptomics, Gene expression, Patch-seq, In situ sequencing, Transgenic animal models

1 Introduction

The mammalian brain carries out a diverse range of functions through complex networks of specialized neurons [1–3]. Neuron types within larger networks and circuits are defined by molecular signatures and patterns of gene expression [4–8]. In recent years, various methods have been developed to unveil neuronal diversity in an objective manner such as RNA sequencing [4, 9]. High-throughput single-cell sequencing studies allow to analyze thousands of genes in thousands of neurons. RNA-seq experiments typically return high-dimensional data and can be analytically demanding to discern what is of biological significance. This can be aided by complementary functional assays to either validate findings within RNA-seq data or to provide some focus to questions to which RNA-seq is applied. Thus, analysis and integration with other techniques is used to identify and understand the

Luis Carrillo-Reid (ed.), *Identification, Characterization, and Manipulation of Neuronal Ensembles*, Neuromethods, vol. 215,
https://doi.org/10.1007/978-1-0716-4208-5_12,

functional properties of different subtypes of neurons identified by sequencing methodologies that make up neuronal ensembles involved in specific behaviors or cognitive functions [7, 8].

1.1 Neuronal Diversity

Since Santiago Ramón y Cajal's pioneering observations over a century ago, neuroscientists have recognized the complexity and abundance of cell types in the brain [10]. Specialized neuron varieties form the foundation for intricate coding within neural circuits. While identifying cell types is pivotal in neural circuit research, defining neuronal categories has been challenging. Criteria for classification span developmental genetic lineage, spatial positioning, morphology, synaptic connections, neurotransmitter contents, electrophysiological features, and functional roles within circuits and behaviors [11]. All these features are determined by sets of genes expressed differentially; the transcriptomic diversity of neurons encompasses the vast array of gene expression patterns among individual neurons, contributing significantly to the functional heterogeneity observed within the nervous system [4]. The emergence of transcriptomics, coupled with the potential for integration with other methodologies, offers the promise of unbiased classification based on gene expression profiles [8, 12–15]. RNA sequencing arises as a powerful tool to unravel brain complexity, allowing researchers to profile individual neurons within circuits and ensembles. This approach has already unveiled numerous populations and subpopulations of neurons, each characterized by specific gene expression patterns that underlie their functional specialization [8].

1.2 Single-Cell RNA Sequencing

In the last decade, the advent of single-cell and single-nuclei RNA sequencing (scRNA-seq and snRNA-seq) has transformed the landscape of neuroscience research, offering insights into the cellular heterogeneity that underlies complex neural networks [4, 8, 12]. Traditional bulk RNA sequencing methods provided averaged gene expression profiles from mixed populations, obscuring the diversity inherent in individual cells [16, 17]. scRNA-seq, however, enables the analysis of gene expression at the single-cell level, displaying detailed cellular diversity, developmental trajectories, and functional states within the nervous system [7, 8]. Unlike other techniques, scRNA-seq enables the dissection of brain structures and microcircuits into their constituent cells, allowing for the unbiased identification and characterization of distinct cell types, subpopulations, and rare cell states [18]. sc-transcriptomics gives the opportunity to dissect and decode the molecular architecture of the brain with exceptional resolution [4]. It is possible to investigate the genetic landscapes of specific neuronal subtypes, glial cells, and other non-neuronal components [19–21]. Moreover, the application of scRNA-seq goes beyond static snapshots of cellular transcriptomes. It allows for the exploration of dynamic processes, such as neuronal differentiation, synaptogenesis, and response to environmental cues [7, 22–24]. This temporal dimension provides a dynamic framework to study cellular transitions and the molecular events that orchestrate the development and function of the nervous system. Despite its transformative potential, scRNA-seq poses unique challenges, including data analysis and the need for specialized computational tools [7, 8]. In this chapter, we describe bimodal Stochastic Neighbor Embedding (biSNE) analysis, a computational resource used for the visualization and analysis of scRNA-seq data. It is an extension of t-SNE (t-Distributed Stochastic Neighbor Embedding), a broadly utilized dimensionality reduction algorithm for visualizing high-dimensional data in a lower-dimensional space. biSNE objective is to handle the challenges that arise because of the inherent bimodality in scRNA-seq data, which often arises due to the presence of both highly expressed and lowly expressed genes. The bimodality can complicate the interpretation of the data and affect the performance of standard dimensionality reduction techniques like t-SNE. These analytical methods help to harness the full potential of scRNA-seq in uncovering the cellular and molecular intricacies that govern the brain's molecular landscapes in health and disease.

1.3 Spatially Defined Transcriptomics

Neuronal ensembles emerge from complex connections and interactions among neurons, frequently influenced by cell positioning [25, 26]. However, during sc-RNAseq, tissue disaggregation leads to the loss of information regarding cell location and connectivity. Advancing beyond traditional transcriptomic profiling, spatially defined transcriptomic techniques like Slide-seq and StarMap solve this problem. These methodologies allow for the concurrent visualization and molecular profiling of gene expression within intact tissue sections, preserving spatial information for a deeper understanding of neuronal circuitry and function [25, 27, 28]. Slide-seq includes the spatial labeling of brain slices by attaching uniquely barcoded beads to acquire RNA fragments. The beads act as flags for positional information, allowing for the reconstruction of gene expression patterns across the tissue [28, 29]. On the other hand, StarMap uses in situ sequencing to directly read out the RNA barcodes within the tissue, removing the need for subsequent sequencing steps and retaining three-dimensional positional anatomy at cellular resolution [30]. Both methods provide high-resolution spatial maps of gene expression, allowing to discern the molecular diversity of cells within their anatomical context [31]. These technologies offer insights into the spatial organization of cellular populations, revealing spatially restricted gene expression patterns and facilitating the exploration of complex tissue architecture at the single-cell level.

1.4 Multimodal Characterization: Patch-seq

The primary function of neurons is to communicate through electrical and chemical signals, with their physiological features determined by the channels, receptors, and signaling pathways encoded by their gene patterns. Patch-clamp electrophysiology has been a pillar in neuroscience, widely used to study neuronal properties, firing patterns and synaptic connectivity in brain circuits. Understanding the biological meaning of transcriptomic configuration of neurons requires correlating this information with physiological characteristics [15, 32]. To achieve the integration of both transcriptomic patterns and physiology, a combined approach was implemented: Patch-seq [8, 13, 14]. Patch-seq, short for patch-clamp and single-cell RNA sequencing, involves recordings of electrical activity using patch-clamp technology with a modified protocol. Subsequently, the contents of the patched neuron, including its RNA, are aspirated and subjected to single-cell or single-nuclei RNA sequencing. Patch-seq thus enables the correlation of electrophysiological properties, such as action potentials and synaptic currents, with the unique transcriptomic signature of the recorded neuron. Moreover, neurons that are recorded can be filled with biocytin for subsequent morphological reconstruction, enabling the precise analysis of their anatomical structure and location [13]. This powerful convergence of functional and molecular information provides a deeper understanding into the cellular diversity, connectivity, and functional roles of distinct neuronal populations within complex neural circuits [8].

Together these techniques are advancing our knowledge about the molecular landscape of neuronal circuits and its correlation with brain function, leading to the development of directed technologies to characterize, manipulate, and modify specific cell types in health and in neurological conditions.

2 Methods

2.1 Dissection of the Region of Interest

In our sequencing efforts, our primary focus was directed towards the Thalamic Reticular Nucleus (TRN), an apparently homogeneous nucleus exclusively comprised of GABAergic neurons, serving as the primary source of inhibition to the thalamus [33–36]. Leveraging single-nuclei RNA sequencing (RNAseq) enabled an unbiased identification of gradients of gene expression in the neuronal population [8]. The application of transcriptomics provides the initial layer of information crucial for adopting a genes-to-function approach.

Utilizing a transgenic mouse line expressing Pvalb-tdTomato and specifically targeting TRN neurons, we conducted fluorescence-aided micro-dissection of tdTomato-positive TRN tissues under a stereo fluorescence microscope (Fig. 1a) [37]. Mice aged 13–15 weeks postnatal were anesthetized with isoflurane,

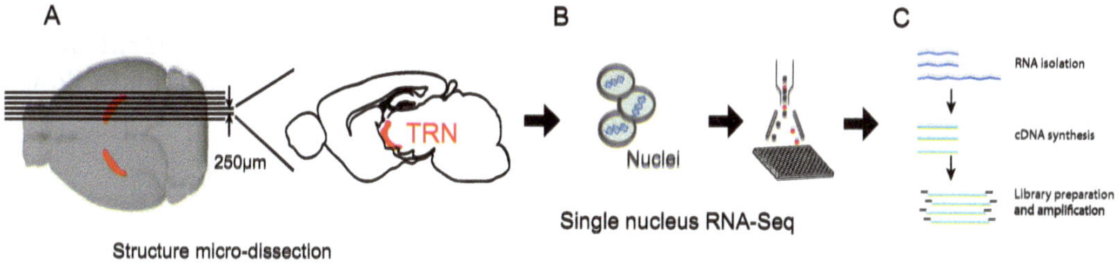

Fig. 1 Schematics of TRN dissection. (**a**) Dorsal view of the brain depicting TRN location (structure in red-*TdTomato+*). (**b**) Single-nuclei sorting before sequencing. Tissue is processed to disaggregate nuclei, which are then individually sorted into wells (**c**) RNAseq working flow

followed by decapitation for tissue harvesting. The brains were promptly removed and immersed in an ice-cold cutting solution. This solution, with a composition in mM of 194 sucrose, 30 NaCl, 4.5 KCl, 1.2 NaH2PO4, 0.2 CaCl2, 2 MgCl2, 26 NaHCO3, and 10 D-(+)-glucose, was saturated with 95% O2 and 5% CO2, maintaining a pH of 7.4 and an osmolality of 320–340 mOsm/L. A series of sagittal slices, each 250 µm thick, covering the entire TRN section, was prepared using a Leica Microsystems VT1200 S slicer. These slices were then placed under a fluorescent stereomicroscope for microdissection. The dissected TRN regions were preserved in ice-cold RNA later (Sigma-Aldrich, #R0901) and subsequently stored at −80 °C after an overnight incubation at 4 °C.

2.2 Single-Nuclei Isolation and FACS

Tissue samples were placed on ice for 5 min in 1 ml lysis buffer (10 mM β-glycerophosphate (pH 7.0), 2 mM MgCl2, 1 mM PMSF, 1 mM β-mercaptoethanol, 1% Tween-20). 1 mL ddH2O was added and lysate was kept on ice for 5 min before cell lysates were homogenized with Dounce homogenizer (Sigma)—20 times with pestle A, followed by 10 times with pestle B. 2 mL equilibration buffer (120 mM β-glycerophosphate pH 7.0, 2 mM MgCl2, 1 mM PMSF, 1 mM β-mercaptoethanol, 50% glycerol, RNAse inhibitor 1 U/µL) was added and nuclei were centrifuged (1000 g for 10 min at 4 °C) using a sucrose gradient (lower: 500 mM sucrose, 2 mM MgCl2, 25 mM KCl, 65 mM β-glycerophosphate pH 7.0, 20% glycerol, 1 mM PMSF, 1 mM β-mercaptoethanol; upper: 340 mM sucrose, 2 mM MgCl2, 25 mM KCl, 65 mM β-glycerophosphate (pH 7.0), 20% glycerol, 1 mM PMSF, 1 mM β-mercaptoethanol); finally nuclei were filtered through a 35 um cell strainer (Corning, NY, Falcon, #352235). The number and quality of purified nuclei were controlled using bright-field microscopy. We used Vybrant® DyeCycle™ Violet Stain (Thermo Fisher Scientific, #V35003) to avoid the spectrum overlapping with tdTomato. Also, in one of the batches, neuronal nuclei were enriched by using NeuN antibody staining [8]. To perform NeuN antibody staining, we incubated purified nuclei in

500 μL Phosphate Buffered Saline (PBS) with 0.5% Bovine Serum Albumin (BSA) (Sigma-Aldrich, #A9418) and 0.2 U/μL RNase inhibitor (Clontech/TaKaRa, #2313A) at 4 °C for 15 min followed by adding 1 μL Anti-NeuN antibody conjugated with Alexa Fluor®488 (EMD Millipore, #MAB377x) and incubating at 4 °C for another 15 min. We removed the staining buffer after spinning down in 500× *g* for 5 min at 4 °C, resuspended the nuclei pellet in buffer as described, and labeled nuclei with Vybrant® DyeCycle™ Violet Stain. We used an Astrios flow cytometry sorter (Beckman Coulter) to sort one nucleus into each well of 96-well plates containing 5 μL TCL buffer (Qiagen, #1031576) with 1% 2-Mercaptoethanol (b-ME) in each well. We set fluorescence activated cell sorting (FACS) gating on forward scatter plot, side scatter plot, and fluorescent channels to include only Violet$^+$ or Violet$^+$ NeuN$^+$ (for neuronal nuclei) (Fig. 1).

2.3 Single Nucleus RNA Library Construction and Sequencing

Single nucleus RNA was first purified using RNAClean XP beads (Beckman Coulter, IN, Agencourt RNAClean XP, #A63987) at 2.2× beads to sample volume ratio. Single nucleus-derived cDNA libraries were generated following a modified Smart-seq2 method [38]. The protocol did not use unique molecular identifiers (UMI). Beads were eluted into 4 μL elution mix made of 1 μL RT primer (10 μm), 1 μL dNTP mix (10 mM each, Thermo Fisher Scientific, #R0191), 1 μL RNAse inhibitor diluted at 1:10 in water (final 1 U/μL), and 1 μL H2O. Eluted samples were incubated at 72 °C for 3 min and immediately placed on ice. Each sample was added with 7 μL reverse transcription (RT) mix made of 0.75 μL H2O, 0.1 μL Maxima RNase-minus RT (Thermo Fisher Scientific, Maxima Reverse Transcriptase, #EP0752), 2 μL 5× Maxima RT buffer, 2 μL Betaine (Sigma Aldrich, 5 M, #B0300), 0.9 μL MgCl2 (Sigma Aldrich, 100 mM, #M1028), 1 μL TSO primer (10 μm), and 0.25 μL RNase inhibitor (40 U/μL). The RT reaction was incubated at 42 °C for 90 min and followed by 10 cycles of 50 °C for 2 min and 42 °C for 2 min and then heat inactivated at 70 °C for 15 min. Samples were then amplified with an addition of 14 μL polymerase chain reaction (PCR) mix made of 1 μL H2O, 0.5 μL ISPCR primer (10 μm), and 12.5 μL KAPA HiFi Hot-Start ReadyMix (KAPA Biosystems, MA, #KK2602). The PCR reaction was performed as follows: 98 °C for 3 min; 21 cycles of 98 °C for 15 s, 67 °C for 20 s, and 72 °C for 6 min; and final extension at 72 °C for 5 min. PCR product was purified using AMPure XP (Beckman Coulter, Agencourt AMPure XP, #A63880) twice and eluted in TE buffer (Thermo Fisher Scientific, #AM9849). Purified cDNA libraries were analyzed on Agilent 2100 Bioanalyzer (Agilent, CA, Agilent High Sensitivity DNA Kit, 4 #5067–4626) and quantified using Picogreen (Thermo Fisher Scientific, Quant-iT PicoGreen dsDNA Assay Kit, #P11496) on a plate reader (Biotek, Synergy H4, wavelength at 485 nm, 528 nm with 20 nm bandwidth).

Sequencing libraries were prepared using Nextera XT kit (Illumina, CA, #FC-131-1024) as described previously [24]. Single nucleus cDNA libraries were sequenced on an Illumina NextSeq 500 to an average depth of 632,169 reads (Fig. 1). Primer sequences: RT primer (Integrated DNA Technologies), /5BiosG/AAGCAGTGG TATCAACGCAGAGTACTTTTTTTTTTTTTTTTTTTTTTTTTTTT TTTTTVN; TSO primer (Exiqon), AAGCAGTGGTATCAACG CAGAGTACrGrG+G; ISPCR primer (Integrated DNA Technologies), /5BiosG/AAGCAGTGGTATCAACGCAG*A*G*T.

Single nucleus RNA library construction was performed. RNA was first purified using Agencourt RNAClean XP beads (Beckman Coulter, #A63987) at 2.2× beads to sample volume ratio. PCR product was purified using AMPure XP beads (Beckman Coulter, Agencourt AMPure XP, #A63880) twice at 0.7× volume ratio and eluted in EB buffer (Qiagen, #19086). Purified cDNA was quality checked on the Agilent 2100 BioAnalyzer with high sensitivity DNA Kit (Agilent, #5067–4626) and quantified using Quant-iT PicoGreen dsDNA Assay Kit (Thermo Fisher Scientific, #P7589) on the EnVision 2104 Multilabel Reader (Perkin Elmer). Sequencing libraries were prepared using a Nextera XT kit (Illumina, #FC-131-1024) with 0.075 ng cDNA from each nucleus and 1/4 of the standard reaction volumes. Single nucleus RNA-Seq libraries were sequenced on an Illumina NextSeq 500 with average depth of >1 million reads per nucleus.

2.4 Initial Processing of Sequencing Reads

The genome mapping rate was estimated using TopHat 2.0.10 [39], aligning reads to the mouse mm10 UCSC genome with default settings. Gene expression estimation utilized RSEM v1.2.8 [40] on alignments generated by Bowtie2 [41] with UCSC gene annotation, also using default parameters. Transforming TPM estimates to log-space involved applying the function log (TPM + 1). Genes were considered detected if their transformed expression levels equaled or exceeded 3 (1.1 in log (TPM + 1) scale). Libraries failing to meet the criteria of less than 40% genome mapping rate or less than 1500 detected genes were excluded from analysis [8].

2.5 Analysis of Nuclei Clusters

Clustering analysis categorizes nuclei into groups based on shared similarities, where nuclei within the same group exhibit greater similarity compared to those in different groups. The effectiveness of this grouping can be assessed using the Dunn index [42]

$$\mathrm{DB} = \frac{\min_{1 \le i \le j \le n} d(i,j)}{\max_{1 \le k \le n} d'(k)}$$

where $d(i, j)$ represents the intergroup distance between group i and j and $d'(k)$ represents the intragroup distance of group d

We predict that the coherent structure of transcriptomes among cells exhibiting high similarity results in observations that reside within a low-dimensional structure in the high-dimensional measurement space [43]. Accordingly, data points corresponding to cells within the same group would align along a continuous and seamless low-dimensional structure, while those from distinct groups would occupy separate structures. We restrict the distances utilized in Dunn index calculations to the low-dimensional structure and define the distance $d'(k)$ as

$$\widehat{\Phi}_{pq} = \text{argminmax}\left\{d_{mn} \mid d_{mn} \in \Phi_{pq}\right\}$$

$$d'(k) = \max\left\{d_{mn} \mid d_{mn} \in \bigcup_{p,q} \widehat{\Phi}_{pq}\right\}$$

where p, q, m, and n are data points belonging to the group k, d_{mn} denotes the pairwise distance of data points m and n, and Φ_{pq} represents a path connecting p and q through data points belonging to the group k. We define the distance $d(i, j)$ correspondingly to $d'(k)$ and limit p, q, m, and n to be data points belonging to the union of the groups i and j.

Here, we present a pipeline of techniques for obtaining nuclei clusters. Initially, we normalize the data and estimate false negatives to mitigate their impact on the calculation of d_{mn}. Subsequently, we employ modified PCA and tSNE [44] to map the low-dimensional structure to a 2-D space, where d_{mn} and Φ_{pq} in the 2-D space correspond to their high-dimensional counterparts. This mapping transforms each low-dimensional structure into dense data clouds in the 2-D space, facilitating cell grouping through density clustering [45]. This nonlinear mapping proves particularly advantageous for datasets characterized by highly varied scales of $d'(k)$ across different cell groups and affected by significant noise in the original high-dimensional space. Finally, we identify cell sub-clusters within each cell cluster using the biSNE algorithm. The PCA-tSNE, biSNE, and density clustering methods are applied hierarchically to each cell cluster to obtain clusters at a finer level. In each iteration, the Dunn index, utilizing the defined local distances $d'(k)$, can be used to assess the quality of the clustering assignment.

2.6 Normalization

Each library of single nuclei was prepared individually. Biases happen among libraries due to unavoidable differences in lysis efficiency, priming rate at RT, amplification efficiency during the initial PCR, the equalization for tagmentation, and ratios in the final sequencing pooling [36]. While various experimental methods have been devised to improve biases, such as incorporating spike-ins or applying unique molecular identifiers, these approaches primarily address bias introduced after the initial PCR step, with a significant portion of bias occurring prior to this stage. We hypothesize that cells of the same type should exhibit high expression levels of a tightly regulated set of genes with minimal "real" intercellular variability [7]. For example, ribosomal and cytoskeletal genes in stem cells or housekeeping genes in dendritic cells have been used previously to normalize single-cell sequencing data [46]. However, there is no consensus on a set of housekeeping genes for brain cells comprising mature neurons, immature neurons, and glia cells. To normalize cells, we devised a computational normalization procedure based on Bland-Altman (MA) plots and density estimation (Fig. 2). Our procedure normalizes one cell with respect to another so that genes belonging to this gene set show no significant differential expression on average [7]. By leveraging a small set of highly expressed and minimally variable genes, our approach offers robustness against noise, as measurements of highly expressed genes are resistant to sampling noise, and minimally variable measurements are unlikely to be affected by substantial noise. Additionally, low intercellular variance allows for the application of simple statistical models, such as the Gaussian model, to yield accurate estimates. Similar rationale underlies previously described normalization methods such as TMM [47] and DESeq [48]. However, these methods are tailored for population RNA-Seq data and are empirically found to be incompatible with single-cell data. A modified DESeq normalization, which accounts for the prevalent occurrence of false negatives in single-cell data, demonstrates comparable performance to our procedure. We initially address the scenario of two cells and subsequently demonstrate how to generalize to a set of arbitrary size. To identify the gene set for normalization, we compute differences and averages of log-transformed expression levels for each gene between a given pair of cells and depict the distribution of differences by averages on an MA plot. Subsequently, gene density in this distribution is estimated [49], and genes within the most densely plotted regions are selected. We calculate a scaling factor as the average of the log expression differences of the selected genes, and the second cell is normalized with respect to the first cell by dividing gene expressions of the second cell by the scaling factor. Specifically, the log expression difference of gene j between two cells is given by

$$r_{12\_j} = \log(e_{2j}) - \log(e_{1j})$$

And average of log expression of gene j is given by

$$a_{12\_j} = [\log(e_{1j}) + \log(e_{2j})]/2$$

where e_{ij} denotes the expression level of gene j in cell i. Gene j is selected into the gene set $\mathbb{S}_j$, if r_{12_j} and a_{12_j}, coordinates of gene j, are within the region having density above the top 70 percentile in the MA plot.

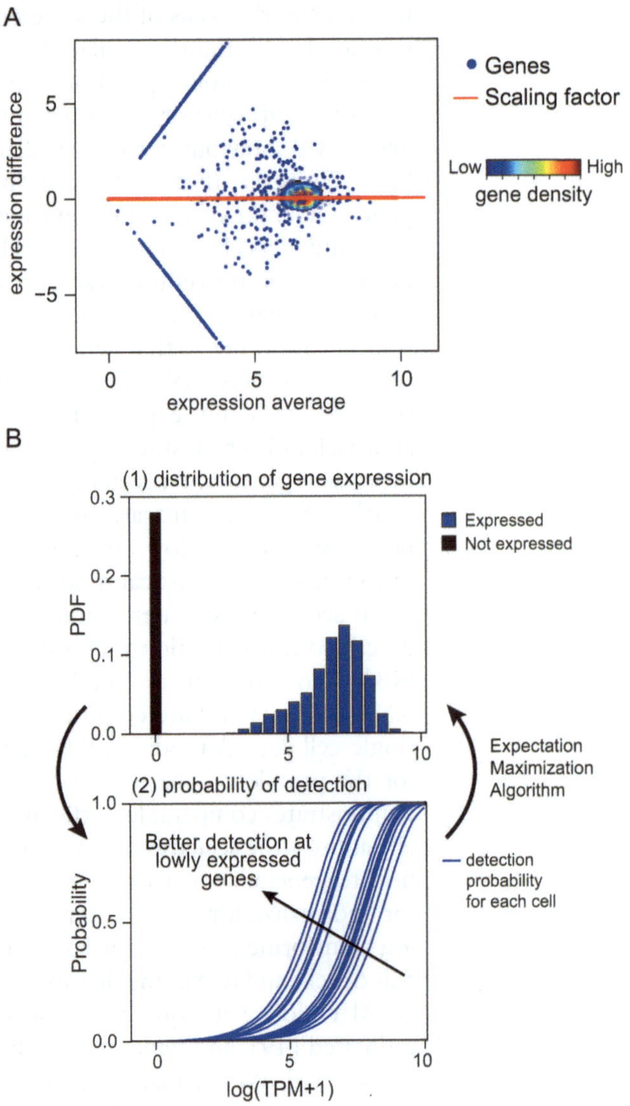

Fig. 2 Computational methods. (**a**) Density MA plot normalization method. MA plot showing the average log (X axis) versus the log ratio (Y axis) of TPM expression of all genes between two single nuclei. High-density region marked by a color scale. Genes within the colored density region are used to calculate the scaling factor between libraries for normalization. (**b**) Illustration of false negative estimation method. An expectation maximization algorithm alternates between estimation of gene expression distribution per gene (top) and the probability of detection (bottom) per cell. Top: histogram shows estimated distribution of expression of an example gene, PDF: probability density function. Bottom: each blue curve represents the probability of successfully detecting genes expressed at different levels in each cell

The scaling factor is obtained by

$$s = \sum_{j, j \in \mathbb{S}_J} r_{12_j} / |\mathbb{S}_J|$$

where $|S|$ indicates the cardinality of a S. Then the second cell is normalized as

$$e'_{2j} = e_{2j} / s$$

To normalize single cells of different types, cells are initially clustered into distinct groups, each comprising cells of similar types, ensuring that normalization aligns with the assumption of uniform cell types. Subsequently, normalization is executed for each group individually. Within each group, scaling factors are computed for each cell relative to several reference cells, selected based on the number of detected genes, such as cells with several genes detected near the 80th percentile. While any reference cell may be influenced by incorrect measurements to varying extents, employing multiple reference cells mitigates the impact of these errors during normalization.

Specifically, for a given group of cells $\{i \mid i \in \mathbb{C}_g \text{ and } g \in \mathrm{G}\}$, a set of cells that have number of genes detected above 80th percentile, are selected as reference cells $\{r \mid r \in \mathbb{C}_{gr} \text{ and } \mathbb{C}_{gr} \subset \mathbb{C}_g\}$. The scaling factor S_{ir} for each cell i with respect to each reference cell r is calculated. To relate S_{ir} obtained with different reference cells, we solve the optimization problem

$$\{\widehat{a}_r | r \in \mathbb{C}_{gr}\} = \text{argmax} \sum_{i \in \mathbb{C}_g} \text{Var}\left[\log(S_{ir}) - \log(a_r)\right],$$

And scaling factors are estimated as

$$s_i = \underset{r \in \mathbb{C}_{gr}}{\text{median}} \left(\frac{s_{ir}}{\widehat{a}_r} \right)$$

To normalize cells from different groups, we use group scaling factors estimated for each group aggregate that are acquired by averaging all cells within the same cluster. Cells from the same group are normalized using their group scaling factor.

Specifically, for each group $g \in \mathrm{G}$, the group aggregate is calculated as

$$e_{gj} = \sum_{i \in \mathbb{C}_g} e'_{ij} \ ,$$

where e_{gj} denotes the expression level of gene j in group g and e'_{ij} is the normalized expression level of gene g in cell i. Multiple reference group aggregates are selected for the estimation of group scaling factors.

2.7 Estimation of Missed Detection Probability

Single-nuclei transcriptome libraries are amplified from extremely small input materials. Consequently, we anticipate that transcripts with low expression levels may go undetected (false negatives) [7]. The likelihood of such undetected transcripts increases for those with lower expression levels and in libraries of inferior quality. These false negatives pose challenges to various analyses, such as undermining the normal distribution assumption essential for the commonly used Student's t-test, rendering the statistical test unreliable. Moreover, false negatives complicate the identification of genes exhibiting bimodal expression patterns, such as cell type-specific markers. Previous studies addressed these false negatives by integrating cell quality and gene expression estimations [50, 51], utilizing parametric estimation of gene expression distribution. However, the distribution of gene expression cannot be easily captured by a single parametric function. In contrast, our approach employs a Bayesian method to assess the likelihood of an observed zero measurement being a missed detection, based on non-parametric estimation of gene expression distribution [7, 8]. Our method hinges on two key observations: (a) Detection rates correlate with expression levels, where higher gene expression increases detection likelihood and (b) detection rates are influenced by library quality, with genes more likely to be detected in high-quality libraries. We model these observations as follows:

- Prior distributions: distributions of expression levels for each gene in cells of the same type
- Sampling probabilities: detection probabilities at different expression levels for each cell

For each observed $e_{ij} = 0$ of gene j in cell i, we then estimate the posterior distribution for two mutually exclusive hypotheses that e_{ij} is a missed detection or that gene j is not expressed in cell i.

Specifically, the distribution of expression level of gene j is calculated as mixture of two distributions.

The first one is the probability that gene j is not expressed

$$p_j(x = 0) = \frac{\sum_{i \in \{e_{ij} = 0\}} 1}{\sum_i 1}$$

where x denotes the true expression level. The second one is a conditional distribution of expression levels of gene j given that gene j is expressed. This distribution is estimated using a KDE (kernel density estimation)-based method [49] using gene expression levels e_{ij} from cells i, $\{i | eij > 0\}$. Combining two parts yields

$$p_j(x) = p_j(x = 0) + \left[1 - p_j(x = 0)\right] p_{j\_KDE^{(x)}},$$

where x denotes the expression level.

The detection probability (1—dropout probability) for a cell i is modeled using a geometric distribution parameterized by β_i, as it captures the Poisson sampling process, mechanism underlying detection stochasticity

$$\Lambda(x,\beta_i) = 1 - e^{-t}, t = \beta_i \begin{bmatrix} 1 \\ x \end{bmatrix} = \beta_{i0} + \beta_{i1} x$$

$$0 \leq \Lambda(x,\beta_i) \leq 1$$

where x denotes the expression level.

Given observed data, the expected value of the log-likelihood function is given by

$$E[L] = \sum_{j\in\{e_{ij}>0\}} \log(1 \cdot \Lambda(e_{ij},\beta_i)) + \sum_{j\in\{e_{ij}=0\}} \sum_x p_j(x) \log\left(p_j(x)(1-\Lambda(x,\beta_i))\right)$$

In each iteration, the log-likelihood function is maximized using gradient descent.

$$\widehat{\beta}_i = \underset{\beta_i}{\text{argmax}}\ \mathrm{E}[L]$$

$$\frac{\partial \mathrm{E}[L]}{\partial \beta_i} = \sum_{j\in\{e_{ij}>0\}} \frac{1}{\Lambda(e_{ij},\beta_i)} \frac{\partial \Lambda(e_{ij},\beta_i)}{\partial \beta_i} + \sum_{j\in\{e_{ij}=0\}} \sum_x p_j(x) \frac{1}{1-\Lambda(x,\beta_i)} (-1) \frac{\partial \Lambda(x,\beta_i)}{\partial \beta_i}$$

Because $\Lambda(x, \beta)$ is constrained to be non-negative, its derivative is modified with a rectifier so that $\Lambda(x, \beta)$ is differentiable for any x,

$$h(x) = \frac{\log(\exp(x \cdot N) + 1)}{N}$$

where N is a large number

$$\frac{\partial \Lambda}{\partial \beta} \approx \frac{\partial h}{\partial \Lambda} \frac{\partial \Lambda}{\partial \beta} = \frac{1}{1 + \exp(-\Lambda(x,\beta) \cdot N)} \cdot e^{-t} \begin{bmatrix} 1 \\ x \end{bmatrix}$$

Then the distribution of expression levels are updated by

$$p(e_{ij} = 0 | x_{ij} > 0) = \sum_x p_j(x) \left(1 - \Lambda\left(x, \widehat{\beta}_i\right)\right)$$

$$p(x_{ij} = 0 | e_{ij} = 0) = \frac{p_j(x_{ij} = 0)}{p_j(x_{ij} = 0) + p(e_{ij} = 0 | x_{ij} > 0)}$$

$$p_j(x=0) = \frac{\sum_{j\in\{e_{ij}=0\}} p(x_{ij}=0|e_{ij}=0)}{\sum_{j\in\{e_{ij}=0\}} p(x_{ij}=0|e_{ij}=0) + \sum_{j\in\{e_{ij}>0\}} 1}$$

$$p_j(x) = p_j(x=0) + \left[1 - p_j(x=0)\right] p_{j\_KDE}(x),$$

where $p\ (x_{ij} = 0|e_{ij} = 0)$ denotes the probability that gene j is not expressed in cell i.

We implemented an expectation-maximization (EM) algorithm that alternates between performing an expectation step for L and a maximization step for searching the maximizer $\widehat{\boldsymbol{\beta}}_i$ of $\mathrm{E}[L]$.

The probability $p\ (x_{ij} = 0|e_{ij} = 0)$ is included in calculations of summary statistics and distances to weight zero measurements. The higher the probability, the more likely that an observed zero represents a truly unexpressed gene in a cell and the more we weight the contribution of the zero. Conversely, the lower the probability, the higher the chance that it is false negative and the lower we weight its contribution in an analysis.

Specifically, we weight summary statistics, Euclidean distance, Pearson correlation coefficient, and cosine similarity in the following ways.

I. The weighted gene expression mean:

$$u_j = \sum_i e_{ij} w_{ij} \Big/ \sum_i w_{ij},$$

where

$$w_{ij} = \begin{cases} p(x_{ij}=0|e_{ij}=0) & \text{if } e_{ij}=0 \\ 1 & \text{if } e_{ij}>0 \end{cases}$$

II. The weighted Euclidean distance between two cells x, y:

$$w_j = w_{xj} w_{yj}$$

$$d_{xy} = \frac{\sum_j \left(e_{xj} - e_{yj}\right)^2 w_j}{\sum_j w_j}$$

III. The weighted Pearson correlation coefficient between two cells x, y:

$$\widehat{\mathbf{e}}_{\mathbf{x}} = \mathbf{e}_{\mathbf{x}} - u_x, \quad \widehat{\mathbf{e}}_{\mathbf{y}} = \mathbf{e}_{\mathbf{y}} - u_y$$

$$S_{xy} = \sum_j \widehat{e}_{xj} \widehat{e}_{yj} w_j, \quad S_{xx} = \sum_j \widehat{e}_{xj}^2 w_j, \quad S_{yy} = \sum_j \widehat{e}_{yj}^2 w_j$$

$$\rho_{xy} = \frac{S_{xy}}{\sqrt{S_{xx}S_{yy}}}.$$

IV. The weighted cosine similarity is calculated in a similar way except no data centering.

V. The weighted Euclidean distance between two cells *x*, *y* under a linear transformation of linear combinations of genes, $Y = XA$, where X is an i x j matrix and A is a j x k transformation matrix, is given by

$$w_j = w_{xj} w_{yj}$$

$$d_{xy} = \sum_k \left(\frac{\sum_j a_{jk}\left(e_{xj} - e_{yj}\right) w_j}{\sum_j w_j} \right)^2.$$

VI. The weighted Pearson correlation coefficient between two cells *x*, *y* under a linear transformation of linear combinations of genes as above is given by

$$u_x = \frac{1}{|K|} \sum_k \frac{\sum_j a_{jk} e_{xj} w_j}{\sum_j w_j}, \quad u_y = \frac{1}{|K|} \sum_k \frac{\sum_j a_{jk} e_{xj} w_j}{\sum_j w_j}$$

$$\widehat{e}_{xk} = \frac{\sum_j a_{jk} e_{xj} w_j}{\sum_j w_j} - u_x, \quad \widehat{e}_{yk} = \frac{\sum_j a_{jk} e_{xj} w_j}{\sum_j w_j} - u_y$$

$$S_{xy} = \sum_k \widehat{e}_{xk} \widehat{e}_{yk}, \quad S_{xx} = \sum_k \widehat{e}^2_{xk}, \quad S_{yy} = \sum_k \widehat{e}^2_{xk}$$

$$\rho_{xy} = \frac{S_{xy}}{\sqrt{S_{xx}S_{yy}}}.$$

VII. The weighted cosine similarity is calculated similarly as the weighted correlation coefficient except no data centering.

VIII. The weighted covariance between two genes under a linear transformation of linear combinations of genes as above is given by

$$u_{xk} = \frac{\sum_i \sum_j a_{jk} \widehat{e}_{ij} w_{ij}}{\sum_i \sum_j w_{ij}} = \frac{\sum_j a_{jk} \sum_i \widehat{e}_{ij} w_{ij}}{\sum_i \sum_j w_{ij}} = 0,$$

$$\operatorname{cov}\left(k, k'\right) = \frac{\sum_i \left(\sum_j a_{jk} \widehat{e}_{ij} w_{ij}\right) \left(\sum_j a_{jk'} \widehat{e}_{ij} w_{ij}\right)}{\sum_i \left(\sum_j w_{ij}\right)^2}.$$

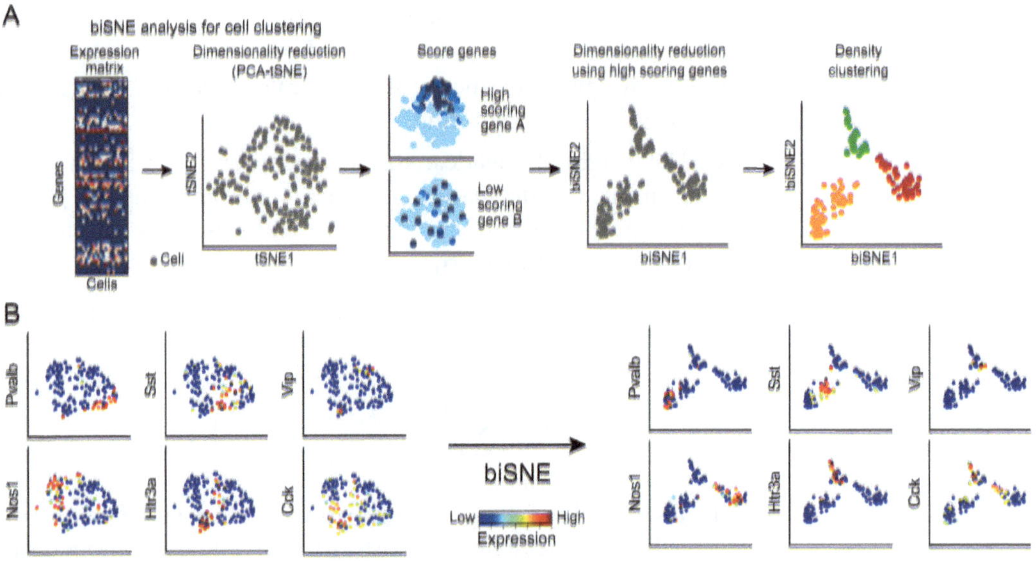

Fig. 3 BiSNE algorithm. (**a**) BiSNE algorithm. Left to right: BiSNE takes as input an expression matrix of genes (rows) across nuclei (or cells, columns). It generates a 2-D plot of nuclei, by dimensionality reduction using PCA followed by tSNE nonlinear embedding, and then scores each gene by their expression across the 2-D plot, such that genes expressed in nuclei in proximity on the 2-D plot (dark blue points, top) are high scoring, whereas those expressed in nuclei scattered across the plot (dark blue points, bottom) are low scoring. Next, it takes an expression matrix of only high scoring genes (heatmap, genes (rows) across all nuclei (columns)) and repeats the dimensionality reduction. BiSNE is followed by density clustering (colored, bottom left). (**b**) Expression of marker genes across 2-D embedded nuclei before and after biSNE. Shown is a panel of the same tSNE 2-D embedding of example GABAergic nuclei, with each panel colored by the expression of a marker genes (denoted on the left). Left: using PCA-tSNE only. Right: using biSNE

2.8 PCA and tSNE

To project cells into a two-dimensional space, we first employ principal component analysis (PCA) to reduce the original data to lower linear dimensions, retaining the most significant variance determined by the largest eigenvalue gap. Subsequently, we compute the cosine distance among cells within the PCA-reduced dimensional space. Lastly, we use t-distributed Stochastic Neighbor Embedding (tSNE) [44, 52, 53] with the cosine distance to further visualize cells in two dimensions, where the Euclidean distances of closely projected cells reflect their cosine distances (Fig. 3a). The cosine distance is contingent upon the angle between two vectors defined by gene expressions in the high-dimensional space. It is favored in our analysis over Euclidean distance and correlation distance due to its greater robustness to noise compared to Euclidean distance and its invariance under rotational transformations such as PCA.

2.8.1 Weighted PCA

The PCA analysis is typically conducted using either singular value decomposition (SVD) or eigenvalue decomposition (EVD) on the covariance matrix, a process that scales quadratically with the number of genes. Considering the substantial number of genes in our dataset, directly performing SVD or EVD on the extensive covariance matrix implies significant computational expenses. To derive the principal components or the transformation matrix A while factoring in weights, we initially center the original data matrix E across genes to obtain $\widehat{E}$, where e_{ij} is the expression level of gene j in cell i. Next, we perform SVD on centered data matrix $\widehat{E}$ to get A^*. We calculate the weighted covariance matrix C_w on $\widehat{E}$ under the linear transformation defined by the matrix A^*. We then perform SVD or EVD on C_w to get A.

2.8.2 tSNE with Cosine Distance

We adapted the conventional tSNE method to enable dimensionality reduction using weighted cosine similarity. The original tSNE approach maps data in a nonlinear manner to a low-dimensional space, wherein the Euclidean distances between neighboring data points in the low-dimensional space reflect the distances between these neighboring data points, or local distances, in the high-dimensional space. tSNE operates on a distance matrix, which delineates all pairwise distances in the high-dimensional space. Prior to employing tSNE, we convert the weighted cosine similarity to cosine distance by examining the associations between the two measures across the closest data points. Specifically, given a cell and its gene expression measurements denoted by a n dimensional vector $\mathbf{x}$, the measurements of its neighbor $\mathbf{y}$ are modeled as

$$\mathbf{y} = k(\mathbf{x} + \mathbf{d}),$$

where k is a scaling factor and $\mathbf{d}$ denotes the distance between $\mathbf{x}$ and $\mathbf{y}$. Under the null hypothesis that $\mathbf{x}$ and $\mathbf{y}$ are measured from two cells of the same type, $\mathbf{d}$ is drawn from a Gaussian distribution with zero mean and variance σ. Our goal is to estimate the distance magnitude $|\mathbf{d}|$, given the measured angle ø between $\mathbf{x}$ and $\mathbf{y}$.

Geometrically, the vector $\mathbf{d}$ lies on a hypersphere defined by radius $|\mathbf{d}|$. The volume and surface area of a hypersphere of dimension n (n-sphere) has the following properties:

$$Sn = (n+1)Vn + 1$$

$$dSn = (n+1)dVn + 1$$

the volume element is

$$dV_{n+1} = |\mathbf{d}|^2 \sin^{n-1}(\emptyset_1)\sin^{n-2}(\emptyset_1)\cdots\sin(\emptyset_{n-1})\mathrm{d}|\mathbf{d}|\mathrm{d}\emptyset_1\mathrm{d}\emptyset_2\cdots\mathrm{d}\emptyset_n$$

$$= \sin^{n-1}(\emptyset_1)\mathrm{d}\emptyset_1\cdot g(|\mathbf{d}|, \emptyset_2, \cdots, \emptyset_n)$$

$$\mathrm{d}S_n = \sin^{n-1}(\emptyset_1)\mathrm{d}\emptyset_1\cdot(n+1)g(|\mathbf{d}|, \emptyset_2, \cdots, \emptyset_n)$$

the probability of drawing $\mathbf{d}$ in a n-sphere of radius $|\mathbf{d}|$ with an angle ø from $\mathbf{x}$ scales as $sin^{n-1}(ø)$.

When n is large, most of $\mathbf{d}$ lie perpendicular to $\mathbf{x}$; thus, there exists a unique mapping between $|\mathbf{d}|$ and ø.

$$\cos(\emptyset) = \frac{1}{\sqrt{(1 + |\mathbf{d}|)^2 + 1}}$$

$$|\mathbf{d}| = \sqrt{\frac{1}{\cos^2(\emptyset)} - 1}$$

2.9 Differential Gene Expression and Pathway Analysis

We use an adjusted Welch's t-test for identifying differentially expressed genes. We applied weights in the calculation of summary statistics, such as sample mean, sample variance, and effective degrees of freedom, used in Welch's t-test.

Specifically, to find the significance level of gene j between cells in group X and cells in group Y,

$$n_{xj} = \sum_{i, i \in X} w_{ij}, \quad n_{yj} = \sum_{i, i \in Y} w_{ij},$$

$$u_{xj} = \sum_{i, i \in X} e_{ij} w_{ij} / n_{xj}, \quad u_{yj} = \sum_{i, i \in Y} e_{ij} w_{ij} / n_{yj},$$

$$S_{xj} = \sum_{i, i \in X} (e_{ij} - u_{xj})^2 w_{ij} / (n_{xj} - 1), S_{xj}$$

$$= \sum_{i, i \in Y} (e_{ij} - u_{yj})^2 w_{ij} / (n_{yj} - 1),$$

t statistics $t_j = \frac{u_{xj} - u_{yj}}{\sqrt{\frac{S_{xj}}{n_{xj}} + \frac{S_{yj}}{n_{yj}}}}$, degrees of freedom

$$v_j \approx \frac{(S_{xj}/n_{xj} + S_{yj}/n_{yj})}{S_{xj}^2 / \left[n_{xj}^2 (n_{xj} - 1)\right] + S_{yj}^2 / \left[n_{yj}^2 (n_{yj} - 1)\right]}.$$

The false discovery rate (FDR) is calculated for each differentially expressed gene in multiple hypothesis testing using the Benjamini and Hochberg procedure [54].

2.10 Density Clustering and Selection of the Number of Clusters

We used a density-based clustering method [35] to partition cells embedded in the 2-D space. The method searches cluster centers that are characterized by two quantities: (1) high local density ρ_i and (2) large distance δ_i from points of higher density, which are centers of other clusters. We unify the two quantities into a single metric by taking the product of the two quantities, $S_i = \rho_i \cdot \delta_i$.

To select cluster centers, we rank each data points by their si in descending order. For a given n, the number of desired clusters, we select the top ranked n cluster centers and perform the cluster assignment as described previously [45]. To evaluate the quality of the clustering, we calculate the Dunn index for each n with $d(i, j)$ and $d'(k)$ defined as local distances. The calculation of the Dunn index can be operated in $O(N^3)$, where N is the number of total data points.

Algorithm

Identification of maximum steps on shortest paths (MaxStep)

Input: pairwise distance of data points (D)

Output: the pairwise shortest link (D')

```
D'← D
n← # of data points
for k← 1 to n do
 for i← 1 to n-1 do
 for j← i+1 to n do
 D'(i,j)← min(D'(i,j),max(D'(i,k),D'(k,j)))
 end
 end
end
return D'
```

Algorithm

Calculation of the Dunn index defined on local distances (DunnLocal)

Input: pairwise distance of data points in the 2-D embedding (D), clustering assignment (Cl)

Output: the Dunn index (θ)

```
cl_uiq← unique(Cl)
n← # of cl_uiq
d'_k← empty array with a length of n
d_ij← empty matrix with a size of (n, n)
for i← 1 to n do
 ii← index of data whose clustering assignment is cl_uiq
(i)
 d'k(i)← max(MaxStep(D(ii,ii)))
end
for i← 1 to n-1 do
 for j← i+1 to n do
 ii← index of data whose clustering assignment is either
cl_uiq(i) or cl_uiq(j)
 d_ij(i, j)← max(MaxStep(D(ii,ii)))
 end
end
θ← min(d_ij)/ max(d'_k)
return θ
```

2.11 BISNE

Cells situated close together in the tSNE mapping exhibit co-expression of a specific set of genes not expressed by distant cells. This gene set may serve to differentiate various cell subpopulations. The genes within this set are co-expressed among cells grouped closely together, indicating localized expression patterns within the tSNE mapping (Fig. 3b).

2.11.1 Statistics for Scoring Expression Patterns

Motivated by this observation, we use two different statistics to identify genes with significantly localized expression patterns in the tSNE mapping and then perform PCA-tSNE using the union of these identified genes to cluster cells.

Moran's I

Moran's I [55] scores correlation between a measurement on a set of mapping positions and pairwise distances of these mapping positions. Given tSNE coordinates, the Moran's I for gene k is given by

$$I(k) = \frac{\sum_i \sum_j Q_{ij}(e_{ik} - u_k)(e_{jk} - u_k) w_{ik} w_{jk} / \sum_i \sum_j Q_{ij} w_{ik} w_{jk}}{\sum_i (e_{ik} - u_k)^2 w_{ik} / \sum_i w_{ik}}$$

where Q_{ij} denotes the pairwise similarity transformed from d_{ij}, the Euclidean distances between cell i and j in the tSNE mapping. We obtain Q_{ij} from d_{ij} using the Gaussian function,

$$Q_{ij} = \frac{1}{\sigma\sqrt{2\pi}} exp\left(-\frac{d_{ij}^2}{2\sigma^2}\right)$$

We choose σ to set the minimal size of localized expressed pattern, as $d_{ij} \approx \sigma$ weights around 60% and $d_{ij} \approx 2\sigma$ weights around 13.5%.

The statistical significance of the pattern of gene k is tested by converting $I(k)$ to a z score,

$E[I] = -1/(N-1)$, where N is the length of $\mathbf{e_k}$

$$V[I] = \frac{1}{S_0^2(N^2-1)(N^2 S_1 - N S_2 + 3S_0^2)} - E[I]^2$$

$$S_0 = 2\sum_i \sum_j Q_{ij}, \quad S_1 = 2\sum_i \sum_j Q_{ij}^2, \quad S_2 = 4\sum_i \left(\sum_j Q_{ij}\right)^2$$

$$z = -\frac{I - E[I]}{\sqrt{V[I]}}.$$

Moran's I uses gene expression levels in its calculation. When identifying marker genes, only the information about whether a gene is expressed or not is necessary. We use a modified Moran's I on binarized gene expression levels.

Specifically, we binarize gene expression level by a threshold,

$$\hat{e}_{ij} = \begin{cases} 1 & \text{if } e_{ij} > 3\text{TPM} \\ 0 & \text{if } e_{ij} \leq 3\text{TPM} \end{cases}$$

We then calculate the modified Moran's I by

$$I(k) = \frac{\sum_i \sum_j Q_{ij} (\hat{e}_{ik} - \hat{u}_k)(\hat{e}_{jk} - \hat{u}_k) w_{ik} w_{jk}}{\sum_i \sum_j Q_{ij} w_{ik} w_{jk}}.$$

Moran's I is a global measure. It has biases towards genes that are widely expressed. To reduce false positives, we filtered out genes expressed in more than 80% of cells.

Manhattan Distance and Order Statistics

The Manhattan distance is an alternative to the Euclidean distance in quantifying proximity. The advantage of using Manhattan distance is that x and y coordinates can be tested independently using order statistics. Assume a given set of cells that express gene j and their positions $\mathbf{z}$ on a coordinate Z, $\overline{\mathbf{z}}$ is defined as the normalized z such that $\overline{z}_i = (z_i - \min(\mathbf{z}))/(\max(\mathbf{z}) - \min(\mathbf{z})), i \in \{i | e_{ij} > \text{TPM3}\}$, and $\hat{\mathbf{z}}$ is defined as the ordered list of $\overline{\mathbf{z}}$, such that $\hat{z}_i < \hat{z}_{i+1}$. The range wz is defined as $w_z = \hat{z}_n - \hat{z}_1$.

Assume that $\hat{\mathbf{z}}$ is a vector of i.i.d. samples from a uniform distribution; the significance level p of w_z can be found using order statistics

PDF of $w f_w(w) = n(n-1) \int_{-\infty}^{+\infty} [F(x+w) - F(w)]^{(n-2)} f(x) f(x+w) \mathrm{d}x$

where $f(x)$ and $F(x)$ are PDF and CDF of z

CDF of w$F_w w = w^{n-1}[n - (n-1)w]$, under null hypothesis

$$p_z = F_w(w_z),$$

where PDF is the probability density function and CDF is the cumulative density function. To robustly estimate w in the presence of outliers, the distribution of $\mathbf{z}$ is fitted using the Gaussian distribution with robust estimators of mean and variance [56].

$u_z = \text{median}(\mathbf{z})$

$$S_z = 1.1926 \,\text{median}(\text{median}(|z_i - z_j|))$$

$$p_i = \phi\left(-\left|\frac{z_i - u_z}{S_z}\right|\right),$$

where Φ denotes the CDF of the standard normal distribution. Samples with $p_i < \epsilon$, a predefined threshold, are considered outliers and are excluded from the estimation of w.

A single p value is calculated for each gene by taking the product of p_x and p_y, the p values obtained for x and y coordinates, respectively. It measures the overall significance level of each gene in both coordinates.

2.12 Selection of Significant Genes

For each statistical measure, we rank genes according to their significance. Genes ranked highly are presumed to offer valuable insights for clustering cells, while those ranked lower are more likely to introduce noise that hampers clustering differentiation. We establish a cutoff rank to identify informative genes, selected based on the eigenvalue statistic of random matrices [57]. This metric suggests that the inclusion of a noisy row (gene) in a data matrix would diminish the maximum eigenvalue gap of the matrix. Conversely, the inclusion of an informative row (gene) would augment the maximum eigenvalue gap, as its variance aligns with that of other genes. Thus, the change in the maximum eigenvalue gap indicates a gene's informativeness. Following gene ranking, we initially populate a data matrix with the top-ranked genes and gradually incorporate subsequent genes with lower rankings. With each addition, we assess the change in the maximum eigenvalue gap before and after adding the gene. Additionally, we randomly shuffle measurements of the gene across cells and determine the change in the maximum eigenvalue gap resulting from the inclusion of this permuted gene. Subsequently, we establish a cutoff rank below which there is no discernible difference in the change of the maximum eigenvalue gap between adding a gene and its permuted counterpart. The selection cutoff can also be formally tested using minimum hypergeometric test [58].

Specifically, for a data matrix $E_{1,j-1}$ and a gene j, We form a new matrix

$$E_{1,j} = \begin{bmatrix} E_{1,j-1} \\ e_j, \end{bmatrix}$$

and we obtain the eigenvalues of $E_{1,j-1}\ E^T{}_{1,j-1}$ using weighted SVD. The eigenvalues are normalized and sorted in order

$$\lambda_1 > \lambda_2 > \cdots > \lambda_n, \text{and} \sum_i \lambda_i = 1$$

The distribution density (Marchenko-Pastur distribution) of higher-order eigenvalues can be approximated by a linear function [57], and its cumulative distribution can be approximated by a quadratic polynomial.

The sorted eigenvalues follow the inverse function of the cumulative distribution and are fitted by

$$\lambda_i = f(i) = \alpha_0 + \alpha_1 \sqrt{\frac{i}{n}}, \alpha_0 \text{ and } \alpha_1 \in \mathbb{R}$$

The eigenvalue gap is approximated as

$$\Delta_j = \sum_{i=1}^{n} \left(\lambda_i - \widehat{\lambda}_i\right)$$

For permutation comparison, expression of gene j is permuted,

$$\tilde{e}_j : \tilde{e}_{ij} = e_{i'j} \ , i' \text{ is drawn without replacement from } [1, n]$$

$$\tilde{E}_{1,j} = \begin{bmatrix} E_{1,j-1} \\ \tilde{e}_j \end{bmatrix},$$

where i' denotes randomly permuted cell index. The eigenvalue gap $\tilde{\Delta}_j$ is obtained for the permuted matrix $\tilde{E}_{1,j}$. A cutoff rank is chosen at k, if the change in the eigenvalue gap $\Delta' - \tilde{\Delta}'$ is not significant for genes ranked below k. To combine top genes, we take the union of genes selected by different statistics.

2.13 Clustering of Gene Signatures Using Cross-Correlation

To create gene signatures and account for cell similarity based on the genes they express, we calculate cross-correlations among highly scored genes, considering the proximity of cells expressing these genes. We then convert the correlation coefficient into distances and employ t-SNE and density clustering to cluster these genes. Specifically, spatial cross-correlation between gene k and $k'$is given by

$$I(k, k') = \frac{\sum_i \sum_j Q_{ij}(e_{ik} - u_k)(e_{jk'} - u_k) w_{ik} w_{jk'} / \sum_i \sum_j Q_{ij} w_{ik} w_{jk'}}{\sqrt{\left(\sum_i (e_{ik} - u_{kl})^2 w_{ik} / \sum_i w_{ik}\right) \left(\sum_i (e_{ik'} - u_{kl})^2 w_{ik'} / \sum_i w_{ik'}\right)}}.$$

It has been noted that the range of I is not [−1, 1], unlike Pearson's correlation coefficient. We empirically found that I is positively biased in the tSNE mapping. The positive bias may underestimate the strength of anticorrelation genes having complementary patterns. A scalar transformation of I that has the exact range [−1, 1] has been proposed [59].

$$\tilde{W} = (n\overline{w})^{-1} H^T W H, \quad \text{where } \tilde{W} \text{ is a } (n-1) \times (n-1) \text{ matrix, and } \tilde{w} = \sum_{i,j=1}^{n} w_i j / n^2$$

$H = (h_1, \ldots, h_{n-1})$ is defined on Helmert orthogonal matrix

$$h_i^T = \frac{(1_i^T, -i, 0_{n-i-1}^T)}{\sqrt{i(i+1)}}, \qquad \text{for } i = 1, \ldots, n-1.$$

The scalar transformation of Moran's I is given by

$$I_M = \begin{matrix} [(n-1)I + 1] / [|(n-1)\lambda_{(1)} + 1|] & \text{if } (n-1)I + 1 < 0 \\ [(n-1)I + 1] / [|(n-1)\lambda_{(n-1)} + 1|] & \text{if } (n-1)I + 1 \geq 0, \end{matrix}$$

where $\lambda_{(1)}$ and $\lambda_{(n-1)}$ are the smallest and largest eigenvalues of the matrix $\tilde{W}$.

The calculation of spatial cross-correlation has a computational complexity that scales quadratically with the number of gene and cells as of $O(N^2 M^2)$, where N is the number of cells and M is the number of genes. When the number of cells and the number of genes is large, it becomes impractical to calculate the spatial cross-correlation. However, for clustering genes using tSNE [44], only the information about k-nearest neighbor (knn) data points is necessary, requiring a linear complexity as of $O(N^2MK)$. The data with knn defined on a metric space can be organized using structures such as vantage point (VP) tree [60] for efficient computation. We develop a conversion between spatial correlation coefficient and a metric.

Theorem

For a given similarity $I(k,k'), I(k,k') \in [\rightarrow B, B], B \in \mathbb{R}$ and $B > 0$, define $g(k,k')$

$$g(k,k') = \begin{cases} 0 & \text{if } k = k' \\ \sqrt{a - I(k,k')} & \text{if } k \neq k' \end{cases}$$

with $a > \frac{5}{3}B$, and $g(I(k,k'))$ is a metric.

Proof For $k = k'$, the proof is trivial. For $k \neq k'$,

1. Non-negativity, $g(k,k') = \sqrt{a - I(k,k')} > \sqrt{\frac{2}{3}B} > 0$.
2. Coincidence, $g(k,k') = \sqrt{a - I(k,k')} > 0$
3. Symmetry, $g(k,k') = \sqrt{a - I(k,k')} > 0$
4. Triangle inequality, $g(k,k'') + g(k'',k) \geq 2\sqrt{a - B} > \sqrt{a - (-B)} > g(k,k')$

2.13.1 Identification of TRN Gene Expression Patterns

In the TRN, we analyzed the transcriptomes of 1687 nuclei, which had an average of 1.3 × 106 mapped reads and a median library complexity of 3909 genes, comparable to previous studies. Clustering analysis revealed expected GABAergic cells (Gad2+) and non-neuronal cell types, along with glutamatergic neurons (Slc17a7+), likely originating from adjacent thalamic regions (Fig. 4a). The predominant cluster, enriched for the inhibitory neuronal marker Pvalb, totaling 671 neurons, represents inhibitory neurons of the TRN. However, t-SNE clustering based on the most variable genes showed no clear sub-clusters of Gad2+ Pvalb+ cells, the putative TRN neurons. To capture finer distinctions among closely related TRN neurons, we corrected for batch effects and applied the biSNE clustering algorithm. Interestingly, biSNE mapped cells to an elongated continuum, with no obvious demarcation separating the neurons into distinct classes (Fig. 4b). The continuum, referred to as a transcriptomic gradient, is characterized by the shift in the number of expressed genes specific to the "Spp1" profile versus the "Ecel1" profile. Cells expressing both Spp1 and Ecel1 or neither appeared to be of similar quality, holding

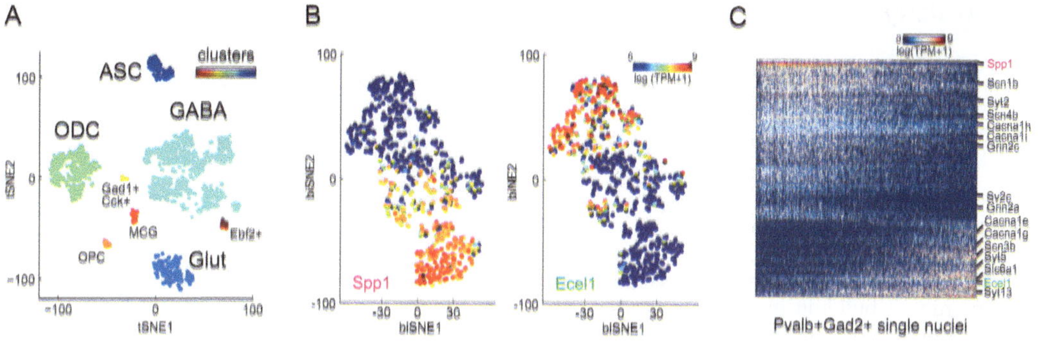

Fig. 4 Single-nuclei RNA sequencing reveals transcriptomic gradient of TRN neurons. (**a**) t-SNE embedding shows nucleus clusters, indicated by colors. GABAergic cells (GABA), oligodendrocytes (ODC), glutamatergic cells (Glut), oligodendrocyte precursor cells (OPC), microglia (MCG), astrocytes (ASC). $n = 1687$ nuclei. (**b**) biSNE embedding of Gad2+ and Pvalb+ single nuclei, colored by Spp1 (left) or Ecel1 (right) expression level, showing two extremes marked by Spp1 or Ecel1, respectively. $n = 671$ nuclei. (**c**) Heat map showing Spp1- and Ecel1-associated transcriptional programs. Some genes relevant to electrophysiological properties are highlighted on the side. Columns: single nuclei; Rows: genes

intermediate positions along the gradient, suggesting a functional spectrum between Spp1+ and Ecel1+ profiles (Fig. 4). Thus, our data suggest that TRN molecular diversity is characterized by a single transcriptomic gradient composed of a gradual mixture of two distinct transcription programs marked by highly expressed Spp1+ or Ecel1+ profiles. This data generated a comprehensive single-cell resolution atlas of TRN revealing its cellular heterogeneity connected to functionally distinct TRN subnetworks [8].

3 Conclusions

Single-cell and single-nuclei RNAseq techniques have revolutionized our understanding of the molecular dynamics within neuronal networks. Development of specific analysis such as biSNE is needed to get clear information of gene expression profiles and separation of cell types circumventing variations inherent to experiments. The biSNE algorithm aims to accurately represent the bimodal structure of the scRNA-seq data in the lower-dimensional embedding space. It achieves this by considering the distribution of gene expression levels across cells and accounting for the bimodal nature of the data during the dimensionality reduction process. When combined with complementary methodologies, they facilitate the exploration of gene expression patterns across diverse behavioral states. This integration offers valuable insights into how various cells collaborate to orchestrate complex cognitive functions, addressing fundamental questions in neuroscience.

Acknowledgments

We are grateful for the assistance of Marcela Palomero-Rivero and Uri Nimrod Ramirez-Jarquin. This work was supported by UNAM-PAPIIT IN216023 granted to VGL-H.

References

1. Yuste R, Cossart R, Yaksi E (2024) Neuronal ensembles: building blocks of neural circuits. Neuron. https://doi.org/10.1016/j.neuron.2023.12.008
2. Carrillo-Reid L, Yuste R (2020) Playing the piano with the cortex: role of neuronal ensembles and pattern completion in perception and behavior. Curr Opin Neurobiol 64. https://doi.org/10.1016/j.conb.2020.03.014
3. Walsh JJ, Christoffel DJ, Malenka RC (2023) Neural circuits regulating prosocial behaviors. Neuropsychopharmacology 48(1):79–89. https://doi.org/10.1038/s41386-022-01348-8
4. Saunders A et al (2018) Molecular diversity and specializations among the cells of the adult mouse brain. Cell 174(4):1015–1030.e16. https://doi.org/10.1016/j.cell.2018.07.028
5. Peng H et al (2021) Morphological diversity of single neurons in molecularly defined cell types. Nature 598(7879):174–181. https://doi.org/10.1038/s41586-021-03941-1
6. Zeng H, Sanes JR (2017) Neuronal cell-type classification: challenges, opportunities and the path forward. Nat Rev Neurosci 18(9): 530–546. https://doi.org/10.1038/nrn.2017.85
7. Habib N et al (2016) Div-Seq: single-nucleus RNA-Seq reveals dynamics of rare adult newborn neurons. Science 353(6302):925–928. https://doi.org/10.1126/science.aad7038
8. Li Y et al (2020) Distinct subnetworks of the thalamic reticular nucleus. Nature 583(7818): 819–824. https://doi.org/10.1038/s41586-020-2504-5
9. Zeisel A et al (2018) Molecular architecture of the mouse nervous system. Cell 174(4): 999–1014.e22. https://doi.org/10.1016/j.cell.2018.06.021
10. Ramon y Cajal S (1899) La Textura del Sistema Nerviosa del Hombre y los Vertebrados, 1st edn, Moya
11. Huang ZJ, Zeng H (2013) Genetic approaches to neural circuits in the mouse. Annu Rev Neurosci 36:183–215. https://doi.org/10.1146/annurev-neuro-062012-170307
12. Welch JD, Kozareva V, Ferreira A, Vanderburg C, Martin C, Macosko EZ (2019) Single-cell multi-omic integration compares and contrasts features of brain cell identity. Cell 177(7):1873–1887.e17. https://doi.org/10.1016/j.cell.2019.05.006
13. Cadwell CR et al (2016) Electrophysiological, transcriptomic and morphologic profiling of single neurons using patch-seq. Nat Biotechnol 34(2):199–203. https://doi.org/10.1038/nbt.3445
14. Fuzik J et al (2016) Integration of electrophysiological recordings with single-cell RNA-seq data identifies neuronal subtypes. Nat Biotechnol 34(2):175–183. https://doi.org/10.1038/nbt.3443
15. Lipovsek M, Bardy C, Cadwell CR, Hadley K, Kobak D, Tripathy SJ (2021) Patch-seq: past, present, and future. J Neurosci 41(5): 937–946. https://doi.org/10.1523/JNEUROSCI.1653-20.2020
16. Cahoy JD et al (2008) A transcriptome database for astrocytes, neurons, and oligodendrocytes: a new resource for understanding brain development and function. J Neurosci 28(1): 264–278. https://doi.org/10.1523/JNEUROSCI.4178-07.2008
17. Liang WS et al (2008) Altered neuronal gene expression in brain regions differentially affected by Alzheimer's disease: a reference data set. Physiol Genomics 33(2):240–256. https://doi.org/10.1152/physiolgenomics.00242.2007
18. Kulkarni A, Anderson AG, Merullo DP, Konopka G (2019) Beyond bulk: a review of single cell transcriptomics methodologies and applications. Curr Opin Biotechnol 58:129–136. https://doi.org/10.1016/j.copbio.2019.03.001
19. Lake BB et al (2016) Neuronal subtypes and diversity revealed by single-nucleus RNA sequencing of the human brain. Science 352(6293):1586–1590. https://doi.org/10.1126/science.aaf1204

20. Eze UC, Bhaduri A, Haeussler M, Nowakowski TJ, Kriegstein AR (2021) Single-cell atlas of early human brain development highlights heterogeneity of human neuroepithelial cells and early radial glia. Nat Neurosci 24(4): 584–594. https://doi.org/10.1038/s41593-020-00794-1
21. Tasic B et al (2018) Shared and distinct transcriptomic cell types across neocortical areas. Nature 563(7729):72–78. https://doi.org/10.1038/s41586-018-0654-5
22. Hrvatin S et al (2018) Single-cell analysis of experience-dependent transcriptomic states in the mouse visual cortex. Nat Neurosci 21(1): 120–129. https://doi.org/10.1038/s41593-017-0029-5
23. Nowakowski TJ et al (2017) Spatiotemporal gene expression trajectories reveal developmental hierarchies of the human cortex. Science 358(6368):1318–1323. https://doi.org/10.1126/science.aap8809
24. Lake BB et al (2018) Integrative single-cell analysis of transcriptional and epigenetic states in the human adult brain. Nat Biotechnol 36(1):70–80. https://doi.org/10.1038/nbt.4038
25. Ortiz C, Carlén M, Meletis K (2021) Spatial transcriptomics: molecular maps of the mammalian brain. Annu Rev Neurosci 44:547–562. https://doi.org/10.1146/annurev-neuro-100520-082639
26. Pang JC et al (2023) Geometric constraints on human brain function. Nature 618(7965): 566–574. https://doi.org/10.1038/s41586-023-06098-1
27. Liu Y et al (2020) High-spatial-resolution multi-omics sequencing via deterministic barcoding in tissue. Cell 183(6):1665–1681.e18. https://doi.org/10.1016/j.cell.2020.10.026
28. Ståhl PL et al (2016) Visualization and analysis of gene expression in tissue sections by spatial transcriptomics. Science 353(6294):78–82. https://doi.org/10.1126/science.aaf2403
29. Rodriques SG et al (2019) Slide-seq: a scalable technology for measuring genome-wide expression at high spatial resolution. Science 363(6434):1463–1467. https://doi.org/10.1126/science.aaw1219
30. Wang X et al (2018) Three-dimensional intact-tissue sequencing of single-cell transcriptional states. Science 361(6400). https://doi.org/10.1126/science.aat5691
31. Rao A, Barkley D, França GS, Yanai I (2021) Exploring tissue architecture using spatial transcriptomics. Nature 596(7871):211–220. https://doi.org/10.1038/s41586-021-03634-9
32. Shao M, Zhang W, Li Y, Tang L, Hao Z-Z, Liu S (2023) Patch-seq: advances and biological applications. Cell Mol Neurobiol 44(1):8. https://doi.org/10.1007/s10571-023-01436-3
33. Wimmer RD, Schmitt LI, Davidson TJ, Nakajima M, Deisseroth K, Halassa MM (2015) Thalamic control of sensory selection in divided attention. Nature 526(7575): 705–709. https://doi.org/10.1038/nature15398
34. McAlonan K, Cavanaugh J, Wurtz RH (2006) Attentional modulation of thalamic reticular neurons. J Neurosci 26(16):4444–4450. https://doi.org/10.1523/JNEUROSCI.5602-05.2006
35. Sherman SM, Guillery RW (2002) The role of the thalamus in the flow of information to the cortex. Philos Trans R Soc Lond Ser B Biol Sci 357(1428):1695–1708. https://doi.org/10.1098/rstb.2002.1161
36. Pinault D (2004) The thalamic reticular nucleus: structure, function and concept. Brain Res Rev 46(1):1–31. https://doi.org/10.1016/j.brainresrev.2004.04.008
37. Kaiser T, Ting JT, Monteiro P, Feng G (2016) Transgenic labeling of parvalbumin-expressing neurons with tdTomato. Neuroscience 321: 236–245. https://doi.org/10.1016/j.neuroscience.2015.08.036
38. Picelli S, Björklund ÅK, Faridani OR, Sagasser S, Winberg G, Sandberg R (2013) Smart-seq2 for sensitive full-length transcriptome profiling in single cells. Nat Methods 10(11):1096–1098. https://doi.org/10.1038/nmeth.2639
39. Kim D, Pertea G, Trapnell C, Pimentel H, Kelley R, Salzberg SL (2013) TopHat2: accurate alignment of transcriptomes in the presence of insertions, deletions and gene fusions. Genome Biol 14(4):R36. https://doi.org/10.1186/gb-2013-14-4-r36
40. Li B, Dewey CN (2011) RSEM: accurate transcript quantification from RNA-Seq data with or without a reference genome. BMC Bioinformatics 12(1):323. https://doi.org/10.1186/1471-2105-12-323
41. Langmead B, Salzberg SL (2012) Fast gapped-read alignment with Bowtie 2. Nat Methods 9(4):357–359. https://doi.org/10.1038/nmeth.1923
42. Dunn JC (1974) Well-separated clusters and optimal fuzzy partitions. J Cybern 4(1): 95–104. https://doi.org/10.1080/01969727408546059
43. Roweis ST, Saul LK (2000) Nonlinear dimensionality reduction by locally linear

embedding. Science 290(5500):2323–2326. https://doi.org/10.1126/science.290.5500.2323

44. Van Der Maaten L (2014) Accelerating t-SNE using tree-based algorithms. J Mach Lear Res 15(1):3221–3245
45. Rodriguez A, Laio A (2014) Clustering by fast search and find of density peaks. Science 344(6191):1492–1496. https://doi.org/10.1126/science.1242072
46. Shalek AK et al (2013) Single-cell transcriptomics reveals bimodality in expression and splicing in immune cells. Nature 498(7453): 236–240. https://doi.org/10.1038/nature12172
47. Robinson MD, McCarthy DJ, Smyth GK (2010) edgeR: a Bioconductor package for differential expression analysis of digital gene expression data. Bioinformatics 26(1): 139–140. https://doi.org/10.1093/bioinformatics/btp616
48. Love MI, Huber W, Anders S (2014) Moderated estimation of fold change and dispersion for RNA-seq data with DESeq2. Genome Biol 15(12):550. https://doi.org/10.1186/s13059-014-0550-8
49. Botev ZI, Grotowski JF, Kroese DP (2010) Kernel density estimation via diffusion. Ann Stat 38(5). https://doi.org/10.1214/10-AOS799
50. Shalek AK et al (2014) Single-cell RNA-seq reveals dynamic paracrine control of cellular variation. Nature 510(7505):363–369. https://doi.org/10.1038/nature13437
51. Kharchenko PV, Silberstein L, Scadden DT (2014) Bayesian approach to single-cell differential expression analysis. Nat Methods 11(7): 740–742. https://doi.org/10.1038/nmeth.2967
52. Amir ED et al (2013) viSNE enables visualization of high dimensional single-cell data and reveals phenotypic heterogeneity of leukemia. Nat Biotechnol 31(6):545–552. https://doi.org/10.1038/nbt.2594
53. Macosko EZ et al (2015) Highly parallel genome-wide expression profiling of individual cells using nanoliter droplets. Cell 161(5): 1202–1214. https://doi.org/10.1016/j.cell.2015.05.002
54. Benjamini Y, Hochberg Y (1995) Controlling the false discovery rate: a practical and powerful approach to multiple testing. J R Stat Soc 57(1):289–300
55. Moran PAP (1950) Notes on continuous stochastic phenomena. Biometrika 37(1–2): 17–23. https://doi.org/10.1093/biomet/37.1-2.17
56. Rousseeuw PJ, Croux C (1993) Alternatives to the median absolute deviation. J Am Stat Assoc 88(424):1273–1283. https://doi.org/10.1080/01621459.1993.10476408
57. Edelman A, Rao NR (2005) Random matrix theory. Acta Numer 14:233–297. https://doi.org/10.1017/S0962492904000236
58. Eden E, Lipson D, Yogev S, Yakhini Z (2007) Discovering motifs in ranked lists of DNA sequences. PLoS Comput Biol 3(3):e39. https://doi.org/10.1371/journal.pcbi.0030039
59. Maruyama Y (2015) An alternative to Moran's I for spatial autocorrelation
60. Yianilos PN (1993) Data structures and algorithms for nearest neighbor search in general metric spaces. In: fourth annual ACM-SIAM symposium on discrete algorithms (SODA'93). Society for Industrial and Applied Mathematics, p 311–321

Correction to: Identification, Characterization, and Manipulation of Neuronal Ensembles

Luis Carrillo-Reid

Correction to:
Luis Carrillo-Reid (ed.), ***Identification, Characterization, and Manipulation of Neuronal Ensembles,*** **Neuromethods, vol. 215,**
https://doi.org/10.1007/978-1-0716-4208-5

"The chapter 'Fast Neuronal Segmentation of Two-Photon Functional Imaging Recordings Using CITE-On' written by Marco Brondi, Luca Sità, Sebastiano Curreli, Tommaso Fellin was originally published without open access. Following the authors' decision to opt for open access, the copyright of the chapter changed to © The Author(s) 2025 and the chapter is now distributed under the terms of the Creative Commons Attribution 4.0 International License (https://creativecommons.org/licenses/by/4.0/), which permits use, sharing, adaptation, distribution and reproduction in any medium or format, as long as you give appropriate credit to the original author(s) and the source, provide a link to the Creative Commons licence and indicate if changes were made.

The images or other third party material in this chapter are included in the chapter's Creative Commons licence, unless indicated otherwise in a credit line to the material. If material is not included in the chapter's Creative Commons licence and your intended use is not permitted by statutory regulation or exceeds the permitted use, you will need to obtain permission directly from the copyright holder."

"The chapter 'Identification of Neuronal Ensembles from Similarity Maps Using Singular Value Decomposition written by Ricardo Velazquez-Contreras & Luis Carrillo-Reid' was originally published without open access. Following the authors' decision to opt for open access, the copyright of the chapter changed to © The Author(s) 2025 and the chapter is now distributed under the terms of the Creative Commons Attribution 4.0 International

The updated versions of these chapters can be found at
https://doi.org/10.1007/978-1-0716-4208-5_4
https://doi.org/10.1007/978-1-0716-4208-5_5
https://doi.org/10.1007/978-1-0716-4208-5_6

Luis Carrillo-Reid (ed.), *Identification, Characterization, and Manipulation of Neuronal Ensembles*, Neuromethods, vol. 215,
https://doi.org/10.1007/978-1-0716-4208-5_13, © The Author(s) 2025

License (https://creativecommons.org/licenses/by/4.0/), which permits use, sharing, adaptation, distribution and reproduction in any medium or format, as long as you give appropriate credit to the original author(s) and the source, provide a link to the Creative Commons licence and indicate if changes were made.

The images or other third party material in this chapter are included in the chapter's Creative Commons licence, unless indicated otherwise in a credit line to the material. If material is not included in the chapter's Creative Commons licence and your intended use is not permitted by statutory regulation or exceeds the permitted use, you will need to obtain permission directly from the copyright holder."

"The chapter 'Detection of Overlapping Neuronal Assemblies from Activity Recordings of Large Neuronal Populations by Means of Non-orthogonal Low-Dimensional State Spaces written by Sebastián A. Romano & Germán Sumbre' was originally published without open access. Following the authors' decision to opt for open access, the copyright of the chapter changed to © The Author(s) 2025 and the chapter is now distributed under the terms of the Creative Commons Attribution 4.0 International License (https://creativecommons.org/licenses/by/4.0/), which permits use, sharing, adaptation, distribution and reproduction in any medium or format, as long as you give appropriate credit to the original author(s) and the source, provide a link to the Creative Commons licence and indicate if changes were made.

The images or other third party material in this chapter are included in the chapter's Creative Commons licence, unless indicated otherwise in a credit line to the material. If material is not included in the chapter's Creative Commons licence and your intended use is not permitted by statutory regulation or exceeds the permitted use, you will need to obtain permission directly from the copyright holder."

INDEX

Luis Carrillo-Reid (ed.), *Identification, Characterization, and Manipulation of Neuronal Ensembles*, Neuromethods, vol. 215, https://doi.org/10.1007/978-1-0716-4208-5,

MIX
Papier aus verantwortungsvollen Quellen
Paper from responsible sources
FSC® C105338

If you have any concerns about our products,
you can contact us on
ProductSafety@springernature.com

In case Publisher is established outside the EU,
the EU authorized representative is:
Springer Nature Customer Service Center GmbH
Europaplatz 3, 69115 Heidelberg, Germany

Printed by Libri Plureos GmbH
in Hamburg, Germany